VOLUME SIX HUNDRED AND TWO

METHODS IN ENZYMOLOGY

Chemical and Biochemical Approaches for the Study of Anesthetic Function Part A

METHODS IN ENZYMOLOGY

Editors-in-Chief

ANNA MARIE PYLE
Departments of Molecular, Cellular and Developmental Biology and Department of Chemistry
Investigator, Howard Hughes Medical Institute
Yale University

DAVID W. CHRISTIANSON
Roy and Diana Vagelos Laboratories
Department of Chemistry
University of Pennsylvania
Philadelphia, PA

Founding Editors

SIDNEY P. COLOWICK and NATHAN O. KAPLAN

VOLUME SIX HUNDRED AND TWO

METHODS IN ENZYMOLOGY

Chemical and Biochemical Approaches for the Study of Anesthetic Function Part A

Edited by

RODERIC G. ECKENHOFF
University of Pennsylvania Perelman School of Medicine, Philadelphia, PA, United States

IVAN J. DMOCHOWSKI
University of Pennsylvania, Philadelphia, PA, United States

ELSEVIER

ACADEMIC PRESS
An imprint of Elsevier

Academic Press is an imprint of Elsevier
50 Hampshire Street, 5th Floor, Cambridge, MA 02139, United States
525 B Street, Suite 1800, San Diego, CA 92101–4495, United States
The Boulevard, Langford Lane, Kidlington, Oxford OX5 1GB, United Kingdom
125 London Wall, London, EC2Y 5AS, United Kingdom

First edition 2018

ISBN: 978-0-12-812740-7
ISSN: 0076-6879

Publisher: Zoe Kruze
Acquisition Editor: Zoe Kruze
Editorial Project Manager: Alina Cleju
Production Project Manager: Divya KrishnaKumar
Cover Designer: Matthew Limbert

Typeset by SPi Global, India

CONTENTS

Section III
Photolabeling

Section IV
Xenon

CONTRIBUTORS

Victoria Bedell
Perelman School of Medicine at the University of Pennsylvania, Philadelphia, PA, United States

Edward J. Bertaccini
Stanford University School of Medicine, Stanford; Palo Alto VA Health Care System, Palo Alto, CA, United States

Grace Brannigan
Rutgers University, Camden, NJ, United States

Elena Buglo
John P. Hussman Institute for Human Genomics; Dr. John T. MacDonald Foundation, University of Miami, Miami; University of Miami, Coral Gables, FL, United States

Robert S. Cantor
Dartmouth College, Hanover, NH, United States

Vincenzo Carnevale
Institute for Computational Molecular Science, College of Science and Technology, Temple University, Philadelphia, PA, United States

Peter Caroccio
NeuroproteXeon, Orchard Park, NY, United States

Charles W. Carspecken
University of Pennsylvania, Philadelphia, PA, United States

Manuel Covarrubias
Vickie and Jack Farber Institute for Neuroscience, Sidney Kimmel Medical College, Jefferson College of Biomedical Sciences, Thomas Jefferson University, Philadelphia, PA, United States

William P. Dailey
University of Pennsylvania School of Arts and Sciences, Philadelphia, PA, United States

Julia E. Dallman
University of Miami, Coral Gables, FL, United States

Ivan J. Dmochowski
University of Pennsylvania, Philadelphia, PA, United States

Roderic G. Eckenhoff
University of Pennsylvania Perelman School of Medicine, Philadelphia, PA, United States

Victoria S. Fahrenbach
Stanford University School of Medicine, Stanford; Palo Alto VA Health Care System, Palo Alto, CA, United States

Stuart A. Forman
Massachusetts General Hospital; Harvard Medical School, Boston, MA, United States

Eleonora Gianti
Institute for Computational Molecular Science, College of Science and Technology, Temple University, Philadelphia, PA, United States

Peter A. Goldstein
Weill Cornell Medical College, New York, NY, United States

Boris D. Heifets
Stanford University School of Medicine, Stanford, CA, United States

Jérôme Hénin
Laboratoire de Biochimie Théorique, Institut de Biologie Physico-Chimique, CNRS, Paris, France

Paul Hoerbelt
Stanford University School of Medicine, Stanford, CA, United States

Thomas T. Joseph
Perelman School of Medicine, University of Pennsylvania, Philadelphia, PA, United States

Shanker Karunanithi
Queensland Brain Institute, The University of Queensland, St Lucia; School of Medical Science and Menzies Health Institute Queensland, Griffith University Gold Coast Campus, Gold Coast, QLD, Australia

Max B. Kelz
Center for Sleep and Circadian Neurobiology, University of Pennsylvania, Perelman School of Medicine, Philadelphia, PA, United States

Qiansheng Liang
Vickie and Jack Farber Institute for Neuroscience, Sidney Kimmel Medical College, Jefferson College of Biomedical Sciences, Thomas Jefferson University, Philadelphia, PA, United States

Kaitlyn L. Maier
University of Pennsylvania, Perelman School of Medicine, Philadelphia, PA, United States

Daniel Marcato
Institute of Toxicology and Genetics, Karlsruhe Institute of Technology, Karlsruhe, Germany

Mervyn Maze
University of California San Francisco, San Francisco, CA, United States

Andrew McKinstry-Wu
University of Pennsylvania, Philadelphia, PA, United States

Ralf Mikut
Institute of Applied Computer Science, Karlsruhe Institute of Technology, Karlsruhe, Germany

Philip G. Morgan
University of Washington and Seattle Children's Research Institute, Seattle, WA, United States

Sruthi Murlidaran
Rutgers University, Camden, NJ, United States

Ravindra Peravali
Institute of Toxicology and Genetics, Karlsruhe Institute of Technology, Karlsruhe, Germany

Alex Proekt
University of Pennsylvania, Philadelphia, PA, United States

Christian Pylatiuk
Institute of Applied Computer Science, Karlsruhe Institute of Technology, Karlsruhe, Germany

Paul M. Riegelhaupt
Weill Cornell Medical College, New York, NY, United States

Richie Robel
NeuroproteXeon, Orchard Park, NY, United States

Benjamin W. Roose
University of Pennsylvania, Philadelphia, PA, United States

Will Scudder
University of Miami, Coral Gables, FL, United States

Margaret M. Sedensky
Center for Integrative Brain Research, Seattle Children's Research Institute; University of Washington, Seattle, WA, United States

Louise M. Steele
Kent State University at Salem, Salem, OH, United States

Johannes Stegmaier
Institute of Applied Computer Science, Karlsruhe Institute of Technology, Karlsruhe, Germany

Uwe Strähle
Institute of Toxicology and Genetics, Karlsruhe Institute of Technology, Karlsruhe, Germany

Gareth R. Tibbs
Weill Cornell Medical College, New York, NY, United States

Michael Troup
Queensland Brain Institute, The University of Queensland, St Lucia, QLD, Australia

Bruno van Swinderen
Queensland Brain Institute, The University of Queensland, St Lucia, QLD, Australia

Andrzej Z. Wasilczuk
University of Pennsylvania, Perelman School of Medicine, Philadelphia, PA, United States

Kellie A. Woll
University of Pennsylvania Perelman School of Medicine, Philadelphia, PA, United States

Maxwell Wray
University of Miami, Coral Gables, FL, United States

Elaine Yang
Vickie and Jack Farber Institute for Neuroscience, Sidney Kimmel Medical College, Jefferson College of Biomedical Sciences, Thomas Jefferson University, Philadelphia, PA, United States

Serge D. Zemerov
University of Pennsylvania, Philadelphia, PA, United States

Lianteng Zhi
Vickie and Jack Farber Institute for Neuroscience, Sidney Kimmel Medical College, Jefferson College of Biomedical Sciences, Thomas Jefferson University, Philadelphia, PA, United States

Stephan Züchner
John P. Hussman Institute for Human Genomics; Dr. John T. MacDonald Foundation, University of Miami, Miami, FL, United States

PREFACE

"Anesthesia" is a reversible, drug-induced state that most importantly (for the surgeon, at least!) causes an inability to generate a motor response to what would normally be painful stimuli. The general anesthetic state generally includes unconsciousness and amnesia. These features are considered to reside in the central nervous system, and in fact, when absent (e.g., during normal "awake" function), they collectively define consciousness. Thus, the study of general anesthesia overlaps with the study of consciousness—probably the most enigmatic state that exists! It should not be surprising then that the range of experimental and theoretical approaches to understanding the anesthetic state is very large. We have attempted to capture the scope, complexity, and significance of the many approaches to understanding the state of anesthesia in this two-volume set of *Methods in Enzymology*. This is, perhaps, an unusual collection for *MIE*, in that many of the approaches do not fit the term "enzymology," or for that matter, even chemistry or biochemistry. This first volume, *Chemical and Biochemical Approaches for the Study of Anesthetic Function, Part A*, details computational approaches, animal models, photolabeling technique, the "simplest" anesthetic xenon, and electrophysiology approaches. Computational approaches offer windows into understanding anesthetic function at both molecular and systems-level scales, with an increasing number of ion channel protein crystal structures available to guide both computation and experiment. In addition, the appropriate means to phenotype various genetic models is crucially important, as the activity of no single-target molecule can serve as a reliable surrogate for the state of anesthesia. Xenon provides a uniquely powerful single-atom probe to investigate anesthetic–protein and anesthetic–lipid interactions by many experimental approaches. Electrophysiology allows the measurement of electrical phenomena associated with nervous activity, on varying time and size scales, ranging from single-ion channels, to cells and intact organs. We have sought out the recognized experts in each field, to provide a succinct account of what these approaches can contribute and how to employ them most effectively. With such a diverse and intense campaign, one might have thought the mechanism of anesthesia would have succumbed by now! But this could not be further from the truth; despite finding ever-increasing use in modern medicine, anesthesia remains almost as enigmatic as when it was discovered over 170 years ago. Through further deployment and refinement of these approaches, and many more to be developed, we may finally reveal the mechanisms underlying anesthesia, and eventually improve it.

Ivan J. Dmochowski and Roderic G. Eckenhoff

SECTION I

Computational Approaches

CHAPTER ONE

Physical Accuracy Leads to Biological Relevance: Best Practices For Simulating Ligand-Gated Ion Channels Interacting With General Anesthetics

Sruthi Murlidaran, Grace Brannigan[1]
Rutgers University, Camden, NJ, United States
[1]Corresponding author: e-mail address: grace.brannigan@rutgers.edu

Contents

Abstract

Efforts to detect binding modes of general anesthetics (GAs) for pentameric ligand-gated ion channels (pLGICs) are often complicated by a large number of indicated sites, as well as the challenges of ranking sites by affinity and determining which sites are occupied at clinical concentrations. Physics-based computational methods offer a powerful route for determining affinities of ligands to isolated binding sites, but preserving accuracy is essential. This chapter describes a step-by-step approach to multiple methods for identifying candidate sites and quantifying binding affinities and also discusses limitations and common pitfalls.

Methods in Enzymology, Volume 602
ISSN 0076-6879
https://doi.org/10.1016/bs.mie.2018.02.001

1. INTRODUCTION

Pentameric ligand-gated ion channels (pLGICs) are widely studied anesthetic targets, but present numerous challenges for structural characterization even in the apo state. EC_{50} of general anesthetics (GAs) such as sevoflurane and propofol for pLGICs such as the GABA(A) receptor fall in the 100 mM to 1 μM range, and these low to moderate affinities introduce a high rate of false positives into most approaches. The ability to isolate binding sites offers an advantage to computational approaches, but one of the most common computational approaches, automated docking, is particularly unreliable on its own for interactions of GAs with pLGICs.

We have developed and refined a robust computational approach for identifying candidate binding sites, determining which binding sites are occupied at clinical concentrations, ranking them according to affinity, and determining the microscopic origins of differences in affinities. This approach considers all atoms of the system explicitly (GA, salt, lipids, water, and protein) and relies on the rigorous physics-based methods of Molecular Dynamics Simulation and Alchemical Free Energy Perturbation.

Our general process essentially involves two components: a screening or discovery phase to identify possible binding modes that might be occupied at clinical concentrations, and a quantification phase in which binding affinities are measured. This manuscript describes the steps to set up the necessary calculations and provides examples of the possible pitfalls in this process. This article is written with the CHARMM all-atom force field (Brooks et al., 2009) and the NAMD molecular dynamics software (Phillips et al., 2005) in mind, but the general approach is not force field or software specific. In this chapter we focus on a model of an $\alpha_1\beta_3\gamma_2$ GABA(A) receptor in a phosphatidylcholine (POPC) bilayer, interacting with the GAs sevoflurane and propofol. The chapter assumes a general familiarity with the structure of pLGICs as well as the principles underlying classical molecular dynamics simulation.

2. MD SIMULATION INVOLVING pLGICs

Setting up MD simulations follows a series of steps that are common for any pentameric channel with/without ligand. The CHARMM-GUI membrane builder website (Jo, Lim, Klauda, & Im, 2009), a simulation input generator, accepts a protein structure file as input to embed the protein in a well-packed lipid bilayer with water molecules and neutralizing counter-ions on either side. For a system involving a pLGIC like the GABA(A) receptor, some of the important steps include:

- Choosing membrane builder from the input options
- Setting up chain names (for nonprotein molecules, like the anesthetic, the chain names should match the name mentioned in the topology files)
- Adding terminal patch groups to the protein N-terminus and C-terminus
- Preserving hydrogen coordinates
- Adding disulfide bonds
- Specifying/preserving protonation states, based on the desired pH of simulation
- Choosing the right alignment of protein in the membrane. For a pentameric channel, choosing the option to align the "first principle axis along *Z*" would align the channel along *Z* axis, with the lipid in *X*–*Y* plane. Further, aligning the TMD region of protein with lipid involves translocating the protein until the pore center is at the box origin. It is advisable to check the orientation of the channel after this step, by clicking on the "view structure" option.
- Choosing the type of lipids for the bilayer and setting water thickness. We typically use the homogenous composition of POPC lipids and choose the default option of 1.5 layers in CHARMM-membrane builder for choosing the size of the lipid membrane. There are ≈250 lipid molecules in GABA(A) receptor system and a water thickness of 20 Å is maintained at the top and bottom of the protein.
- Adding salt; we typically use 150 m*M* of neutralizing NaCl.

The total number of atoms is typically between 140,000 and 200,000.

3. DISCOVERY OF CANDIDATE SITES

We use two main approaches for identifying candidate binding modes: (1) spontaneous binding of GAs during equilibrium MD, also known as "flooding" and (2) docking followed by refinement using equilibrium MD (Fig. 1). The former allows a blind, unbiased search for spontaneously occupied binding modes without targeting any particular region of the receptor, but is computationally very expensive, while the latter is primarily suitable when a specific region of the receptor is of interest. We consider each in turn.

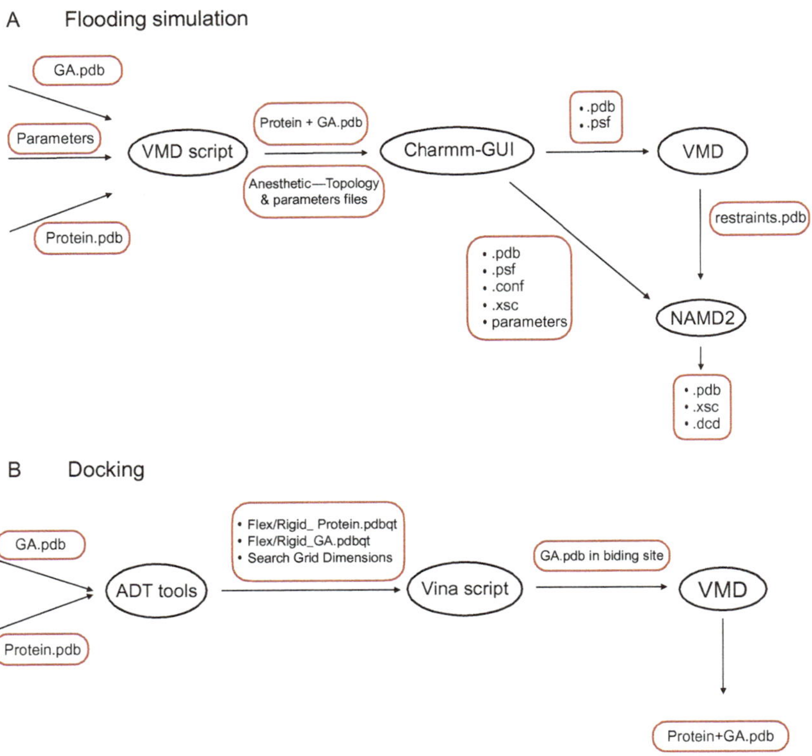

Fig. 1 Flowchart describing the sequential steps involved in performing (A) flooding simulation and (B) docking using AutoDock and MD simulation.

3.1 Flooding Simulation

Flooding involves placing a high concentration of GA in the water surrounding the pLGIC (or, in some cases, the membrane) and allowing the GA to spontaneously partition into the membrane and protein binding sites over a simulation time that usually lasts 400 ns to 2 μs. It is most effective for small, relatively soluble ligands, and we have used this approach with isoflurane and nicotinic acetylcholine receptors and the prokaryotic homolog, GLIC (Brannigan, LeBard, Hénin, Eckenhoff, & Klein, 2010) as well as sevoflurane interacting with GABA(A) receptors. It has also been used with ethanol interacting with glycine receptors (Murail, Wallner, Trudell, Bertaccini, & Lindahl, 2011) and interactions of GAs with voltage-gated channels (Raju, Barber, LeBard, Klein, & Carnevale, 2013).

Prerequisites:

- Protein system in a water-box and lipid membrane (procedure for this is elaborated under the previous section).
- Ligand in PDB format.
- Parameters for ligand that are compatible with the force field used for water, protein, and lipids (see chapter "Molecular mechanics parameterization of anesthetic molecules" by Joseph and Hénin).

Simulation setup:

- Initial coordinates: Simulation should begin with a receptor embedded in a hydrated lipid bilayer with counter-ions and GA molecules randomly distributed in the water. GA molecules can be randomly distributed around the receptor in the desired region of the simulation box using automated tcl scripts and VMD, and then the .pdb file containing both GAs and the receptor can be used as an input into the CHARMM-GUI membrane builder and system can be built as described in Section 2.

  For a typical simulation box, a single GA molecule in aqueous phase corresponds to a concentration greater than 1 m*M*. To obtain a range of candidate sites over a reasonable simulation time, at least 20–100 GAs must be present in the simulation. Over the course of a flooding simulation, therefore, the concentration of GA in the aqueous phase may drop from 100 m*M* down to 0–1 m*M*, as GA molecules partition into the lipid phase. The poor correspondence between concentration in the simulation and in vitro or in vivo systems mandates the use of more sophisticated methods to predict likelihood of occupancy at clinical concentrations, as described subsequently; for a discovery stage, a higher than usual concentration is actually advantageous.

  An example of sevoflurane being used to flood the GABA(A) receptor is shown in Fig. 2, with a sevoflurane-to-lipid ratio of about 1:3, and a sevoflurane-to-water molecule ratio of about 1:430.
- Nondefault NAMD parameters:
  - minimize: This parameter denotes the number of timesteps to minimize the system. Since this simulation involves flooding the system with multiple anesthetic molecules, it is recommended to run longer minimization runs of 50,000–100,000 timesteps.
  - constraints: To avoid major changes to the backbone of the protein, this parameter can be set to "on" and a pdb file flagging the backbone atoms to be constrained under the parameters "conskfile" and "consref." Typically the constraints are on the order of 5 kcal/mol/Å on the C_α atoms of the protein.

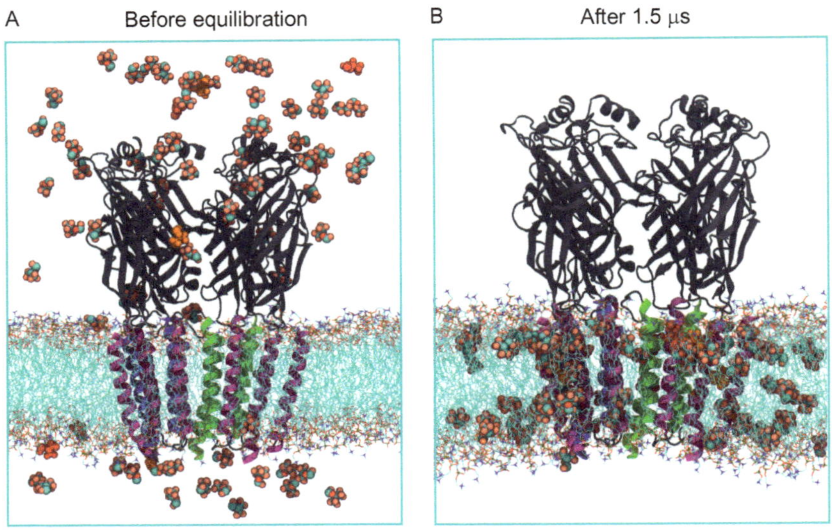

Fig. 2 Cross-sectional view of the sevoflurane flooded GABA(A) receptor system with the TMD aligned along POPC lipid membrane (*colored* by name) and placed in a box of explicit water (molecules not shown, represented as a *blue box*). (A) Initial frame with sevoflurane (*colored* by name) flooded in the water; (B) after 1.5 μs sevoflurane completely localizes in the lipid membrane with some binding the inter- and intrasubunit sites (*red* and *orange*) in the TMD.

Analysis: Analyzing the trajectory of a flooding simulation using VMD, one could:

- Identify microscopic interactions between protein and anesthetic, including hydrogen bonds as described in Section 3.3.
- Visualize the path of entry of ligand into the binding site.
- Estimate a residence time for a site in which the ligand binds and unbinds several times over the course of the simulation.
- Use VMD plug-in, "VOLMAP tool" to create images showing the average density of the ligand in multiple binding sites (Fig. 3B).
- Visualize the competitive binding between lipid and anesthetic as shown in Fig. 3A.
- Identify sites likely to have multiple occupancy.
- Identify other binding sites not recognized by docking software. For instance, intrasubunit sites and multiple occupancy of specific sites were identified in flooding simulations with sevoflurane as shown in Fig. 3A.

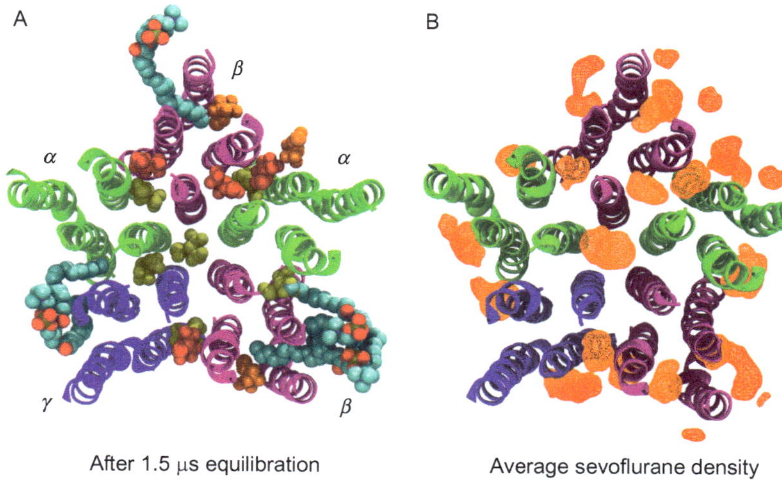

Fig. 3 View from the extracellular domain, of a sevoflurane flooded GABA(A) receptor system. (A) Final frame from the flooding simulation showing sevoflurane occupying the inter, intra and pore sites (*red*, *orange*) overlaid with sevoflurane (*yellow*) docked using AutoDock; flooding simulation also identified lipid (*colored* by name) interference in β intrasubunit site and α–γ, α–β intersubunit sites. (B) Image created using VMD plugin, VOL-MAP tool, to depict the density isosurface (*orange*) averaged over the last 700 ns of the simulation; large mesh represent occupation over majority of the trajectory, whereas smaller mesh represents occupation for lesser time.

Pitfalls: Some of the possible errors while setting up and running simulation involve:

- Incomplete PSF files being generated after GA is added to the protein, with overlapping water molecules.
- Insufficient minimization or equilibration. Simulations must be run sufficiently long for GA molecules to partition into the lipid and access binding sites in the protein that may be deep.

Limitations: The high concentrations required can cause aggregation in GAs with poor solubility (such as propofol). It is also usually not practical to calculate concentration-dependent occupancies because of the high concentrations required; AFEP methods described subsequently are essential for estimating features of dose–response.

3.2 Docking

Our approach usually limits use of automated docking to the generation of reasonably favorable binding modes in a particular region of a protein, as initial configurations for significant refinement via MD simulation. Scoring functions from automated docking are rarely meaningful as absolute values, but relative scores may be appropriate for suggesting the most favorable binding mode to use as an initial configuration. Although docking scores are often presented as binding affinities in units of kcal/mol, the method of assigning these units rarely involves an explicit calculation of interaction differences between bound and apo states. Docking scores *cannot* meaningfully be used to calculate a K_D in its usual meaning as the ligand concentration at which half the sites are occupied.

Prerequisites:

- Receptor model and ligand in PDB format.
- Particular region of the receptor of interest, such as the pore, a given subunit interface in the TMD, or the center of a given subunit.

Softwares used:

- AutoDock Vina (Trott & Olson, 2010) is a docking algorithm that predicts a number of preferred orientations of ligand molecule bound to the protein and ranks them based on approximate scoring functions.
- VMD (Humphrey, Dalke, & Schulten, 1996) is a software that makes it possible to visualize and analyze the protein–anesthetic complex following docking.

Docking setup:

- Initial coordinates: Starting configurations for docking is a PDB structure of the protein and the ligand to be docked with, in the PDB format. In Fig. 4 we have docked sevoflurane and propofol to GABA(A) receptor system.

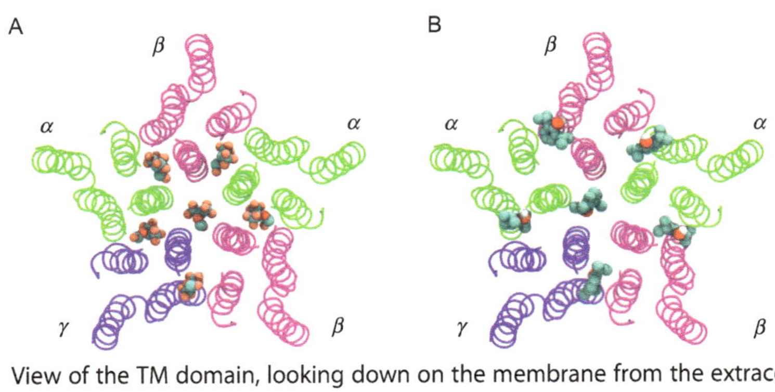

Fig. 4 View of the TM domain, looking down on the membrane from the extracellular region; Sevoflurane (A) and propofol (B) docked to the TMD of GABA(A) receptor. Docking was individually performed at all the intersubunit sites and the pore by making few of the protein residues flexible. For instance, at the α–β site, βMET289 residue side-chain was made flexible while docking to this interface.

- Nondefault parameters: Docking can be performed from command-line prompt with a configuration file containing the following commands:
  - flex: Some side-chains of the channel can be made flexible and this part of the protein can be saved as a separate ".pdbqt" file, while the rigid part of the protein would be the argument for the previous command. Use of this command is to facilitate and focus docking in proximity to these "flex" residues. The software further rotates these "flex" residues to predict more number of orientations of the ligand at the docked site. This parameter is useful when there is prior knowledge regarding the location of possible binding sites, or specific residues present in the binding site. To select flexible residues, one can name the chain and the residue number in the dialog box from the toolbar as shown in Fig. 5B.
  - ligand: Input ligand in ".pdbqt" format. Similar to the receptor, the ligand can be loaded into Autodock tool, in ".pdb" format. Rotatable bonds/torsions must be set for the ligand molecule using the options from toolbar. This specifies the flexibility of the molecule; for propofol all bonds should be left as rigid.
  - Center and size: Docking is most suitable when a particular region of the protein has been identified via other means, and this parameter can be used to limit the search space. For instance, the site of interest might be the intersubunit cavity around $\beta N265$, a residue that has been implicated numerous times for propofol (Chiara et al., 2016; Jurd et al., 2002; Stern & Forman, 2016; Stewart, Pierce, Hotta, Stern, & Forman, 2014) (Fig. 5). AutoDock tools is useful for visualizing the search space overlayed with the protein.
  - exhaustiveness: Docking consists of multiple individual runs starting from random conformations of the protein–ligand complex. The number of these runs depends on the flexibility of the ligand and the protein side-chains and can be controlled by this parameter. We use a value of 15 for this parameter, but also run multiple docking runs with the same parameters for a specific system.
  - nummodes: This parameter specifies the maximum number of binding poses to be generated. We use a value of 20 with multiple individual runs.

Pitfalls: Even with the precautions listed above, the ligand may be inherently unstable in the binding mode for a number of reasons. The two most common are:

- A ligand binding to an unoccupied binding site displaces solvent, and the overall stability depends not just on how the ligand interacts with the

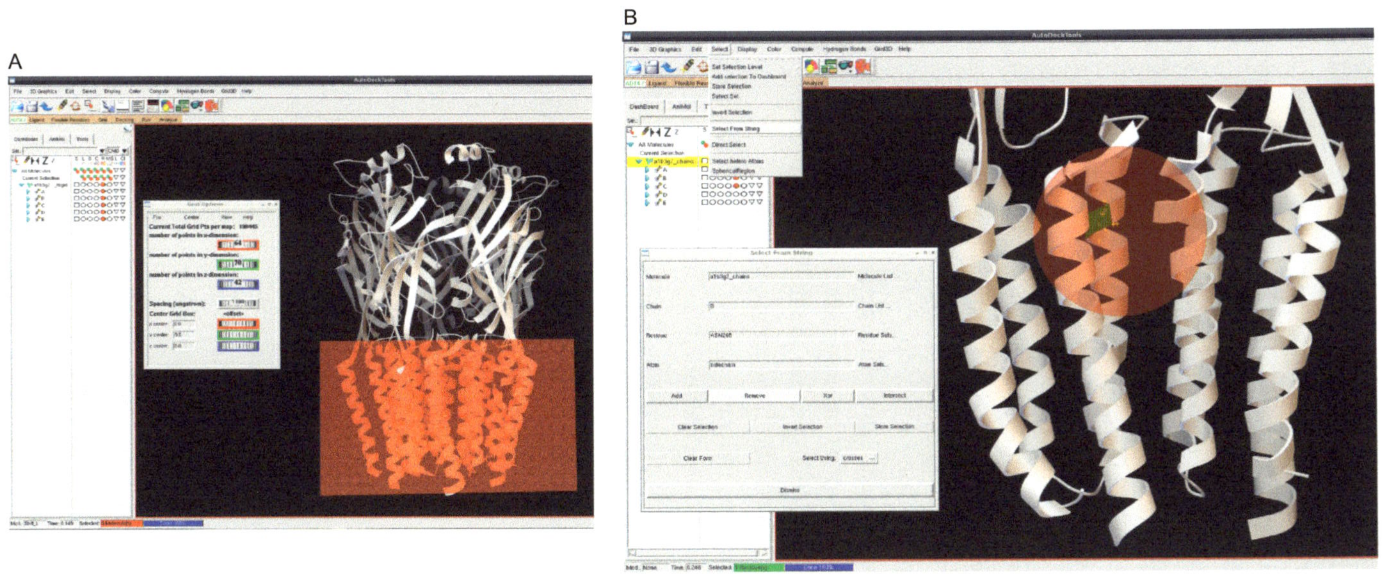

Fig. 5 Screenshot of AutoDock tools screen depicting a dialog box describing (A) the measurements of the grid box over the TMD of the protein and (B) the process of selecting the flexible residues in the protein.

protein residues, but also on how solvent interacts with protein, and how the ligand interacts with solvent. Automated docking algorithms do not take into account the latter in a rigorous way even for aqueous solvent. The situation is even worse when the site fills with lipid acyl chains in the absence of ligand, as GA sites on pLGICs often do; in many cases a GA can compete much more favorably with water than with an acyl chain, and docking programs will overestimate the favorability of such a binding mode.

As a result, it is not uncommon for GAs to migrate far enough from a docked location even in shorter (~100 ns long) MD simulations that contact residues are entirely different between the initial and final frame. For example, AutoDock typically returns a binding mode for sevoflurane that is particularly far from the extracellular domain (Fig. 6), but in MD simulation, sevoflurane reliably migrates to a site much more similar to that for propofol. Rigorous calculation of affinities using AFEP identifies the latter binding mode as much more favorable than the initial docked conformation.

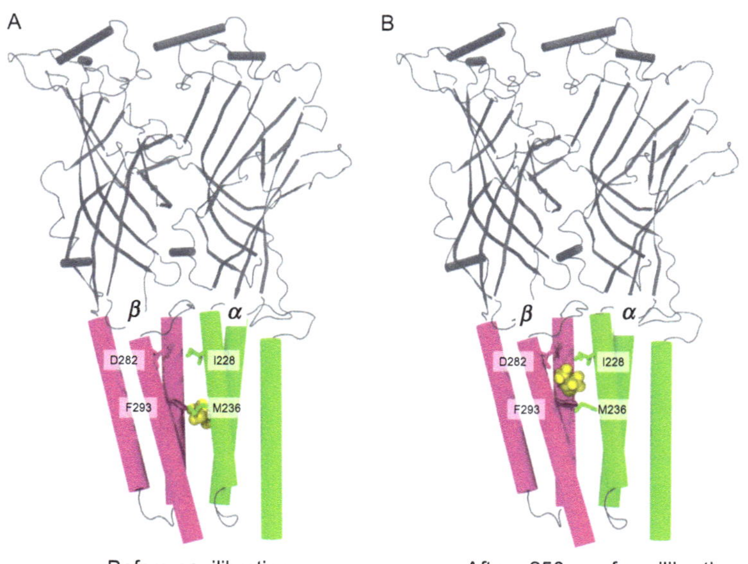

Fig. 6 Side view of the sevoflurane in the β–α interface of the GABA(A) receptor system; two images depicting the docked (A) and equilibrated conformation (B) of the system. Equilibration of the docked conformation allows sevoflurane to reorient itself at a higher affinity site.

- The docking algorithm usually uses flexible ligands, with an energy associated with rotation around bonds. Configurations for propofol in which the hydroxyl clashes with the isopropyl group do not actually have unfavorable dihedral angles and many docking programs will not identify this clash as unfavorable, returning a propofol conformation in which the hydroxyl is overlapping the isopropyl (Fig. 7). This can introduce an instability or rapid unbinding in the MD simulation (Fig. 8A and B), depending on how equilibration is carried out. For this reason, we strongly recommend removing ligand flexibility when docking with propofol.

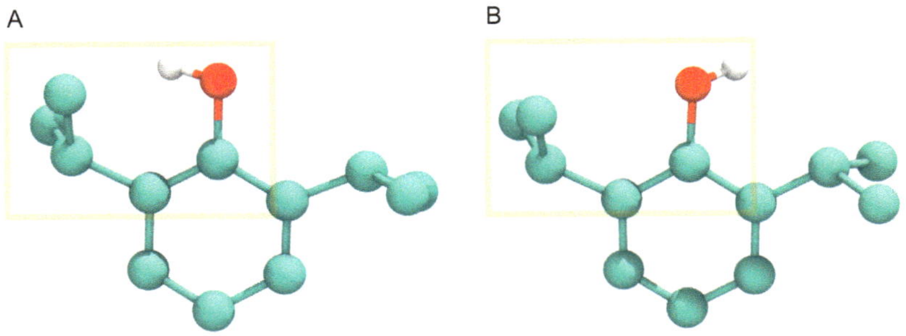

Fig. 7 Comparison between an unfavorable (A) and favorable low energy (B) conformation of propofol.

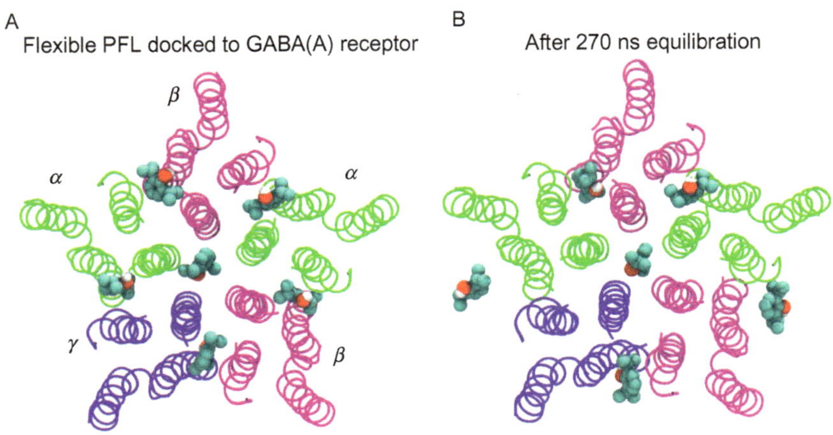

Fig. 8 View of the TM domain, looking down on the membrane from the extracellular region; MD of PFL-GABA(A) receptor system, images depict the position of propofol at initial and final frames of the trajectory. Unfavorable conformation of propofol led to expulsion of the ligand in the course of the simulation.

3.3 MD Refinement and Characterization of Specific Interactions

Several of the limitations of automated docking can be overcome by refinement with unbiased MD simulation, and if possible, it is almost always advisable to do so before drawing conclusions about microscopic interactions or calculating binding affinity.

Prerequisites:

- Docked protein–anesthetic complex
- Parameters for the anesthetic

Simulation setup:

- Initial coordinates: Starting configurations for the simulation would be anesthetics docked to binding sites in the protein, generated using docking software. The protein–anesthetic complex can be used as an input to the CHARMM-GUI membrane builder website (Jo et al., 2009) to build the system.
- Nondefault parameters: restartfreq, dcdfreq, xstFreq, outputEnergies, and outputTiming: These parameters denote the number of timesteps between which each output is generated. By default, the configuration files generated by CHARMM-membrane builder has these parameters set to 125 or 500 timesteps. A frequency of 5000 timesteps is ideal for further analysis unless a much lower frequency is required.

Analysis: Running the simulation for a considerable amount of time (≈200–500 ns), one has to analyze the dynamics before setting up FEP simulations. Initial analysis includes:

- Checking the stability/mobility of the ligand in the binding site. This can be done by tracking the location of the center of mass (COM) of the ligand in the site throughout the trajectory, using tcl scripts in VMD. Dispersion in the COM is essential to estimate prior to constructing spherical restraints for the ligand while setting up FEP calculations.
- Visualizing microscopic interaction between ligand and its environment. For example, on loading the trajectory into VMD, one can make following selections for a ligand with resname PFL:
  - "(protein and within three of (resname SEV)) or (resname PFL)." Making the representation as H-Bonds would allow us to view the hydrogen bonds formed between protein and the ligand (Fig. 9).
  - "(protein or lipid or water) and same residue as within four of (resname PFL)". This selection would let us see the interactions among the protein residues, water, lipids, and the anesthetic, in the

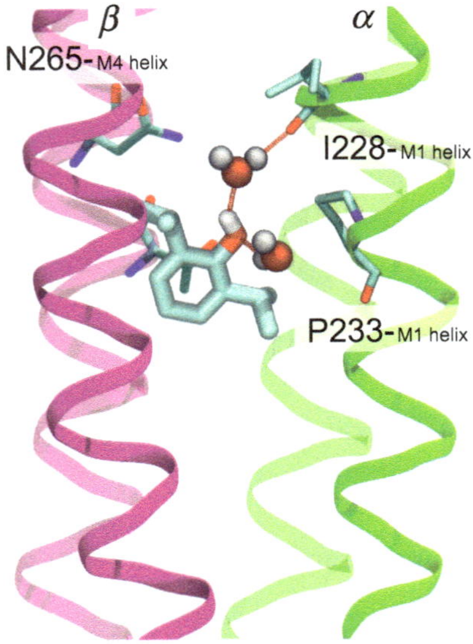

Fig. 9 Depiction of propofol interacting with water and protein side-chains at the $\beta^{-}-\alpha^{+}$ intersubunit site, in a snapshot from MD simulation.

binding site. Analyzing these interactions may provide insight into the affinity (obtained through AFEP) of the anesthetic to a specific site, as in Woll et al. (2016).

- Identifying higher affinity sites, compared to the sites identified through docking. In the case of sevoflurane docked to a GABA(A) receptor, the low specificity and smaller size of the anesthetic results in greater mobility in the binding site, thus allowing it to explore/move to higher affinity sites as shown in Fig. 6.
- Ensure the bound anesthetic is in its favorable/low-energy conformation. In case of docking to GABA(A) system, docking with an unfavorable conformation of propofol (Fig. 7) led to immediate expulsion of the molecule as the simulation began. This is described by depicting few snap-shots from the simulation in Fig. 8, and this issue can be overcome by removing propofol flexibility during the docking process.

Pitfalls: Some of the common pitfalls encountered during setup of a simulation are:

- Not adding necessary disulfide bonds while setting up the system in CHARMM-GUI.
- Inaccurate protonation states of amino acids at pH 7.

- Not confirming the proper alignment of the protein in the lipid membrane.
- Mismatch in names of the GA atoms in the pdb file and topology file.
- Ligand should have a separate chain name different from the rest of the protein, to be recognized as HETATM by CHARMM-GUI.

Limitations: Inability to witness lipid mixing or domain formation around the receptor in a mixed membrane. This would require extensively long simulations or can be achieved through coarse-graining simulations.

4. CALCULATION OF ABSOLUTE AND RELATIVE BINDING AFFINITIES

Following the identification of a stable binding mode for the anesthetic, the binding affinity can be rigorously calculated using a theoretically exact simulation technique called Alchemical Free Energy Perturbation (AFEP) (Fig. 10) that relies on the equation introduced by Zwanzig (1954) for calculating the difference in Helmholtz free energy between two states, X and Y:

$$\Delta A_{X\rightarrow Y} = -RT\ \ln\left\langle e^{-(H_Y(\mathbf{r})-H_X(\mathbf{r}))/RT}\right\rangle_X \tag{1}$$

where R is the gas constant, T is temperature, $H(\mathbf{r})$ is the Hamiltonian for state X or state Y. In a binding free energy application, X would represent the bound state, Y would represent the apo state, and $\Delta A_{X\rightarrow Y} \sim \Delta G_{X\rightarrow Y}$ because ligand binding will not significantly change the system volume.

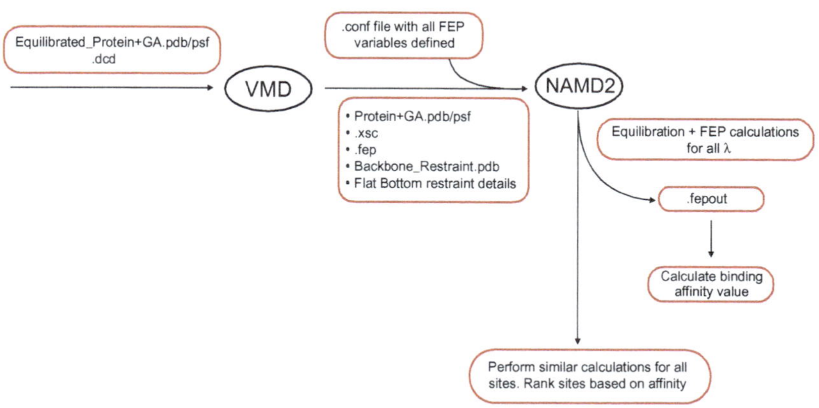

Fig. 10 Flowchart describing the sequential steps involved in calculating binding affinity of anesthetics.

In practice (Beveridge & DiCapua, 1989; Mark, 2002) for convergence purposes this is carried out by summing over a series of windows in which the interaction between the ligand and all other atoms of its environment are gradually decreased to zero:

$$H_{\lambda_i} = H_{\text{env}} + H_{\text{lig}} + (1-\lambda_i)H_{\text{env-lig}}, \tag{2}$$

where $\lambda = 0$ is equivalent to the bound state X, $\lambda = 1$ is the unbound state Y, $0 <= \lambda_i < 1$ for all i and $\lambda_i < \lambda_{i+1}$. The free energy change for "decoupling" the ligand from its environment is

$$\Delta G_{\text{env}} = -RT\sum_i \ln\left\langle e^{-(H_{\lambda_{i+1}}(\mathbf{r})-H_{\lambda_i}(\mathbf{r}))/RT}\right\rangle_{\lambda_i} = -RT\sum_i \ln\left\langle e^{-(\lambda_i-\lambda_{i+1})H_{\text{env-lig}}/RT}\right\rangle_{\lambda_i}. \tag{3}$$

Treating the coupled ligand as the bound state requires that the fully coupled ensemble (at $\lambda = 0$) includes no states with ligand outside the binding site. This assumption can breakdown for moderate affinity ligands like GAs, but can be resolved by making the requirement of localization to the binding site explicit, via a restraint potential on the ligand center of mass.

A flat-well potential that vanishes within the binding site but is very high outside the binding site accomplishes this with minimal need for correction, and also alleviates convergence problems for windows close to $\lambda = 1$ (fully decoupled). The standard binding affinity is

$$\Delta G^0 = \Delta G_{\text{site}} - \Delta G_{\text{solv}} - k_B T \ln(V_{\text{site}}/V^0) \tag{4}$$

where ΔG_{site} is free energy of decoupling from environment of the protein binding site, ΔG_{bulk} is the free energy of decoupling ligand from bulk solvent (solvation free energy), V_{site} is the volume accessible to the center of mass of the bound ligand (as specified by Eq. (5)), and $V^0 = 1661\text{Å}^3$ is the accessible volume per molecule in a 1 M solution. Although this method is theoretically exact and all degrees of freedom are considered automatically by the method, the success and inherent challenge of the technique relies on obtaining reasonable convergence of each average in Eq. (3).

Prerequisites:

- Starting configuration from MD simulation.
- PDB files indicating the atoms/ligands to be unbound/decoupled.
- Restraint files indicating the type of restraints to be applied to the movement of the ligand during the simulation.

- Solvation free energy of the anesthetic. This involves, decoupling the ligand from bulk solvent in the absence of protein. This is required to calculate the free energy cost of moving the ligand from bulk solvent to a binding site in the protein.

Simulation setup:

- Initial coordinates: Starting coordinates should be taken from the output of a fully interacting equilibrium MD simulation, as described in the previous section.
- Nondefault parameters:
  - alchEquilSteps: The number of steps at the beginning of each alchemical window that is excluded in the cumulative average used to calculate $\Delta\Delta G_i \equiv -RT \ln \left\langle e^{-(\lambda_i-\lambda_{i+1})H_{\text{env-lig}}/RT} \right\rangle_{\lambda_i}$ for window i, to allow the system to adjust to the new value of λ. We typically use values of 25,000–50,000 steps to balance the need for equilibration time with need for actual samples. It affects on-the-fly analysis but not the actual trajectory, and different values can be used in postprocessing.
  - alchElecLambdaStart: The value of λ for which the electrostatics should be entirely switched off, through the soft-core potential necessary for FEP calculations. We use the value of 0.5 recommended by the NAMD User's guide (Git et al., 2017).
  - alchVdwLambdaEnd: The value of λ for which the van der Waals interaction should be entirely switched off, through the soft-core potential necessary for FEP calculations. We use the value of 1.0 recommended by the NAMD User's guide (Git et al., 2017).
  - alchDecouple: This parameter specifies whether either intermolecular interactions of the anesthetic are turned "off" (decoupling) or both inter- and intramolecular interactions are turned "off" (annihilation). By default this parameter is set to "off", which chooses the annihilation option. We use the decoupling method, i.e., set the parameter "on" as the intramolecular annihilation free energies simply get canceled when the solvation free energy is subtracted.
  - alchLambda/alchLambda2: Every configuration file would have different values for λ, denoting the progress of the perturbation. We typically use windows with $\lambda_{i+1} - \lambda_i = 0.05$ for λ between 0 and 0.8 and $\lambda_{i+1} - \lambda_i = 0.025$ for λ between 0.8 and 1.0. These values can be modified during reruns to improve sampling.

- Additional files:
  - Restraint file: We use flat-bottom spherical restraints for the inter and intrasubunit sites with the following potential:

$$U_{rest}(\vec{r}_{COM}) = \begin{cases} \frac{k(\vec{r}_{COM} - R)^2}{2}, & |\vec{r_0} - r_{COM}| > R \\ 0, & |\vec{r_0} - r_{COM}| \leq R \end{cases} \quad (5)$$

where $\vec{r}_{COM}$ is the ligand center of mass, $\vec{r_0} = \langle \vec{r}_{COM} \rangle$ in the fully coupled and bound state, and $R = \max(|\vec{r_0} - r_{COM}|)$ in the fully coupled and bound state. Both $\vec{r_0}$ and R can be determined from equilibrium MD simulations and determine the value $V_{\text{site}} = 4/3\pi R^3$.

Typical values of k are 5 kcal/mol/Å. For pore sites we adjust Eq. (5) to represent a cylindrical site by applying separate potentials on the vertical and radial coordinates. In NAMD, these restraints can be implemented using tclForces or the collective variables module, both of which require an additional file. For simple geometries, using tclForces tends to be computationally faster and is implemented using the parameter "tclForces" in the configuration file.
  - FEP file: PDB file used to denote atoms that are to be decoupled. This can be generated using VMD, where the "beta" column in the PDB file can be flagged with number "−1.00" for outgoing atoms and "1.00" for incoming atoms.

Analysis:

- Multiple FEP runs with same or different starting configurations can be performed to check the consistency of the results.
- Perform a recoupling run (ligand-unbound state to bound state) followed by decoupling and combining the results using BAR or SOS estimator to obtain more accurate binding energy value.
- Understanding the affinity values obtained. The standard free energy of binding is related to the dissociation constant K_D via,

$$\Delta G^0 = RT \ln K_D \quad (6)$$

Any K_D value that is smaller (stronger) than the EC_{50} of a GA indicates the site may be essential to action of the GA. Previously we used measured K_D for isoflurane in the GLIC pore to argue that inhibition by isoflurane likely occurred via pore block rather than an allosteric mechanism (LeBard, Hénin, Eckenhoff, Klein, & Brannigan, 2012); although there is still no available crystal structure for isoflurane bound to GLIC, crystal structures released several years later confirmed pore block as the dominant mechanism for a homologous prokaryotic channel, ELIC (Chen et al., 2015). Similarly, we were able to rank K_D for propofol interacting with $GABA_A$r subunit interfaces, with sites containing α and β subunits having $K_D < EC_{50}$ but sites with γ subunits having $K_D > EC_{50}$, consistent with results using photolabeling with azipropofol and a click agent in neurons. Correlating these results with observations of interactions from equilibrium MD explained the surprising result that propofol had a particularly low affinity for the site with the most polar residues, due to competition with water (Woll et al., 2016).

Pitfalls:

- Insufficient sampling to properly calculate the necessary averages is a primary pitfall (Pohorille, Jarzynski, & Chipot, 2010); it can be assessed by examining $\Delta\Delta G$ per window by plotting the last column of the ". fepout" output file from the FEP simulation (Fig. 11). Converged

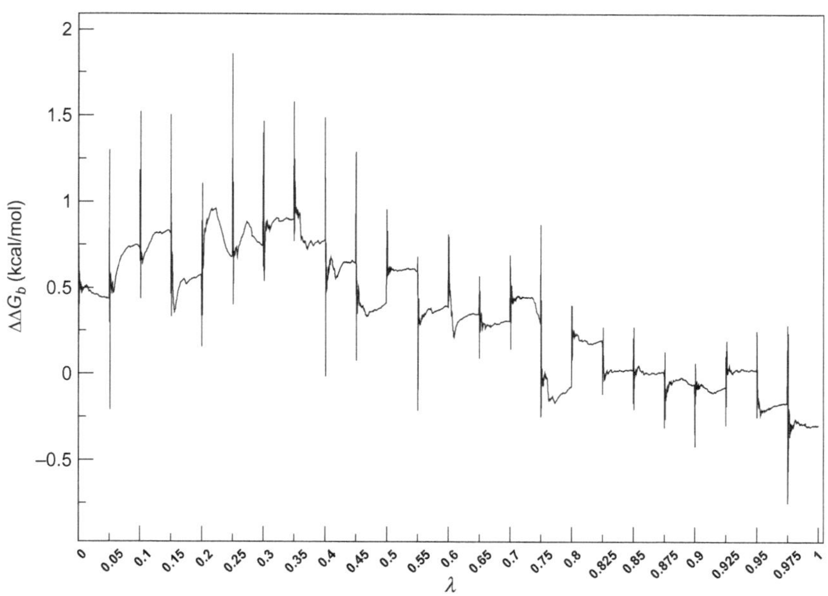

Fig. 11 Sample data set for $\Delta\Delta G_i$ variations per window. Curves that plateau (as at $\lambda =$ 0.825–0.85) indicate convergence, while curves that still change rapidly by the end of the window (as at $\lambda = 0.20$–0.25) indicate a need for extending the calculation for that window or dividing the window into two.

windows will flatten out by the end of the window, while abrupt changes indicate a need for:
  - Extending the calculation for that window.
  - Further dividing $\delta\lambda$ into multiple separate windows.
  - Removing some of the early values from the average (equivalent to increasing alchEquilSteps)
- Insufficient total simulation time to sufficiently equilibrate decoupled states. One way to assess this is tracking the rehydration of the binding site following the decoupling of the ligand, by comparing the number of solvent molecules with a hydrated site in an apo simulation. Increasing the overall simulation time is necessary if the site does not become fully solvated, and usually adding this time to later, mostly decoupled windows works best.
- Neglecting the analytical correction for the standard state in Eq. (4).
- Applying restraints with inaccurate values for R or $\vec{r}_0$. Analyzing unbiased MD simulations of the protein–ligand complex for each potential site is critical for doing this properly.
- Beginning AFEP using a configuration generated by docking without performing sufficient equilibration. In the case of sevoflurane docked to a lower site in the TMD (Fig. 6A), FEP run resulted in an affinity in the 0.1 *M* range, while AFEP runs with well-equilibrated conformations (Fig. 6B) yielded affinities in the 20–200 m*M* range, on the order of EC_{50}.

5. SUMMARY

We have presented a step-by-step approach to identifying candidate binding sites and quantifying and ranking affinities of volatile and injected GAs with pLGICs. While automated docking calculations do have limited usefulness, the realistic parameterization and explicit water and lipids used in molecular dynamics simulation is particularly important for characterizing interactions of GAs with binding sites in pLGICs. For identifying multiple candidate sites, flooding simulations provide a useful balance between classical simulation that is straightforward to interpret and conceptualize in analogy with experiments and a realistic environment and set of interactions. Alchemical free energy perturbation calculations can provide high-accuracy estimates for affinities if carried out carefully, but require skill and attention to convergence.

REFERENCES

Beveridge, D. L., & DiCapua, F. M. (1989). Free energy via molecular simulation: Applications to chemical and biomolecular systems. *Annual Review of Biophysics and Biophysical Chemistry*, *18*(1), 431–492. https://doi.org/10.1146/annurev.bb.18.060189.002243.

Brannigan, G., LeBard, D. N., Hénin, J., Eckenhoff, R. G., & Klein, M. L. (2010). Multiple binding sites for the general anesthetic isoflurane identified in the nicotinic acetylcholine receptor transmembrane domain. *Proceedings of the National Academy of Sciences of the United States of America*, *107*(32), 14122–14127. https://doi.org/10.1073/pnas.1008534107. http://www.pnas.org/content/early/2010/07/20/1008534107.

Brooks, B. R., Brooks, C. L., Mackerell, A. D., Nilsson, L., Petrella, R. J., Roux, B., … Karplus, M. (2009). CHARMM: The biomolecular simulation program. *Journal of Computational Chemistry*, *30*(10), 1545–1614. https://doi.org/10.1002/jcc.21287.

Chen, Q., Kinde, M. N., Arjunan, P., Wells, M. M., Cohen, A. E., Xu, Y., & Tang, P. (2015). Direct pore binding as a mechanism for isoflurane inhibition of the pentameric ligand-gated ion channel ELIC. *Scientific Reports*, *5*(1), 13833. https://doi.org/10.1038/srep13833. http://www.nature.com/articles/srep13833.

Chiara, D. C., Jounaidi, Y., Zhou, X., Savechenkov, P. Y., Bruzik, K. S., Miller, K. W., & Cohen, J. B. (2016). General anesthetic binding sites in human $\alpha 4\beta 3\delta\gamma$-aminobutyric acid type-A receptors ($GABA_AR$). *Journal of Biological Chemistry*, *291*, 26529–26539. https://doi.org/10.1074/jbc.M116.753335.

Git, V., Bernardi, R., Bhandarkar, M., Bhatele, A., Bohm, E., Brunner, R., … Zhu, F. (2017). *NAMD user's guide*. http://www.ks.uiuc.edu/Research/namd/2.12/ug/.

Humphrey, W., Dalke, A., & Schulten, K. (1996). VMD: Visual molecular dynamics. *Journal of Molecular Graphics*, *14*(1), 33–38. 27–38. http://www.ncbi.nlm.nih.gov/pubmed/8744570.

Jo, S., Lim, J. B., Klauda, J. B., & Im, W. (2009). CHARMM-GUI Membrane Builder for mixed bilayers and its application to yeast membranes. *Biophysical Journal*, *97*(1), 50–58. https://doi.org/10.1016/j.bpj.2009.04.013.

Jurd, R., Arras, M., Lambert, S., Drexler, B., Siegwart, R., Crestani, F., … Rudolph, U. (2002). General anesthetic actions in vivo strongly attenuated by a point mutation in the $GABA_A$ receptor $\beta 3$ subunit. *The FASEB Journal*, *17*(2), 250–252. https://doi.org/10.1096/fj.02-0611fje. http://www.ncbi.nlm.nih.gov/pubmed/12475885.

LeBard, D. N., Hénin, J., Eckenhoff, R. G., Klein, M. L., & Brannigan, G. (2012). General anesthetics predicted to block the GLIC pore with micromolar affinity. *PLoS Computational Biology*, *8*(5), e1002532. https://doi.org/10.1371/journal.pcbi.1002532.

Mark, A. E. (2002). Free energy perturbation calculations. In *Encyclopedia of computational chemistry*. Chichester, UK: John Wiley & Sons, Ltd. https://doi.org/10.1002/0470845015.cfa010.

Murail, S., Wallner, B., Trudell, J. R., Bertaccini, E., & Lindahl, E. (2011). Microsecond simulations indicate that ethanol binds between subunits and could stabilize an open-state model of a glycine receptor. *Biophysical Journal*, *100*(7), 1642–1650. https://doi.org/10.1016/j.bpj.2011.02.032.

Phillips, J. C., Braun, R., Wang, W., Gumbart, J., Tajkhorshid, E., Villa, E., … Schulten, K. (2005). Scalable molecular dynamics with NAMD. *Journal of Computational Chemistry*, *26*(16), 1781–1802. https://doi.org/10.1002/jcc.20289. http://www.pubmedcentral.nih.gov/articlerender.fcgi?artid=2486339&tool=pmcentrez&rendertype=abstract.

Pohorille, A., Jarzynski, C., & Chipot, C. (2010). Good practices in free-energy calculations. *The Journal of Physical Chemistry B*, *114*(32), 10235–10253. https://doi.org/10.1021/jp102971x.

Raju, S. G., Barber, A. F., LeBard, D. N., Klein, M. L., & Carnevale, V. (2013). Exploring volatile general anesthetic binding to a closed membrane-bound bacterial voltage-gated sodium channel via computation. *PLoS Computational Biology*, *9*(6), e1003090. https://doi.org/10.1371/journal.pcbi.1003090.

Stern, A. T., & Forman, S. A. (2016). A cysteine substitution probes β3h267 interactions with propofol and other potent anesthetics in $\alpha 1\beta 3\gamma 2l$ γ-aminobutyric acid type A receptors. *Anesthesiology*, *124*(1), 89–100. https://doi.org/10.1097/ALN.0000000000000934. http://www.pubmedcentral.nih.gov/articlerender.fcgi?artid=PMC4681607. http://www.ncbi.nlm.nih.gov/pubmed/26569173.

Stewart, D. S., Pierce, D. W., Hotta, M., Stern, A. T., & Forman, S. A. (2014). Mutations at beta N265 in γ-aminobutyric acid type A receptors alter both binding affinity and efficacy of potent anesthetics. *PLoS ONE*, *9*(10), e111470. https://doi.org/10.1371/journal.pone.0111470.

Trott, O., & Olson, A. J. (2010). AutoDock Vina: Improving the speed and accuracy of docking with a new scoring function, efficient optimization, and multithreading. *The Journal of Computational Chemistry*, *31*(2), 455–461. https://doi.org/10.1002/jcc.21334.

Woll, K. A., Murlidaran, S., Pinch, B. J., Henin, J., Wang, X., Salari, R., … Eckenhoff, R. G. (2016). A novel bifunctional alkylphenol anesthetic allows characterization of GABAA receptor subunit binding selectivity in synaptosomes. *The Journal of Biological Chemistry*, *291*(39), 20473–20486. https://doi.org/10.1074/jbc.M116.736975.

Zwanzig, R. W. (1954). High-temperature equation of state by a perturbation method. I. Nonpolar gases. *The Journal of Chemical Physics*, *22*(8), 1420–1426. https://doi.org/10.1063/1.1740409.

CHAPTER TWO

Computational Approaches to Studying Voltage-Gated Ion Channel Modulation by General Anesthetics

Eleonora Gianti, Vincenzo Carnevale[1]

Institute for Computational Molecular Science, College of Science and Technology, Temple University, Philadelphia, PA, United States

[1]Corresponding author: e-mail address: vincenzo.carnevale@temple.edu

Contents

Abstract

Voltage-gated ion channels (VGICs) are responsible for the propagation of electrical signals in excitable cells. Small-molecule modulation of VGICs affects transmission of action potentials in neurons and thus can modulate the activity of the central nervous system. For this reason, VGICs are considered key players in the medically induced state of general anesthesia. Consistently, VGICs have been shown to respond to several general anesthetics. However, in spite of extensive electrophysiological characterizations, modulation of VGICs by anesthetics is still only partially understood. Among the challenging aspects are the presence of multiple binding sites and the observation of paradoxical effects, i.e., evidence, for the same channel, of inhibition and potentiation. In this

Methods in Enzymology, Volume 602
ISSN 0076-6879
https://doi.org/10.1016/bs.mie.2018.01.002

context, molecular simulations emerged in the recent past as the tool of choice to complement electrophysiology studies with a microscopic picture of binding and allosteric regulation. In this chapter, we describe the most effective computational techniques to study VGIC modulation by general anesthetics. We start by reviewing the VGIC conduction cycle, the corresponding set of channel conformations, and the approaches used to model them. We then review the most successful strategies to identify binding sites and estimate binding affinities.

1. VOLTAGE-GATED ION CHANNELS

1.1 Structure and Function

Voltage-gated ion channels (VGICs) are pore-forming membrane proteins responsible for electrical signaling in cells. As such, VGICs are key players in many electrically mediated cellular functions, including neuronal signaling, muscle cell contraction, cell secretion, hormone regulation, etc. (Miller, 1992).

As electrochemical gradients across cell membranes vary in response to a variety of stimuli, VGICs transition through different states in their activation cycle (also known as *gating* cycle), progressing from the resting (closed) to the activated (open) states, before becoming inactivated. Opening of transmembrane (TM) pores of VGICs results in the selective passage of ions (such as Na^+, Ca^{2+}, and K^+) through otherwise impermeable cell membranes, upon the opening of one or several structural elements in the channel pore known as *gates* (i.e., *gating* process).

VGICs constitute a large superfamily (~150 members) of structurally related ion channels (Yu & Catterall, 2004) including channels selective for Na^+ (Nav), Ca^{2+} (Cav), and K^+ (Kv), in addition to the more promiscuous cation-selective channels (HCN) (Stock, Souza, & Treptow, 2013). Less restrictive classifications also list calcium-activated potassium channels, cyclic nucleotide-modulated ion channels, transient receptor potential channels, inward-rectifying potassium channels, and two-pore potassium channels among the superfamily members (Yu & Catterall, 2004).

All VGICs share a common architecture characterized by a homotetrameric (for prokaryotic channels) or pseudotetrameric (for eukaryotic channels) assembly, as shown with the structure of mammalian Nav (Fig. 1). In eukaryotic VGICs, an α subunit, a large polypeptide chain of approximately 2000 amino acids, arranges into four adjacent domains (DI–DIV), surrounding a central, membrane-spanning channel pore.

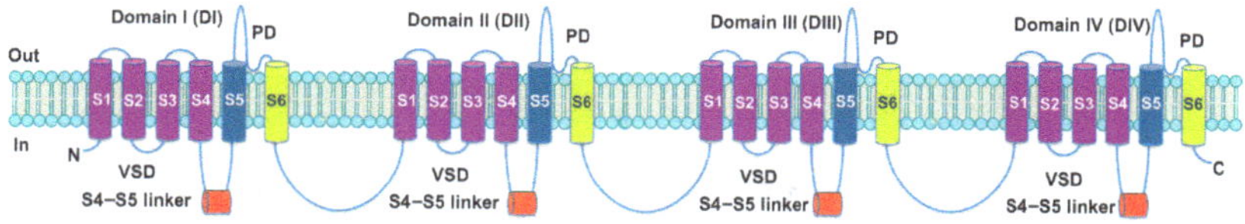

Fig. 1 Nav channel architecture. Topology of the human Nav channel α subunits. These proteins consist of four homologous domains DI, DII, DIII, and DIV connected by intracellular linkers. Each domain contains six transmembrane helix segments (S1 through S6). Segments S1–S4 form the voltage-sensing domain (VSD). A movement of S4 leads to channel opening in response to membrane depolarization. Segments S5, S6 and the connecting loop form the pore domain (PD).

Each domain is constituted by six TM helical segments (S1–S6) in addition to a loop connecting S5 to S6. Of these segments, S1–S4 form the voltage-sensor domain (VSD), a regulatory element that "senses" the changes in the TM potential. Instead, S5 and S6, along with the extracellular pore loops (P-loops), contribute to forming the channel pore (or pore domain, PD) and the selectivity filter (SF). While the SF is responsible for ion selectivity, it is across the PD that ion conduction pathways are established. In addition to the α subunit, one or more β subunits can associate to the structure of an ion channel and exert important functions related to the channel modulation (Brackenbury & Isom, 2011).

1.2 Conduction Cycle and Functional States

The conduction cycle (or activation cycle) of a VGIC is regulated by changes in the membrane potential (or transmembrane voltage, ΔV) of a cell. Progressing through the conduction cycle correlates with conformational changes of the channel structure, resulting in distinct functional states.

The process of activation of a VGIC is triggered by changes in the transmembrane voltage (depolarization by positive potentials), perceived by the channel through its four VSDs (Figs. 1 and 2). These structural elements are constituted by TM helices S1–S4, of which S4 contains a number (3–7) of positively charged residues (generally arginine residues engaged in interactions with residues of opposite charge) that are sensitive to the changes in the surrounding electric field (Vargas et al., 2012). Thus, in response to membrane depolarization, the salt bridge pattern of resting VSDs rearranges quite dramatically, while S4 translates along the normal to the lipid bilayer transitioning from inward (facing the cytoplasmic side of the cell membrane) to outward (facing the extracellular side) configurations.

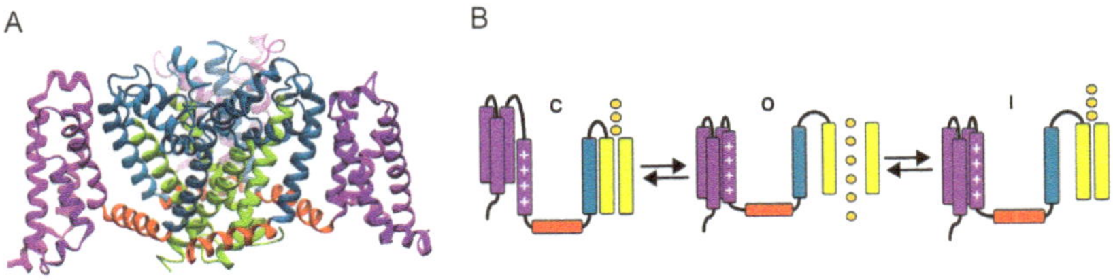

Fig. 2 Cartoon representation of the structure and mechanism of voltage-gated sodium channels. The first four transmembrane (TM) helices (S1–S4 from each subunit) constitute the voltage-sensor domain (*purple*). The pore domain, instead, comprises two TM helices, S5 and S6 (*blue* and *yellow*, respectively). The pore and voltage-sensor domains are connected by the linker domain (*red*). (B) The gating mechanism of VGSCs requires a coordinated motion of the voltage-sensor and pore domains. Under depolarizing conditions, the positively charged S4 helix translates along the bilayer normal causing the splaying of the S6 helix bundle. The channel thus transitions from the closed (C) to the open state (O). The latter is unstable and decays to the inactivated state (I), a nonconductive state characterized by an activated voltage-sensor domain and a closed pore domain. *Adapted from Carnevale, V. & Klein, M.L. (2017). Small molecule modulation of voltage gated sodium channels. Current Opinion in Structural Biology, 43, 156–162, https://doi.org/10.1016/j.sbi.2017.02.002.*

In doing so, S4 visits five distinct transition states (Fig. 3), which were identified either from experimental structures or by computational studies (Delemotte, Kasimova, Klein, Tarek, & Carnevale, 2015; Delemotte, Tarek, Klein, Amaral, & Treptow, 2011; Tarek & Delemotte, 2013). Despite several groups having proposed distinct physical models of the molecular mechanisms of VSD activation (Bezanilla, 2000; Catterall, 1986; Guy & Seetharamulu, 1986; Jiang, Ruta, Chen, Lee, & MacKinnon, 2003; Tiwari-Woodruff, Lin, Schulteis, & Papazian, 2000), they all support the idea that rearrangement of the salt bridge patterns in the VSD is a key to achieve activation of a VGIC (Catterall, 2010; DeCaen, Yarov-Yarovoy, Sharp, Scheuer, & Catterall, 2009; Yarov-Yarovoy et al., 2012).

As a result of outward movements of VSDs, VGICs also produce so-called *gating currents*—i.e., transient currents of small entities (~dozens of elementary charges) (Armstrong & Bezanilla, 1977). Accordingly, an estimate of the amount of charge transported by a channel during activation is provided by the *gating charge*, obtained by integrating the related gating current (Delemotte et al., 2015, 2011; Khalili-Araghi et al., 2010; Tarek & Delemotte, 2013). Finally, cooperative motion of S4 segments in the four VSDs of a channel induces the S4–S5 linkers to adopt a "tense" configuration that, in turn, weakens the linker S6 interactions, as shown by Jensen

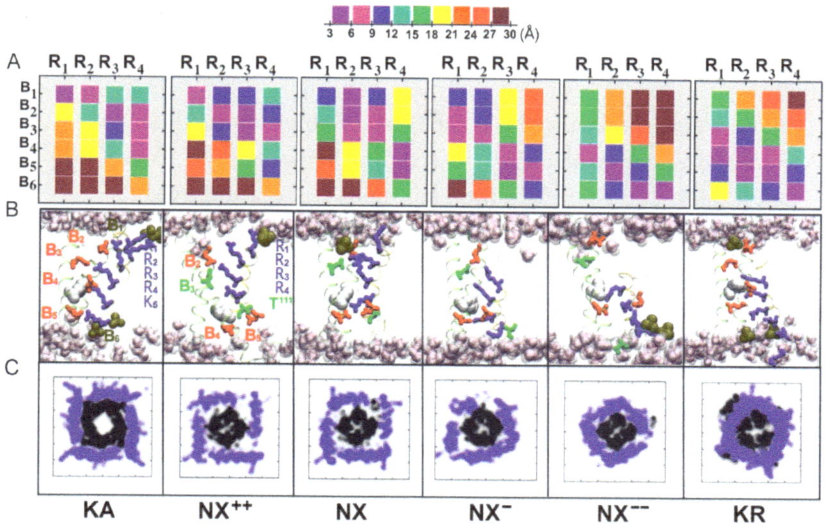

Fig. 3 Voltage-sensor conformations of NavAb. (A) Distance matrices M_{ij} (NavAb, VSD) mapping the domain electrostatic interactions between the S4 basic residues (numbered R_1–R4) and their binding sites [numbered B_1–B_6 for (*upper*) PO_4^-, E^{32}, N^{49}, E^{59}, and D^{80} and (*lower*) PO_4^-, respectively] in each of the VSD conformations NX^{++}, NX, NX^-, and NX^{--}. For the purpose of comparison, the reference, distance matrices M_{ij} (Kv1.2, VSD) for Kv1.2 in the KA and KR conformations are also depicted; Kv1.2 binding sites are numbered B_1–B_6 for (*upper*) PO_4^-, E^{183}, E^{226}, E^{236}, and D^{259} and (*lower*) PO_4^-, respectively. Distance matrices are computed by averaging over the four independent subunits. (B) Snapshots of the VSD for the NavAb and Kv1.2 conformations, highlighting the position of the S4 basic residues (*blue sticks*) and the salt bridges/hydrogen bonds they form with the acidic/polar residues (*red/green sticks*) of other VSD segments or with lipid phosphate moieties (*ocher*). Note that the NavAb "catalytic center," formed by B_4, B_5, and F^{56} (*white*; F^{233} in Kv1.2), is occupied by the S4 residues R^2, R^3, R^4, and T^{111}. (C) Atomic density plots for the selected NavAb and Kv1.2 conformations, depicting the positions of the S4–S5 linkers (*blue*) and S6 helices (*black*) at the membrane plane placed at the intracellular entrance of the channel. The S4–S5 linker constrains the bundle-crossing region of S6 in the conformations NX^- and NX^{--}, thereby prohibiting pore opening in these structures. *Adapted from Amaral, C., Carnevale, V., Klein, M.L., & Treptow, W. (2012). Exploring conformational states of the bacterial voltage-gated sodium channel NavAb via molecular dynamics simulations. Proceedings of the National Academy of Sciences of the United States of America, 109, 21336–21341, https://doi.org/10.1073/pnas.1218087109.*

et al. in their study of activation/deactivation pathways of VGICs (Jensen et al., 2012). Perturbation of these contacts induces opening of the pore gates and allows hydration of the pore domain, a prerequisite to achieve channel activation and ion transport (Amaral, Carnevale, Klein, & Treptow, 2012).

In summary, while progressing through the conduction cycle, VGICs undergo major state transitions and therefore can adopt the following functional states:

1. *Resting state* (the channel is closed): At resting potentials (hyperpolarized membrane) the probability of a VGIC being open is low. The pore is dehydrated and therefore the channel is nonconductive.
2. *Activated state* (the channel is open): Membrane depolarization activates a VGIC by leading to an open-channel conformation that induces full hydration of the pore lumen and allows movement of ions along biological membranes across the electrochemical gradient. From a thermodynamic standpoint, the activated state is quite unstable and therefore decays to the more stable inactivated state.
3. *Inactivated state* (the channel is nonconductive). The VSDs of inactivated channels are activated (S4 is in the outward configuration); however, the pore is closed. Inactivation is a very important process to maintain the electrical activity of excitable cells within physiological ranges, and it is fundamental to allow the action potential to be generated and transmitted. Indeed, inactivation is regulated through a positive feedback: in the case of Na^+ channels, for example, membrane depolarization causes the channel to open allowing Na^+ ions influx which, in turn, further depolarizes the membrane. Inactivation regulates this process by preventing a harmful depletion of the transmembrane potential. While activation is thought to occur through a common mechanism to all VGICs (Kasimova, Granata, & Carnevale, 2016; Palovcak, Delemotte, Klein, & Carnevale, 2014), it has been postulated that inactivation is achieved through subfamily-specific mechanisms. In Na^+ channels, fast inactivation occurs within the time frame of milliseconds and is followed by slow inactivation that takes seconds to minutes to complete. When the inactivation process is completed, the pore domain becomes closed and fully impermeable to ions (Ahern, 2013; Armstrong, 1973; Armstrong & Bezanilla, 1977; Bezanilla, 1977).

1.3 Modulation by General Anesthetics

General anesthesia refers to the medical practice of administering one or more general anesthetic agents that induce reversible loss of consciousness accompanied by immobility, unresponsiveness, loss of reflective functions and, most importantly, of pain sensation. Hence, for nearly two centuries, anesthesia has allowed practicing otherwise intolerable—and yet lifesaving—surgical procedures. Despite being a long-standing practice, general anesthesia remains far from being a fully appreciated medical topic, as the molecular mechanisms underpinning general anesthesia still remain to be understood.

Current general anesthetics are drugs that can be used to induce anesthesia. Structurally, they are all quite hydrophobic, small in size (often fragment-like compounds), and invariantly uncharged. The absence of common substructures and molecular recognition features makes it practically pointless using classical structure–activity relationship (SAR) approaches to study general anesthetics.

Another layer of complexity arises when searching for the physiological targets that general anesthetics might bind to—i.e., general anesthetic targets. The most important effects of general anesthesia involve excitability of the central nervous system (CNS); on that account, ion channels emerge as legitimate anesthetic targets (Hemmings et al., 2005; Krasowski & Harrison, 1999; Miller, 2002).

It is well known that general anesthetics influence the activity and the function of many ion channels (Franks, 2006, 2008; Herold & Hemmings, 2012; Kopp Lugli, Yost, & Kindler, 2009; Lioudyno et al., 2013; Woll et al., 2017). In particular, gamma-aminobutyric acid (GABA) and *N*-methyl-D-aspartate receptors that mediate synaptic transmission became the first established anesthetic targets. More recently, a large body of evidence contributed to validating VGICs as additional physiological targets of general anesthesia (Covarrubias, Barber, Carnevale, Treptow, & Eckenhoff, 2015). Remarkably, VGICs are found in many excitable tissues, including the brain where they participate in regulating the membrane potential and action potential—all actions that are clearly relevant to general anesthetic pathways.

Electrophysiological experiments indicate that halogenated inhalational anesthetics, one of the largest classes of general anesthetic drugs (including desflurane, halothane, isoflurane, etc.), modulate several VGIC subfamilies, spanning both mammalian and prokaryotic Kv and Nav channels (Covarrubias et al., 2015). It is, however, not so clear how such modulation effects are achieved on a case-by-case basis and many questions remain unanswered.

For example, it is known that most halogenated inhalational anesthetics bind and inhibit K-Shaw2 channel by stabilizing its closed state (Barber, Liang, & Covarrubias, 2012; Harris, Shahidullah, Ellingson, & Covarrubias, 2000). Confusingly, sevoflurane, another halogenated inhalational anesthetics, is shown to potentiate the channel's conductance. This positive modulation effect seems to be associated with distinct mechanisms, involving two different sets of binding sites (one localized in the S4–S5 linker region and the other presumably located across the pore). As for Nav channels, independent studies show that halogenated inhalational anesthetics promote inactivation by stabilizing the inactivated channel (Barber, Carnevale, Klein, Eckenhoff, & Covarrubias, 2014; Duch, Rehberg, & Vysotskaya, 1998; Herold, Nau, Ouyang, & Hemmings, 2009; OuYang & Hemmings, 2007; Ouyang, Herold, & Hemmings, 2009; Ouyang, Jih, Zhang, Correa, & Hemmings, 2007). All things considered, a complex picture emerges where general anesthetics may act through multiple molecular pathways and bind to multiple binding sites on ion channels (Barber et al., 2014); accordingly, multiple classes of ion channels can serve as general anesthetic targets.

More generally, VGICs constitute valuable targets for other biomedical applications. Many drugs currently in the market act on one or more channels of the subfamily: local anesthetics, anticonvulsants, and antiarrhythmic drugs are among them. In addition, genetic mutations are known that impair the function of VGICs dramatically, causing severe diseases in humans, collectively known as *channelopathies*. For example, mutations in the genes encoding for subtypes of Nav have been recently shown to affect the structure and biophysical properties of these channels leading to conditions, such as epilepsy, cardiac arrhythmia, and chronic pain (de Lera Ruiz & Kraus, 2015).

Current approaches to ion channel drug discovery need to face a number of challenges, including the absence of nontraditional binding pockets typical of soluble proteins and enzymes (e.g., the protein kinase binding pocket), limited or absent surface for anchoring small-molecule ligands, explicit inclusion of water and/or lipids to the binding environment, multiple conformational states involved in binding events, the presence of multiple binding events that might cooperatively modulate the channel's function, allosteric regulation, etc. Recent advances in computational approaches together with unprecedented progress in the field of high-performance computing have enabled the use of computer simulation to study the structure and function of VGICs, thereby offering a way to overcome experimental challenges and limitations (Allen, Andersen, & Roux, 2006; Allen, Kuyucak, & Chung, 1999; Bernèche & Roux, 2001; Bernsel et al., 2008; Chen, Buyan, & Corry, 2017; Corry, 2016; Domene, 2016; Domene & Furini, 2009; Lindahl & Sansom, 2008; Medovoy, Perozo, & Roux, 2016; Olsen et al., 2014; Roux, Allen, Bernche, & Im, 2004; Treptow & Klein, 2012; Vargas et al., 2012; Vemparala, Domene, & Klein, 2010).

2. STRUCTURAL APPROACHES TO STUDYING FUNCTIONAL STATES

Recent advances in experimental approaches to structure determination have contributed to the field a growing number of high-quality structures of ion channels in one or more gating states, employing X-ray crystallography, nuclear magnetic resonance (NMR), and cryoelectron microscopy (Carnevale & Klein, 2017; Catterall, 2014; Cournia et al., 2015; de Lera Ruiz & Kraus, 2015).

Nonetheless, given the large number of transition states and intermediate substates associated with the process of gating, direct modeling of any ion channel should account for all its relevant conformations. This observation is particularly relevant when studying modulation of VGICs by general anesthetics, as the hypothesis that the process involves multiple binding events and distinct mechanisms has received recently received experimental support (Barber et al., 2014; Covarrubias et al., 2015; Kinde et al., 2016). When the full spectrum of gating states is not entirely covered by the appropriate conformational transition, molecular modeling techniques offer valid solutions to structure determination.

When sequence and structure information of closely related homologs (so-called molecular templates) is available, homology-modeling techniques are a preferable choice. To this end, computer programs like Modeller (Eswar et al., 2006), Prime (Jacobson et al., 2004), Catalyst (http://accelrys.com/), and Rosetta (Uchime et al., 2012) are generally used to generate initial structure models that can be further refined by molecular dynamics (MD) simulation. This combined approach is also a popular option for experimental structure refinement, whenever poor resolution of local density maps prevents accurate structure determination. Additionally, it is almost routinely applied to generate solutions for building unresolved loops and/or refining the orientation of single amino acid side chains. Conversely, when no suitable templates are available (sequence similarity <30%), ab initio or de novo protein structure predictions represent viable alternatives (Floudas, Fung, McAllister, Mönnigmann, & Rajgaria, 2006).

2.1 Generating Structural Models of Functional States

In an effort to understand the molecular determinants responsible for binding events of ligands to the pore of ion channels—i.e., pore blockers—Gianti et al. used the human Hv1 proton channel, a pH sensor involved in many biological functions and a channel with an exquisitely simple architecture, as a prototype to study structurally more intricate VGICs from both biophysical and drug discovery perspectives (Gianti, Delemotte, Klein, & Carnevale, 2016).

Unlike most ion channels, Hv1 channel is constituted by a dimer of homotetrameric structures, each functioning as both the VSD and pore domain. Hence, outward movements of the S4 segment trigger Hv1 activation. Once achieved, the pore gates open and proton currents can be established through the channel lumen. Starting from the only available experimental structure of Hv1 (Takeshita et al., 2014), solved in a closed/intermediate state, Gianti et al. used homology modeling followed by extensive MD simulations to explore all conformations relevant to the Hv1 gating process (namely, the resting, the intermediate, and the fully activated states). Careful structural analyses of the generated states revealed that only the fully activated state retains the features to allow proton conduction. The authors were also able to identify a number of structural substates associated with each primary state, a result that well correlates with an exceptional VSD flexibility observed for Hv1 by experimental evidence (Li et al., 2015). A fully equilibrated structure of the activated Hv1 channel was then used in a structure-based drug design implementation that revealed binding modes of 2-guanidinobenzimidazole (2GBI) derivatives (Hong, Kim, & Tombola, 2014), the only chemical class of Hv1 pore blockers known to date. Furthermore, given the lack of experimental structures of the activated Hv1, direct modeling of the fully activated state, obtained by combining classical and QM/MM molecular dynamics simulations, was instrumental to shed light on the mechanism of proton transport in this very peculiar channel (van Keulen et al., 2016). A number of convincing examples can be mentioned that highlight the significance of using computational approaches to generate multiple conformational states for direct modeling of ion channels in general and, in particular, VGICs (Amaral et al., 2012; Raju, Barber, LeBard, Klein, & Carnevale, 2013; Shen et al., 2015; Treptow & Klein, 2012; Vemparala et al., 2010).

3. DETECTION OF BINDING SITES AND ACCESS PATHWAYS

3.1 "Coarse" Molecular Docking Approaches

Molecular docking is a very popular computer-aided drug design method for predicting preferred binding orientations or poses—the binding modes—of one molecule (typically a small-molecule ligand) to another molecule (a macromolecule, generally a protein or a nucleic acid), in a process mimicking the formation of a stable complex. Docking is a two-stage procedure: first is the placement step, where a docking search algorithm is used to generate putative binding modes. This is generally achieved by searching for all the alternative orientations of ligand conformers that can "fit" one or more receptor-binding pockets. Second, a scoring function (mathematical equations that provide the likelihood for a ligand structure to bind to a receptor conformation) is generally used to rank all the generated poses according to the predicted strength of their noncovalent interactions, i.e., binding affinity. A number of software tools for molecular docking are currently available, some of which are commercially available, while others are free to the general public or restricted to academic users (Claußen, Buning, Rarey, & Lengauer, 2001; Friesner et al., 2004, 2006; Halgren et al., 2004; Jones, Willett, Glen, Leach, & Taylor, 1997; Morris et al., 2009).

Most docking programs account for full conformational flexibility of ligands within the docking sites, while treating the protein receptor as a rigid body. Clearly, this is a preferable choice for virtual screening (VS) applications, requiring *fast* methods to evaluate up to several million compounds for their ability to bind to biological targets of interest. In this case, VS protocols that combine ligand-based approaches (a fast, although less accurate, step to prefilter *hits* from very large databases of compounds) with receptor-based techniques (a more accurate step to evaluate selected hits by using docking and/or molecular dynamics free energy perturbation (FEP) calculations) can be very successful (Dixon, Smondyrev, Knoll, et al., 2006; Dixon, Smondyrev, & Rao, 2006; Grant, Gallardo, & Pickup, 1996; Rush, Grant, Mosyak, & Nicholls, 2005; Zauhar, Gianti, & Welsh, 2013; Zauhar, Moyna, Tian, Li, & Welsh, 2003).

Nonetheless, when fine-tuning the binding affinity of a ligand is the priority, the rigid receptor approximation becomes a major drawback in the pursuit of accuracy and reliability of the proposed binding modes. In this case, approaches that explicitly take into account receptor flexibility, even at the expense of calculation speed, seem to offer an obvious solution to the dilemma.

A variety of approaches are available that incorporate partial or full flexibility of the receptor in molecular docking applications (Armen, Chen, & Brooks, 2009; Bottegoni, Kufareva, Totrov, & Abagyan, 2009; Gagnon, Law, & Brooks, 2016; Huang & Zou, 2006; Ravindranath,

Forli, Goodsell, Olson, & Sanner, 2015; Totrov & Abagyan, 2008). An elegant solution to the problem is offered by the induced fit docking (IFD) approach, a multistep procedure that incorporates partial receptor flexibility (Sherman, Day, Jacobson, Friesner, & Farid, 2006). In IFD, one or more amino acid side chains in the binding pocket are mutated to alanine residues to maximize the space available before the ligand placement. Once a number of binding modes is generated, only the top scoring poses are accepted and submitted to a protein refinement step. Then, the mutated side chains are returned to their original amino acid type before a minimization step is performed that removes steric clashes and optimizes newly generated binding complexes. One last step of docking calculation is performed that applies a more accurate scoring function.

General anesthetics are very small in size and very hydrophobic, both features that tend to associate with a large number of putative binding orientations to be explored at any given binding site. For this reason, and because of the intrinsic structural flexibility of VGICs, IFD may offer a practical solution by introducing *local* structural flexibility (limited to the putative binding sites along with shells of immediate residues) without exponentially increasing the time allotted for calculations. Examples of successful use of the IFD for structure-based drug design either alone or in combination with molecular dynamics simulations are available for both globular proteins (enzymes, transcription factors, etc.) and membrane-bound proteins (the Hv1 proton channel) (Clark et al., 2016; Farid, Day, Friesner, & Pearlstein, 2006; Gianti, Delemotte, et al., 2016; Gianti & Zauhar, 2012).

For all proteins in general and for VGICs in particular, given the conformational rearrangements that occur during conduction, receptor flexibility certainly makes a formidable challenge (Allen, Andersen, & Roux, 2004). Hence, full receptor flexibility becomes a preferable choice in docking implementations. This additional level of sophistication arises from the use of the ensemble docking of multiple receptor conformations, also known as *ensemble docking*. In this docking protocol, multiple receptor conformations (experimentally and/or computationally derived) are selected that maximize conformational heterogeneity. Then, the same set of ligands is docked in parallel against each individual receptor. Whether the receptor structures are experimentally determined or computationally derived, ensemble docking has proven to significantly improve the accuracy and the predictive power of docking calculations (Huang & Zou, 2006; Lam, Abagyan, & Totrov, 2018; Lorber & Shoichet, 2008; Totrov & Abagyan, 2008).

Other fruitful strategies to improve the overall quality of docking results include ranking the binding modes according to a consensus of multiple scoring functions or by combining the original docking score with binding energy predictions from molecular mechanics–generalized Born surface area (MMGBSA) (Kollman et al., 2000; Rastelli, Rio, Degliesposti, & Sgobba, 2010) or MD free energy calculations (Alonso, Bliznyuk, & Gready, 2006; Frembgen-Kesner & Elcock, 2006; Gianti, Messick, Lieberman, & Zauhar, 2016; Gianti & Zauhar, 2015). The latter two methodologies both have the advantage of including the solvent into the calculations, either implicitly (MMGBSA) or explicitly (MD).

Binding site (or pocket) identification and druggability assessment is an important prerequisite in any structure-based drug design implementation. Many computational solutions have been proposed to address this task, ranging from dedicated software tools (Dias, Martins, Nguyen, & Gomes, 2017; Halgren, 2007, 2009; Hall, Kozakov, & Vajda, 2012; Kozakov et al., 2015; Ravindranath & Sanner, 2016; Votapka & Amaro, 2013) to protocols that attempt to solve the problem with docking of fragment probes (i.e., fragment-like compounds selected to "probe" a spectrum of interaction types, e.g., hydrogen bond donor/acceptor properties, hydrophobic features, etc.) (Brenke et al., 2009; Gianti, Messick, et al., 2016; Mattos & Ringe, 1996). Here, the entire surface volume available to the receptor is searched for binding hot spots, i.e., locations where most fragments bind more frequently and tightly. Used in combination with orthogonal approaches for site mapping (e.g., SiteMap, Schrödinger, LLC) the fragment-based docking approach has offered a valid strategy for binding site detection in many proteins, including ion channels (Gianti, Delemotte, et al., 2016; Gianti, Messick, et al., 2016). Automated molecular docking calculations have been largely used to generate initial structural complexes of VGICs (Kv channels) and other ion channels with anesthetic drugs, further refined by MD simulations (Barber, Liang, Amaral, Treptow, & Covarrubias, 2011; Liu, Haddadian, Willenbring, Xu, & Tang, 2010; Liu, Willenbring, Xu, & Tang, 2009; Stock et al., 2013; Willenbring, Liu, Mowrey, Xu, & Tang, 2011).

3.2 Molecular Dynamics Flooding Simulations

Despite being a computationally expensive calculation, unlike "coarse" docking, MD flooding simulations unquestionably represent the most promising techniques recently applied to detecting binding sites and access pathways on ion channels (Brannigan, LeBard, Henin, Eckenhoff, & Klein, 2010; Salari, Murlidaran, & Brannigan, 2014; Vemparala et al., 2010; Wang, Roose, Palovcak, Carnevale, & Dmochowski, 2016) and in particular on VGICs (Barber et al., 2014; Kinde et al., 2016; Raju et al., 2013; Vemparala, Domene, & Klein, 2008).

In flooding simulations, the system (generally consisting of protein, ligands, ions, lipids, and water molecules) is simulated (by all atom MD calculations) in the presence of a high concentration of anesthetic ligand molecules in water ($\sim$200 m*M* initial aqueous concentration). Once the system has reached equilibration, all protein degrees of freedom are unrestrained and ligands are allowed to partition into binding sites on both protein and lipid compartments ($\sim$2 m*M* final aqueous concentration). This way, the search for binding sites is completely "unbiased," as no preliminary information is required to "guide" the search for binding hot spots (Salari et al., 2014). Putative binding sites are then detected by prioritizing regions of higher anesthetic concentration (Raju et al., 2013). The flooding approach has been proven particularly useful in the case of the small inhalational anesthetics (Salari et al., 2014; Vemparala et al., 2008).

MD flooding is certainly a more expensive solution than computational site mapping followed by molecular docking. However, unlike molecular docking, in flooding water molecules are explicitly included and therefore the system solvation is accounted for thoroughly. Another net benefit of flooding is that the system flexibility is fully incorporated into the calculations.

3.2.1 Flooding of General Anesthetics on VGICs

The pioneering work on flooding simulations (Vemparala et al., 2008) investigated the effects of the inhalational anesthetic halothane on both open and closed conformations of the KirBac1.1 channel (a prokaryotic homologue of the eukaryotic inward rectifier potassium (Kir) channel) embedded in a lipid bilayer. By suggesting that anesthetic molecules bind preferentially to open-state channels, potentially impairing the motion associated to the process of gating, this study provided a first attempt to understanding the mechanism of modulation of VGICs by anesthetics.

Corroborating studies provided further support for MD flooding simulations as a preferable approach to identify binding sites for general anesthetic drugs on VGICs. By applying flooding simulations to structural models of all the relevant activation states of NaChBac in a lipid bilayer, the bacterial prototype for eukaryotic Nav channels, it was possible to

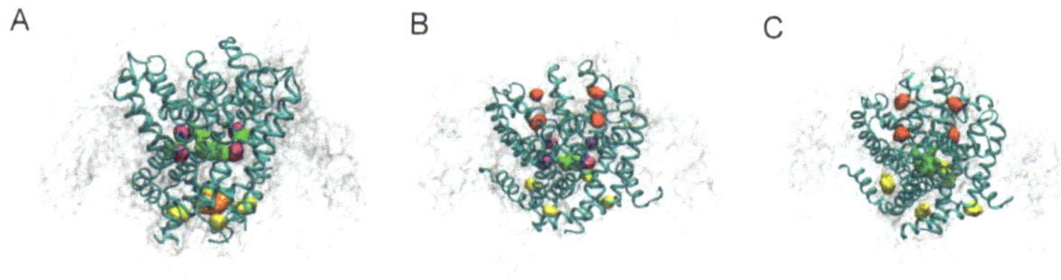

Fig. 4 Putative sevoflurane binding sites in various states of NaChBac: (A) resting/closed, (B) partially activated/closed, and (C) activated/open. The four binding sites identified by clustering analysis are extracellular site (*red*), linker site (*yellow*), activation gate site (*orange*), and cavity/fenestration site (*green/purple*). *Adapted from Barber, A.F., Carnevale, V., Klein, M.L., Eckenhoff, R.G., & Covarrubias, M. (2014). Modulation of a voltage-gated Na + channel by sevoflurane involves multiple sites and distinct mechanisms. Proceedings of the National Academy of Sciences of the United States of America, 111, 6726–6731, https://doi.org/10.1073/pnas.1405768111.*

identify binding sites (Fig. 4) for general inhalational anesthetics, including isoflurane and sevoflurane (Barber et al., 2014; Kinde et al., 2016; Raju et al., 2013). These putative binding spots spanned four distinct locations: (i) the pore domain, likely impeding ion transport; (ii) regions adjacent to the pore, but not on the conduction pathway; (iii) in proximity to the interface between the S4–S5 linker and S6 segment; and (iv) the selectivity filter region. The latter is the only "conformation-selective" site: being only accessible from the extracellular milieu, anesthetic binding is possible only when the channel is activated (the pore domain is open and fully hydrated). The remaining sites are rather "conformation promiscuous." Intriguingly, all binding spots, including the pore site, are indeed accessible even when the pore domain is dewetted and the gates are closed.

Including alternative drug-access pathways—i.e., lateral fenestrations—into the picture provides a rationale to explain the promiscuous binding of general anesthetics to distinct conformations of NaChBac (see Section 3.4 for details), a concept that is likely generalizable to other ion channels.

The anesthetic binding sites mapped on Nav were initially proposed solely on the basis of computational studies guided by MD flooding simulations (Raju et al., 2013).

Remarkably, subsequent work provided experimental evidence confirming and refining the computational predictions of all the binding sites for anesthetic ligands, such as isoflurane and sevoflurane. In a first study, which combined computational studies (MD flooding simulations) with experiments (electrophysiological mutagenesis) and kinetic analyses, it was revealed that the general anesthetic sevoflurane binds to multiple sites on all relevant conformations of NaChBac (Barber et al., 2014). In a second work, site-directed placement of ^{19}F probes in NMR experiments, as guided by MD simulations, allowed for quantification of isoflurane binding to NaChBac (Kinde et al., 2016). Indeed, by saturation transfer difference (STD) NMR spectroscopy it was possible not only to monitor the binding of isoflurane to specific sites of NaChBac but also to rank these sites by their affinity for isoflurane (Fig. 5).

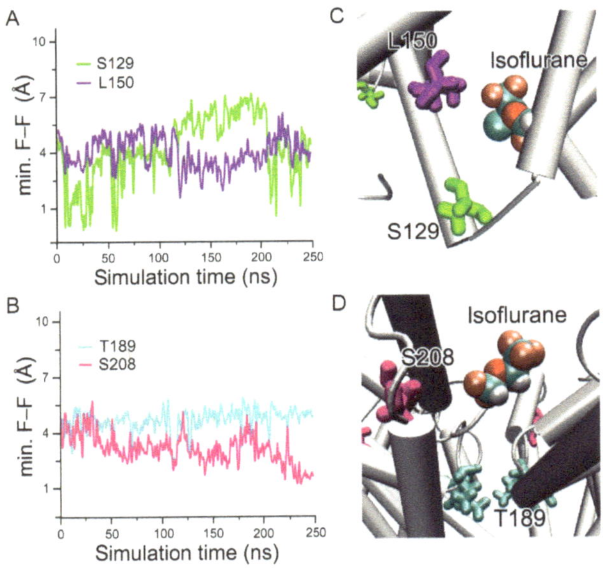

Fig. 5 MD simulations of isoflurane binding to NaChBac. Time evolution of the minimal distance between fluorine of the TET-labeled side chain and the fluorine atom of the nearest isoflurane molecule at (A) the linker sites, S129 (*green*) and L150 (*purple*); (B) the selectivity filter site, T189 (*cyan*); and the extracellular surface site, S208 (*magenta*). Two independent MD simulations were performed and the corresponding time series were evaluated. In both cases, the isoflurane molecules stay close to the side chain of the S129- and S208-labeled positions, in agreement with the maximum STD observed in the NMR experiments. (C and D) Snapshots of the MD simulations showing the proximity of the isoflurane molecule to each of the labeled side chains. The isoflurane molecule is shown in van der Waals representation colored by atom type, and the protein side chain is shown as licorice using the same colors as in (A) and (B). *Adapted from Kinde, M.N., Bondarenko, V., Granata, D., Bu, W., Grasty, K.C., Loll, P.J., et al. (2016). Fluorine-19 NMR and computational quantification of isoflurane binding to the voltage-gated sodium channel NaChBac. Proceedings of the National Academy of Sciences of the United States of America 113, 13762–13767, https://doi.org/10.1073/pnas.1609939113*

The data supported the idea that at least two distinct mechanisms exist for NaChBac modulation by isoflurane: (i) isoflurane can bind to the pore central cavity and block the pore; (ii) isoflurane can bind to and restrict the motion of the S4–S5 linker, thus interfering with S6 motion, a critical step in channel activation. Altogether, these results contributed important understanding of the molecular mechanisms of binding and modulation of Nav channels by general anesthesia.

3.2.2 Flooding of General Anesthetics on Other Ion Channels

Remarkably, flooding simulations have also been applied successfully to study the binding of general anesthetics to a number of ion channels from different subfamilies (other than VGICs) (Vemparala et al., 2010; Wang et al., 2016).

For example, Brannigan et al. mapped multiple binding sites for isoflurane on the nicotinic acetylcholine receptor (nAChR) embedded in lipid bilayers (Brannigan et al., 2010). In their work, the authors show how isoflurane partitions into both the lipid and protein environments. Isoflurane preferentially binds at two locations in the TM domain of nAChR: (i) the pore domain, thus impairing translocation of ions across the lumen; and (ii) three sites at the subunit interfaces, as confirmed by experiments. Additional sites were mapped at centers located on both α subunits, outside the TM domain. In other work by the same group, case studies were presented that discuss the application of flooding simulations for blind identification of anesthetic binding sites on both ion channels and their lipid environments (Salari et al., 2014).

3.3 Binding Mode Hypotheses

Molecular docking offers a valid starting point for exploring putative binding modes of chemical ligands to a binding site on protein receptors and is increasingly used in modern drug discovery approaches. However, as detailed earlier, very high failure rates (up to 70%) are still associated with docking procedures that neglect receptor flexibility (Totrov & Abagyan, 2008). While this approximation may be acceptable for high-throughput VS applications, it is definitely a poor option when it comes to delineating *binding mode hypotheses* for the purpose of understanding and fine-tuning molecular recognition. In the latter case, additional factors must be included in the picture. One and foremost, solvation is an important component of the binding process, and the role of water molecules in ligand binding and stabilization should not be underestimated. This is particularly compelling for ion channels, given that full hydration of the channel lumen is a prerequisite for ion translocation across membrane pores.

Unlike drug binding pockets of soluble proteins, ligand binding sites located in pore domains of ion channels almost invariantly are associated with specific conduction states and show marked hydrophobic character. Characterizing water density fluctuations in pore lumens of ion channels provides a meaningful estimate of hydrophobicity. Large amplitude density fluctuations are the major hallmark of hydrophobicity and are involved in many microscopic wetting/dewetting phenomena (Allen et al., 2004; Baron & McCammon, 2013; Baron, Setny, & Andrew McCammon, 2010; Berne, Weeks, & Zhou, 2009; Chandler, 2005; Hummer, Rasaiah, & Noworyta, 2001; Klebe, 2015; Köfinger, Hummer, & Dellago, 2011; Liu, Huang, Zhou, & Berne, 2005; Mittal & Hummer, 2010; Mondal, Morrone, & Berne, 2013; Ostmeyer, Chakrapani, Pan, Perozo, & Roux, 2013; Pratt & Pohorille, 2002; Rasaiah, Garde, & Hummer, 2008; Vaitheeswaran, Yin, Rasaiah, & Hummer, 2004; Waghe, Rasaiah, & Hummer, 2012; Weingarth et al., 2014; Willard & Chandler, 2008). Accordingly, monitoring solvent fluctuations along ion translocation pathways can be used to reveal the location of binding sites for hydrophobic moieties in the pore domain of ion channels, as reported in the case of the human Hv1 proton channel (Gianti, Delemotte, et al., 2016). In a related work, the same authors discuss how the regions characterized by "overdamped" density fluctuations in the M2 proton channel of the influenza virus contain clusters of waters that can be used to guide drug design (Gianti, Carnevale, DeGrado, Klein, & Fiorin, 2015). In particular, they find that shape similarity between the ligand to be bound and the water molecules to be displaced is crucial to guide hydrophobic scaffold replacement.

3.4 Characterization of Access Pathways

Lateral fenestrations are transient, state-dependent hydrophobic pathways that can assemble at the interface of the channel pore with the adjacent membrane lipids, thus offering alternative access pathways to drug binding. As such, extensive computational efforts have been dedicated to studying lateral fenestration in the context of modulation of VGICs by general anesthetics. In a study dedicated to the modulation of the bacterial NaChBac, by the inhalational general anesthetics isoflurane and sevoflurane, lateral fenestrations were detected at distinct locations and for multiple conformations of the channel. As pointed out earlier, molecular flooding simulations of a computer-generated model of this bacterial channel detected four distinct locations for anesthetic binding. Remarkably, lateral fenestrations associated to sites located at the central pore region and the adjacent interfaces were shown to provide drug access to the channel lumen, even when the central pore adopted closed configurations. Hence, binding events at the fenestration occur solely in closed deactivated channels (Barber et al., 2014; Kinde et al., 2016; Raju et al., 2013).

An independent computational investigation conducted by large-scale simulations of another bacterial Nav channel, NavAb, highlighted the presence of two distinct drug-access pathways to the pore, one of which through lateral fenestrations. These lateral fenestrations were shown to point to a high-affinity pore-blocking site through a minimum free energy pathway (Boiteux et al., 2014).

Remarkably, experimental evidence for lateral fenestrations on the structure of NavAb is also available, clearly showing lipid tails accessing the central lumen of the closed pore (Adler, 2011; Payandeh, Scheuer, Zheng, & Catterall, 2011). Computational studies by Jorgensen et al. (2016) showed that many potassium channel subfamilies share transmembrane fenestrations. By using MD simulations, three potassium channels were investigated to show that these channels all share four main lateral fenestrations of similar nature.

4. PREDICTING BINDING AFFINITY

Several combined computational and experimental investigations revealed that general anesthetics exert an allosteric effect on VGICs, rather than merely occluding the conduction pore (Barber et al., 2014; Kinde et al., 2016; Raju et al., 2013). Allostery describes how certain nonlocal energetic perturbations couple to and alter the equilibrium distribution of conformations that a macromolecule visits. In VGICs, allostery provides a framework for understanding how conformational changes in domains responsible for sensing stimuli couple to channel gate domains that open and close, controlling the flux of permeant ions (Chowdhury & Chanda, 2012). Importantly, the concept of allostery, which subsumes a complex set of molecular mechanisms, is not only useful to characterize the normal functioning of these exquisite nanomachines but also rationalizes their chemical regulation. Pharmacologically relevant ion channel modulators including neurotoxins and endogenous ligands are often allosteric in nature, i.e., they exert their effect at distal locations from the relative binding sites. Thus, their effect can be understood as a reshaping of the free energy landscape, which alters the relative population of the states defining the activation cycle of the channel (Chowdhury & Chanda, 2012; Comitani, Limongelli, & Molteni, 2016; Linder, de Groot, & Stary-Weinzinger, 2013; Tsai & Nussinov, 2014). It is indeed the selective stabilization of specific conformational states that result in the overall modulation of channel's function.

In practice, to quantitatively characterize allosteric effects, one would need a computational method able to predict the binding affinity of a ligand for each distinct configuration of the channel. Although in principle appropriate for this task, molecular dynamics simulations are rarely used to estimate the dissociation constant of a binding reaction. Indeed, a reliable calculation of the latter would require statistically accurate estimates of the rate constants for association and dissociation (k_{on} and k_{off}, respectively) and thus the sampling of several binding and unbinding events. This is often impossible given the limitation intrinsic to even the most powerful current supercomputers, which can determine the time evolution of the simulated system for only tens to hundreds of μs. To overcome these limitations and characterize the equilibrium properties of biomolecular systems, several approaches have been developed. These methods, usually classified under the category of *enhanced sampling techniques*, allow one to glean quantitative insight on the energetics of a process, being it a conformational transition or a binding event. Here we will focus on two of them, namely, FEP and funnel metadynamics, which can be conveniently used to calculate the absolute binding affinity of a small molecule for a protein target.

4.1 FEP Approach

This method is based on a celebrated result from Zwanzig (1954) that relates the free energy difference between two thermodynamics states to an ensemble average easily obtainable via molecular dynamics simulations. The crux of the matter is the exclusive dependence of the free energy estimate on the microstates corresponding to only one of the two thermodynamics states. In principle, one needs to simulate only the bound state (the complex between the channel and the ligand) to obtain its free energy relative to the unbound one (in which the ligand is in solution and at an infinite distance from the target). In practice, for statistical efficiency, the state transformation between the bound and unbound state is partitioned into several intermediate steps through a computational approach called alchemical transformation (Chipot & Pohorille, 2007). Historically, the first calculations of protein–ligand binding affinities were reported by the groups of Jorgensen, McCammon, and Kollman (Bash, Singh, Brown, Langridge, & Kollman, 1987; Jorgensen & Ravimohan, 1985; Kollman, 1993; McCammon, Gelin, & Karplus, 1977; Merz & Kollman, 1989; Wong & McCammon, 1986). Since then, thanks to continuing development of algorithms and computer hardware, FEP has grown into a mature quantitative technique that is now extensively used in pharmaceutical industry to interpret structure–activity relations and to optimize drugs (Wang et al., 2015). Of particular relevance to this context is a recent application of this approach to a series of congeners binding to several GPCRs (Lenselink et al., 2016). In spite of the complexity of the system (GPCRs are large transmembrane proteins), the study showed that accurate affinity predictions could be obtained for different targets in distinct conformations. Most importantly, the method was able to distinguish potent compounds from weak binders.

4.2 Funnel Metadynamics

In contrast to FEP, which uses the nonphysical pathway generated via the alchemical transformation, funnel metadynamics aims at gleaning insight into the physical pathway connecting the bound and the unbound states. In particular, the simulations protocol is appropriately designed so that binding and unbinding events can be frequently sampled, thereby greatly improving the statistical convergence of the calculation. The protocol is based on the metadynamics method, in which a history-dependent potential is added to the regular atom–atom interactions (Laio & Parrinello, 2002). This additional biasing potential discourages the system from visiting repeatedly the same regions of the configurational space. Specifically, by appropriately choosing the biasing potential, one is able to encourage a tightly bound ligand to leave the binding site and explore other accessible conformational states (and thus explore also the unbound state). Crucial for this method is the fact that, asymptotically, the constantly changing biasing potential compensates exactly for the underlying free energy surface. Thus, the free energy surface can be easily calculated from the history-dependent potential at the end of the simulation run. By promoting dissociation events, metadynamics greatly simplifies the problem of estimating reliably k_{off}. However, this approach does not prove useful in promoting the sampling of binding events. Indeed, geometric considerations suggest that the thermodynamic state corresponding to the ligand in solution is constituted by an exceedingly large number of microscopic states. Under these circumstances the metadynamics biasing potential cannot enforce the exploration of the bound state and therefore the estimation of k_{on} remains unreliable, if at all possible. To overcome this limitation, an important variant of this protocol has been recently developed, the so-called funnel metadynamics (Limongelli, Bonomi, & Parrinello, 2013). In this approach, the region of space that can be explored by the ligand in the unbound state is greatly reduced by an appropriate restraining potential, which prevents the ligand from diffusing away from the protein. The effect of this restraining potential on the overall energetics of the system can be analytically estimated and thus removed from the calculation. This approach has been successfully applied to the binding of GABA to a pentameric ligand channel to investigate the entire free energy landscape of ligand binding (Comitani et al., 2016).

5. SUMMARY AND FUTURE PERSPECTIVES

In spite of its medical relevance, the pharmacologically induced state of general anesthesia is still poorly understood at the molecular level. Due to their potential long-term neurotoxicity (Bianchi, et al., 2008; Carnini, Fazen Eckenhoff, & Eckenhoff, 2006; Eckenhoff et al., 2004; Jevtovic-Todorovic et al., 2003; Wang et al., 2009; Wei et al., 2008; Zhang et al., 2012), general anesthetic agents are far from being "perfect" drugs; a detailed understanding of their molecular mechanism(s) of action could enable structure-based drug optimization and thus lead to a mitigation of the most dangerous side effects. However, the large number of protein targets identified and the paradoxical effects often observed (e.g., evidence of both potentiation and inhibition within a single ion channel) makes the investigation of the molecular determinants of general anesthesia an extremely challenging task.

Since their inception, molecular modeling approaches and, in particular, molecular dynamics simulations have been conveniently used to connect the microscopic interactions characterizing a complex molecular system to its emerging macroscopic properties. Molecular simulations are especially useful to generate mechanistic hypotheses: along with relevant molecular events, they predict the outcomes of possible experiments. Accordingly, in the last decade simulations helped to clarify the mechanism of action of general anesthetics by guiding the design and interpretation of a wealth of experimental techniques, ranging from chemical biology to electrophysiology, structural biology, and spectroscopy (Barber et al., 2014; Bertaccini, Yoluk, Lindahl, & Trudell, 2013; Brömstrup, Howard, Trudell, Harris, & Lindahl, 2013; Chiara et al., 2014; Heusser et al., 2013; Liang et al., 2015; Voigt et al., 2015; Woll, Dailey, Brannigan, & Eckenhoff, 2016; Woll, Peng, et al., 2017; Woll, Skinner, et al., 2017; Zhang, Qu, Covarrubias, & Germann, 2013). As a result, we have now quantitative testable models describing the complex modulatory action exerted by general anesthetics on ion channels and, in particular, on VGICs.

These remarkable successes naturally raise a series of questions: What is, ultimately, the predictive power of these computational approaches? Will we ever develop quantitative models relating the chemical features of active compounds to the physiologically relevant end points of general anesthesia, such as reversible loss of consciousness and analgesia? The answer to these questions lies in the possibility of accurately integrating molecular simulations within the framework of multiscale descriptions of the central nervous system. In other words, success in this endeavor will depend on our ability to develop novel theoretical descriptions able to span a wide range of time- and length-scales: from ligand binding to neural network excitability. Recent studies suggest that this ambitious goal might be within reach and that theoretical models of consciousness, although necessarily simplistic, could provide quantitative insights. For instance, Zhou et al. have shown that a percolation model is able to reproduce key spectral features of electroencephalographic recordings of transitions from conscious to unconscious brain activities, thereby providing a manageable working model to study general anesthesia (Zhou, Mowrey, Tang, Xu, 2015). Investigating these neural network dynamics from an ab initio, bottom-up perspective will require the prediction of the kinetics of channel activation and inactivation using molecular dynamics simulations. To this end, computational approaches such as Markov State Models (Chodera, Singhal, Pande, Dill, & Swope, 2007; Mukherjee, Pantelopulos, & Voelz, 2016; Noe, Schutte, Vanden-Eijnden, Reich, & Weikl, 2009; Pande, Beauchamp, & Bowman, 2010; Plattner & Noé, 2015; Schütte & Sarich, 2015; Shukla, Hernández, Weber, & Pande, 2015; Voelz, Elman, Razavi, & Zhou, 2014) are proving extremely effective and hold promise to bridge this gap between the microscopic dynamics of ion channels and long-timescale behavior in neurons.

REFERENCES

Adler, E. M. (2011). Insight into the sodium channel. *Science Signaling*, *4*, ec207. https://doi.org/10.1126/scisignal.4183ec207.

Ahern, C. A. (2013). What activates inactivation? *The Journal of General Physiology*, *142*, 97–100. https://doi.org/10.1085/jgp.201311046.

Allen, T. W., Andersen, O. S., & Roux, B. (2004). On the importance of atomic fluctuations, protein flexibility, and solvent in ion permeation. *The Journal of General Physiology*, *124*, 679–690. https://doi.org/10.1085/jgp.200409111.

Allen, T. W., Andersen, O. S., & Roux, B. (2006). Molecular dynamics—Potential of mean force calculations as a tool for understanding ion permeation and selectivity in narrow channels. *Biophysical Chemistry, 124,* 251–267. https://doi.org/10.1016/j.bpc.2006.04.015.

Allen, T. W., Kuyucak, S., & Chung, S.-H. (1999). Molecular dynamics study of the KcsA potassium channel. *Biophysical Journal, 77,* 2502–2516. https://doi.org/10.1016/S0006-3495(99)77086-4.

Alonso, H., Bliznyuk, A. A., & Gready, J. E. (2006). Combining docking and molecular dynamic simulations in drug design. *Medicinal Research Reviews, 26,* 531–568. https://doi.org/10.1002/med.20067.

Amaral, C., Carnevale, V., Klein, M. L., & Treptow, W. (2012). Exploring conformational states of the bacterial voltage-gated sodium channel NavAb via molecular dynamics simulations. *Proceedings of the National Academy of Sciences of the United States of America, 109,* 21336–21341. https://doi.org/10.1073/pnas.1218087109.

Armen, R. S., Chen, J., & Brooks, C. L. (2009). An evaluation of explicit receptor flexibility in molecular docking using molecular dynamics and torsion angle molecular dynamics. *Journal of Chemical Theory and Computation, 5,* 2909–2923. https://doi.org/10.1021/ct900262t.

Armstrong, C. M. (1973). Destruction of sodium conductance inactivation in squid axons perfused with pronase. *The Journal of General Physiology, 62,* 375–391. https://doi.org/10.1085/jgp.62.4.375.

Armstrong, C. M., & Bezanilla, F. (1977). Inactivation of the sodium channel II. Gating current experiments. *The Journal of General Physiology, 70,* 567–590.

Barber, A. F., Carnevale, V., Klein, M. L., Eckenhoff, R. G., & Covarrubias, M. (2014). Modulation of a voltage-gated Na+ channel by sevoflurane involves multiple sites and distinct mechanisms. *Proceedings of the National Academy of Sciences of the United States of America, 111,* 6726–6731. https://doi.org/10.1073/pnas.1405768111.

Barber, A. F., Liang, Q., Amaral, C., Treptow, W., & Covarrubias, M. (2011). Molecular mapping of general anesthetic sites in a voltage-gated ion channel. *Biophysical Journal, 101,* 1613–1622. https://doi.org/10.1016/j.bpj.2011.08.026.

Barber, A. F., Liang, Q., & Covarrubias, M. (2012). Novel activation of voltage-gated K(+) channels by sevoflurane. *The Journal of Biological Chemistry, 287*(48), 40425–40432. https://doi.org/10.1074/jbc.M112.405787(Epub 2012 Oct 4. PubMed PMID: 23038249; PubMed Central PMCID: PMC3504757).

Baron, R., & McCammon, J. A. (2013). Molecular recognition and ligand association. *Annual Review of Physical Chemistry, 64,* 151–175. https://doi.org/10.1146/annurev-physchem-040412-110047.

Baron, R., Setny, P., & Andrew McCammon, J. (2010). Water in cavity−ligand recognition. *Journal of the American Chemical Society, 132,* 12091–12097. https://doi.org/10.1021/ja1050082.

Bash, P. A., Singh, U. C., Brown, F. K., Langridge, R., & Kollman, P. A. (1987). Calculation of the relative change in binding free energy of a protein-inhibitor complex. *Science, 235,* 574–576.

Berne, B. J., Weeks, J. D., & Zhou, R. (2009). Dewetting and hydrophobic interaction in physical and biological systems. *Annual Review of Physical Chemistry, 60,* 85–103. https://doi.org/10.1146/annurev.physchem.58.032806.104445.

Bernèche, S., & Roux, B. (2001). Energetics of ion conduction through the K+ channel. *Nature, 414,* 73–77. https://doi.org/10.1038/35102067.

Bernsel, A., Viklund, H., Falk, J., Lindahl, E., von Heijne, G., & Elofsson, A. (2008). Prediction of membrane-protein topology from first principles. *Proceedings of the National Academy of Sciences of the United States of America, 105,* 7177–7181. https://doi.org/10.1073/pnas.0711151105.

Bertaccini, E. J., Yoluk, O., Lindahl, E. R., & Trudell, J. R. (2013). Assessment of homology templates and an anesthetic binding site within the γ-aminobutyric acid receptor. *Anesthesiology*, *119*, 1087–1095. https://doi.org/10.1097/ALN.0b013e31829e47e3.
Bezanilla, F. (1977). Inactivation of the sodium channel. I. Sodium current experiments. *The Journal of General Physiology*, *70*, 549–566. https://doi.org/10.1085/jgp.70.5.549.
Bezanilla, F. (2000). The voltage sensor in voltage-dependent ion channels. *Physiological Reviews*, *80*, 555–592.
Bianchi, S. L., Tran, T., Liu, C., Lin, S., Li, Y., Keller, J. M., et al. (2008). Brain and behavior changes in 12-month-old Tg2576 and nontransgenic mice exposed to anesthetics. *Neurobiology of Aging*, *29*, 1002–1010. https://doi.org/10.1016/j.neurobiolaging.2007.02.009.
Boiteux, C., Vorobyov, I., French, R. J., French, C., Yarov-Yarovoy, V., & Allen, T. W. (2014). Local anesthetic and antiepileptic drug access and binding to a bacterial voltage-gated sodium channel. *Proceedings of the National Academy of Sciences of the United States of America*, *111*, 13057–13062. https://doi.org/10.1073/pnas.1408710111.
Bottegoni, G., Kufareva, I., Totrov, M., & Abagyan, R. (2009). Four-dimensional docking: A fast and accurate account of discrete receptor flexibility in ligand docking. *Journal of Medicinal Chemistry*, *52*, 397–406. https://doi.org/10.1021/jm8009958.
Brackenbury, W. J., & Isom, L. L. (2011). Na+ channel? Subunits: Overachievers of the ion channel family. *Frontiers in Pharmacology*, *2*, 53. https://doi.org/10.3389/fphar.2011.00053.
Brannigan, G., LeBard, D. N., Henin, J., Eckenhoff, R. G., & Klein, M. L. (2010). Multiple binding sites for the general anesthetic isoflurane identified in the nicotinic acetylcholine receptor transmembrane domain. *Proceedings of the National Academy of Sciences of the United States of America*, *107*, 14122–14127. https://doi.org/10.1073/pnas.1008534107.
Brenke, R., Kozakov, D., Chuang, G.-Y., Beglov, D., Hall, D., Landon, M. R., et al. (2009). Fragment-based identification of druggable "hot spots" of proteins using Fourier domain correlation techniques. *Bioinformatics*, *25*, 621–627. https://doi.org/10.1093/bioinformatics/btp036.
Brömstrup, T., Howard, R. J., Trudell, J. R., Harris, R. A., & Lindahl, E. (2013). Inhibition versus potentiation of ligand-gated ion channels can be altered by a single mutation that moves ligands between intra- and intersubunit sites. *Structure*, *21*, 1307–1316. https://doi.org/10.1016/j.str.2013.06.018.
Carnevale, V., & Klein, M. L. (2017). Small molecule modulation of voltage gated sodium channels. *Current Opinion in Structural Biology*, *43*, 156–162. https://doi.org/10.1016/j.sbi.2017.02.002.
Carnini, A., Fazen Eckenhoff, M., & Eckenhoff, R. G. (2006). Interactions of volatile anesthetics with neurodegenerative-disease-associated proteins. *Anesthesiology Clinics*, *24*, 381–405. https://doi.org.10.1016/j.atc.2006.02.002.
Catterall, W. A. (1986). Voltage-dependent gating of sodium channels: Correlating structure and function. *Trends in Neurosciences*, *9*, 7–10. https://doi.org/10.1016/0166-2236(86)90004-4.
Catterall, W. A. (2010). Ion channel voltage sensors: Structure, function, and pathophysiology. *Neuron*, *67*, 915–928. https://doi.org/10.1016/j.neuron.2010.08.021.
Catterall, W. A. (2014). Structure and function of voltage-gated sodium channels at atomic resolution: Voltage-gated sodium channels at atomic resolution. *Experimental Physiology*, *99*, 35–51. https://doi.org/10.1113/expphysiol.2013.071969.
Chandler, D. (2005). Interfaces and the driving force of hydrophobic assembly. *Nature*, *437*, 640–647. https://doi.org/10.1038/nature04162.
Chen, R., Buyan, A., & Corry, B. (2017). Voltage-gated sodium channel pharmacology. In D. P. Geraghty & L. D. Rash (Eds.), *Advances in pharmacology* (pp. 255–285). Elsevier.

Chiara, D. C., Gill, J. F., Chen, Q., Tillman, T., Dailey, W. P., Eckenhoff, R. G., et al. (2014). Photoaffinity labeling the propofol binding site in GLIC. *Biochemistry (Moscow), 53*, 135–142. https://doi.org/10.1021/bi401492k.

Chipot, C., & Pohorille, A. (Eds.), (2007). In *Springer series in chemical physics*, Vol. 86. *Free energy calculationsBerlin, Heidelberg: Springer Berlin Heidelberg.*

Chodera, J. D., Singhal, N., Pande, V. S., Dill, K. A., & Swope, W. C. (2007). Automatic discovery of metastable states for the construction of Markov models of macromolecular conformational dynamics. *The Journal of Chemical Physics, 126*, 155101. https://doi.org/10.1063/1.2714538.

Chowdhury, S., & Chanda, B. (2012). Estimating the voltage-dependent free energy change of ion channels using the median voltage for activation. *The Journal of General Physiology, 139*, 3–17. https://doi.org/10.1085/jgp.201110722.

Clark, A. J., Tiwary, P., Borrelli, K., Feng, S., Miller, E. B., Abel, R., et al. (2016). Prediction of protein–ligand binding poses via a combination of induced fit docking and metadynamics simulations. *Journal of Chemical Theory and Computation, 12*, 2990–2998. https://doi.org/10.1021/acs.jctc.6b00201.

Claußen, H., Buning, C., Rarey, M., & Lengauer, T. (2001). FlexE: Efficient molecular docking considering protein structure variations. *Journal of Molecular Biology, 308*, 377–395. https://doi.org/10.1006/jmbi.2001.4551.

Comitani, F., Limongelli, V., & Molteni, C. (2016). The free energy landscape of GABA binding to a pentameric ligand-gated ion channel and its disruption by mutations. *Journal of Chemical Theory and Computation, 12*, 3398–3406. https://doi.org/10.1021/acs.jctc.6b00303.

Corry, B. (2016). Computer simulation of ion channels. In C. Domene (Ed.), *Computational biophysics of membrane proteins* (pp. 161–196). Cambridge: Royal Society of Chemistry. chapter 6.

Cournia, Z., Allen, T. W., Andricioaei, I., Antonny, B., Baum, D., Brannigan, G., et al. (2015). Membrane protein structure, function, and dynamics: A perspective from experiments and theory. *The Journal of Membrane Biology, 248*, 611–640. https://doi.org/10.1007/s00232-015-9802-0.

Covarrubias, M., Barber, A. F., Carnevale, V., Treptow, W., & Eckenhoff, R. G. (2015). Mechanistic insights into the modulation of voltage-gated ion channels by inhalational anesthetics. *Biophysical Journal, 109*, 2003–2011. https://doi.org/10.1016/j.bpj.2015.09.032.

DeCaen, P. G., Yarov-Yarovoy, V., Sharp, E. M., Scheuer, T., & Catterall, W. A. (2009). Sequential formation of ion pairs during activation of a sodium channel voltage sensor. *Proceedings of the National Academy of Sciences of the United States of America, 106*, 22498–22503. https://doi.org/10.1073/pnas.0912307106.

Delemotte, L., Kasimova, M. A., Klein, M. L., Tarek, M., & Carnevale, V. (2015). Free-energy landscape of ion-channel voltage-sensor–domain activation. *Proceedings of the National Academy of Sciences of the United States of America, 112*, 124–129. https://doi.org/10.1073/pnas.1416959112.

Delemotte, L., Tarek, M., Klein, M. L., Amaral, C., & Treptow, W. (2011). Intermediate states of the Kv1.2 voltage sensor from atomistic molecular dynamics simulations. *Proceedings of the National Academy of Sciences of the United States of America, 108*, 6109–6114. https://doi.org/10.1073/pnas.1102724108.

de Lera Ruiz, M., & Kraus, R. L. (2015). Voltage-gated sodium channels: Structure, function, pharmacology, and clinical indications. *Journal of Medicinal Chemistry, 58*, 7093–7118. https://doi.org/10.1021/jm501981g.

Dias, S. E. D., Martins, A. M., Nguyen, Q. T., & Gomes, A. J. P. (2017). GPU-based detection of protein cavities using Gaussian surfaces. *BMC Bioinformatics, 18*, 493. https://doi.org/10.1186/s12859-017-1913-4.

Dixon, S. L., Smondyrev, A. M., Knoll, E. H., Rao, S. N., Shaw, D. E., & Friesner, R. A. (2006). PHASE: A new engine for pharmacophore perception, 3D QSAR model development, and 3D database screening: 1. Methodology and preliminary results. *Journal of Computer-Aided Molecular Design, 20*, 647–671. https://doi.org/10.1007/s10822-006-9087-6.

Dixon, S. L., Smondyrev, A. M., & Rao, S. N. (2006). PHASE: A novel approach to pharmacophore modeling and 3D database searching. *Chemical Biology & Drug Design, 67*, 370–372. https://doi.org/10.1111/j.1747-0285.2006.00384.x.

Domene, C. (2016). *Computational biophysics of membrane proteins*. Royal Society of Chemistry, 274.

Domene, C., & Furini, S. (2009). Examining ion channel properties using free-energy methods. In *Methods in enzymology* (pp. 155–177). Elsevier.

Duch, D. S., Rehberg, B., & Vysotskaya, T. N. (1998). Volatile anesthetics significantly suppress central and peripheral mammalian sodium channels. *Toxicology Letters, 100–101*, 255–263.

Eckenhoff, R. G., Johansson, J. S., Wei, H., Carnini, A., Kang, B., Wei, W., et al. (2004). Inhaled anesthetic enhancement of amyloid-beta oligomerization and cytotoxicity. *Anesthesiology, 101*, 703–709.

Eswar, N., Webb, B., Marti-Renom, M. A., Madhusudhan, M. S., Eramian, D., Shen, M., et al. (2006). Comparative protein structure modeling using Modeller. In A. Bateman, W. R. Pearson, L. D. Stein, G. D. Stormo, & J. R. Yates (Eds.), *Current protocols in bioinformatics* (pp. 5.6.1–5.6.30). Hoboken, NJ: John Wiley & Sons, Inc.

Farid, R., Day, T., Friesner, R. A., & Pearlstein, R. A. (2006). New insights about HERG blockade obtained from protein modeling, potential energy mapping, and docking studies. *Bioorganic & Medicinal Chemistry, 14*, 3160–3173. https://doi.org/10.1016/j.bmc.2005.12.032.

Floudas, C. A., Fung, H. K., McAllister, S. R., Mönnigmann, M., & Rajgaria, R. (2006). Advances in protein structure prediction and de novo protein design: A review. *Chemical Engineering Science, 61*, 966–988. https://doi.org/10.1016/j.ces.2005.04.009.

Franks, N. P. (2006). Molecular targets underlying general anaesthesia: Molecular targets underlying general anaesthesia. *British Journal of Pharmacology, 147*, S72–S81. https://doi.org/10.1038/sj.bjp.0706441.

Franks, N. P. (2008). General anaesthesia: From molecular targets to neuronal pathways of sleep and arousal. *Nature Reviews Neuroscience, 9*, 370–386. https://doi.org/10.1038/nrn2372.

Frembgen-Kesner, T., & Elcock, A. H. (2006). Computational sampling of a cryptic drug binding site in a protein receptor: Explicit solvent molecular dynamics and inhibitor docking to p38 MAP kinase. *Journal of Molecular Biology, 359*, 202–214. https://doi.org/10.1016/j.jmb.2006.03.021.

Friesner, R. A., Banks, J. L., Murphy, R. B., Halgren, T. A., Klicic, J. J., Mainz, D. T., et al. (2004). Glide: A new approach for rapid, accurate docking and scoring. 1. Method and assessment of docking accuracy. *Journal of Medicinal Chemistry, 47*, 1739–1749. https://doi.org/10.1021/jm0306430.

Friesner, R. A., Murphy, R. B., Repasky, M. P., Frye, L. L., Greenwood, J. R., Halgren, T. A., et al. (2006). Extra precision glide: Docking and scoring incorporating a model of hydrophobic enclosure for protein – ligand complexes. *Journal of Medicinal Chemistry, 49*, 6177–6196. https://doi.org/10.1021/jm051256o.

Gagnon, J. K., Law, S. M., & Brooks, C. L. (2016). Flexible CDOCKER: Development and application of a pseudo-explicit structure-based docking method within CHARMM: Adding receptor flexibility improves protein-ligand docking within CDOCKER. *Journal of Computational Chemistry, 37*, 753–762. https://doi.org/10.1002/jcc.24259.

Gianti, E., Carnevale, V., DeGrado, W. F., Klein, M. L., & Fiorin, G. (2015). Hydrogen-bonded water molecules in the M2 channel of the influenza a virus guide the binding preferences of ammonium-based inhibitors. *The Journal of Physical Chemistry. B, 119*, 1173–1183. https://doi.org/10.1021/jp506807y.

Gianti, E., Delemotte, L., Klein, M. L., & Carnevale, V. (2016). On the role of water density fluctuations in the inhibition of a proton channel. *Proceedings of the National Academy of Sciences of the United States of America, 113*, E8359–E8368. https://doi.org/10.1073/pnas.1609964114.

Gianti, E., Messick, T. E., Lieberman, P. M., & Zauhar, R. J. (2016). Computational analysis of EBNA1 "druggability" suggests novel insights for Epstein-Barr virus inhibitor design. *Journal of Computer-Aided Molecular Design, 30*, 285–303. https://doi.org/10.1007/s10822-016-9899-y.

Gianti, E., & Zauhar, R. J. (2012). Modeling androgen receptor flexibility: A binding mode hypothesis of CYP17 inhibitors/antiandrogens for prostate cancer therapy. *Journal of Chemical Information and Modeling, 52*, 2670–2683. https://doi.org/10.1021/ci3002342.

Gianti, E., & Zauhar, R. J. (2015). An SH2 domain model of STAT5 in complex with phospho-peptides define "STAT5 Binding Signatures." *Journal of Computer-Aided Molecular Design, 29*, 451–470. https://doi.org/10.1007/s10822-015-9835-6.

Grant, J. A., Gallardo, M. A., & Pickup, B. T. (1996). A fast method of molecular shape comparison: A simple application of a Gaussian description of molecular shape. *Journal of Computational Chemistry, 17*, 1653–1666. https://doi.org/10.1002/(SICI)1096-987X(19961115)17:14<1653::AID-JCC7>3.0.CO;2-K.

Guy, H. R., & Seetharamulu, P. (1986). Molecular model of the action potential sodium channel. *Proceedings of the National Academy of Sciences of the United States of America, 83*, 508–512.

Halgren, T. (2007). New method for fast and accurate binding-site identification and analysis. *Chemical Biology & Drug Design, 69*, 146–148. https://doi.org/10.1111/j.1747-0285.2007.00483.x.

Halgren, T. A. (2009). Identifying and characterizing binding sites and assessing druggability. *Journal of Chemical Information and Modeling, 49*, 377–389. https://doi.org/10.1021/ci800324m.

Halgren, T. A., Murphy, R. B., Friesner, R. A., Beard, H. S., Frye, L. L., Pollard, W. T., et al. (2004). Glide: A new approach for rapid, accurate docking and scoring. 2. Enrichment factors in database screening. *Journal of Medicinal Chemistry, 47*, 1750–1759. https://doi.org/10.1021/jm030644s.

Hall, D. R., Kozakov, D., & Vajda, S. (2012). Analysis of protein binding sites by computational solvent mapping. In R. Baron (Ed.), *Computational drug discovery and design* (pp. 13–27). New York: Springer New York.

Harris, T., Shahidullah, M., Ellingson, J. S., & Covarrubias, M. (2000). General anesthetic action at an internal protein site involving the S4-S5 cytoplasmic loop of a neuronal K(+) channel. *The Journal of Biological Chemistry, 275*(7), 4928–4936(PubMed PMID:10671530).

Hemmings, H. C., Akabas, M. H., Goldstein, P. A., Trudell, J. R., Orser, B. A., & Harrison, N. L. (2005). Emerging molecular mechanisms of general anesthetic action. *Trends in Pharmacological Sciences, 26*, 503–510. https://doi.org/10.1016/j.tips.2005.08.006.

Herold, K. F., & Hemmings, H. C. J. (2012). Sodium channels as targets for volatile anesthetics. *Frontiers in Pharmacology, 3*, 50. https://doi.org/10.3389/fphar.2012.00050.

Herold, K. F., Nau, C., Ouyang, W., & Hemmings, H. C. (2009). Isoflurane inhibits the tetrodotoxin-resistant voltage-gated sodium channel Nav1.8. *Anesthesiology, 111*, 591–599. https://doi.org/10.1097/ALN.0b013e3181af64d4.

Heusser, S. A., Howard, R. J., Borghese, C. M., Cullins, M. A., Broemstrup, T., Lee, U. S., et al. (2013). Functional validation of virtual screening for novel agents with general anesthetic action at ligand-gated ion channels. *Molecular Pharmacology*, *84*, 670–678. https://doi.org/10.1124/mol.113.087692.

Hong, L., Kim, I. H., & Tombola, F. (2014). Molecular determinants of Hv1 proton channel inhibition by guanidine derivatives. *Proceedings of the National Academy of Sciences of the United States of America*, *111*, 9971–9976. https://doi.org/10.1073/pnas.1324012111.

Huang, S.-Y., & Zou, X. (2006). Ensemble docking of multiple protein structures: Considering protein structural variations in molecular docking. *Proteins: Structure, Function, and Bioinformatics*, *66*, 399–421. https://doi.org/10.1002/prot.21214.

Hummer, G., Rasaiah, J. C., & Noworyta, J. P. (2001). Water conduction through the hydrophobic channel of a carbon nanotube. *Nature*, *414*, 188–190. https://doi.org/10.1038/35102535.

Jacobson, M. P., Pincus, D. L., Rapp, C. S., Day, T. J. F., Honig, B., Shaw, D. E., et al. (2004). A hierarchical approach to all-atom protein loop prediction. *Proteins: Structure, Function, and Bioinformatics*, *55*, 351–367. https://doi.org/10.1002/prot.10613.

Jensen, M. O., Jogini, V., Borhani, D. W., Leffler, A. E., Dror, R. O., & Shaw, D. E. (2012). Mechanism of voltage gating in potassium channels. *Science*, *336*, 229–233. https://doi.org/10.1126/science.1216533.

Jevtovic-Todorovic, V., Hartman, R. E., Izumi, Y., Benshoff, N. D., Dikranian, K., Zorumski, C. F., et al. (2003). Early exposure to common anesthetic agents causes widespread neurodegeneration in the developing rat brain and persistent learning deficits. *The Journal of Neuroscience*, *23*, 876–882.

Jiang, Y., Ruta, V., Chen, J., Lee, A., & MacKinnon, R. (2003). The principle of gating charge movement in a voltage-dependent K+ channel. *Nature*, *423*, 42–48. https://doi.org/10.1038/nature01581.

Jones, G., Willett, P., Glen, R. C., Leach, A. R., & Taylor, R. (1997). Development and validation of a genetic algorithm for flexible docking. *Journal of Molecular Biology*, *267*, 727–748. https://doi.org/10.1006/jmbi.1996.0897.

Jorgensen, C., Darré, L., Oakes, V., Torella, R., Pryde, D., & Domene, C. (2016). Lateral fenestrations in K+-channels explored using molecular dynamics simulations. *Molecular Pharmaceutics*, *13*, 2263–2273. https://doi.org/10.1021/acs.molpharmaceut.5b00942.

Jorgensen, W. L., & Ravimohan, C. (1985). Monte Carlo simulation of differences in free energies of hydration. *The Journal of Chemical Physics*, *83*, 3050–3054. https://doi.org/10.1063/1.449208.

Kasimova, M. A., Granata, D., & Carnevale, V. (2016). Voltage-gated sodium channels: Evolutionary history and distinctive sequence features, Voltage-gated sodium channels. In *Current topics in membranes* (pp. 261–286). Elsevier.

Khalili-Araghi, F., Jogini, V., Yarov-Yarovoy, V., Tajkhorshid, E., Roux, B., & Schulten, K. (2010). Calculation of the gating charge for the Kv1.2 voltage-activated potassium channel. *Biophysical Journal*, *98*, 2189–2198. https://doi.org/10.1016/j.bpj.2010.02.056.

Kinde, M. N., Bondarenko, V., Granata, D., Bu, W., Grasty, K. C., Loll, P. J., et al. (2016). Fluorine-19 NMR and computational quantification of isoflurane binding to the voltage-gated sodium channel NaChBac. *Proceedings of the National Academy of Sciences of the United States of America*, *113*, 13762–13767. https://doi.org/10.1073/pnas.1609939113.

Klebe, G. (2015). Applying thermodynamic profiling in lead finding and optimization. *Nature Reviews Drug Discovery*, *14*, 95–110. https://doi.org/10.1038/nrd4486.

Köfinger, J., Hummer, G., & Dellago, C. (2011). Single-file water in nanopores. *Physical Chemistry Chemical Physics*, *13*, 15403. https://doi.org/10.1039/c1cp21086f.

Kollman, P. (1993). Free energy calculations: Applications to chemical and biochemical phenomena. *Chemical Reviews*, *93*, 2395–2417. https://doi.org/10.1021/cr00023a004.

Kollman, P. A., Massova, I., Reyes, C., Kuhn, B., Huo, S., Chong, L., et al. (2000). Calculating structures and free energies of complex molecules: Combining molecular mechanics and continuum models. *Accounts of Chemical Research, 33*, 889–897.

Kopp Lugli, A., Yost, C. S., & Kindler, C. H. (2009). Anaesthetic mechanisms: Update on the challenge of unravelling the mystery of anaesthesia. *European Journal of Anaesthesiology, 26*, 807–820. https://doi.org/10.1097/EJA.0b013e32832d6b0f.

Kozakov, D., Grove, L. E., Hall, D. R., Bohnuud, T., Mottarella, S. E., Luo, L., et al. (2015). The FTMap family of web servers for determining and characterizing ligand-binding hot spots of proteins. *Nature Protocols, 10*, 733–755. https://doi.org/10.1038/nprot.2015.043.

Krasowski, M. D., & Harrison, N. L. (1999). General anaesthetic actions on ligand-gated ion channels. *Cellular and Molecular Life Sciences: CMLS, 55*, 1278–1303. https://doi.org/10.1007/s000180050371.

Laio, A., & Parrinello, M. (2002). Escaping free-energy minima. *Proceedings of the National Academy of Sciences of the United States of America, 99*, 12562–12566. https://doi.org/10.1073/pnas.202427399.

Lam, P. C.-H., Abagyan, R., & Totrov, M. (2018). Ligand-biased ensemble receptor docking (LigBEnD): A hybrid ligand/receptor structure-based approach. *Journal of Computer-Aided Molecular Design, 32*, 187–198. https://doi.org/10.1007/s10822-017-0058-x.

Lenselink, E. B., Louvel, J., Forti, A. F., van Veldhoven, J. P. D., de Vries, H., Mulder-Krieger, T., et al. (2016). Predicting binding affinities for GPCR ligands using free-energy perturbation. *ACS Omega, 1*, 293–304. https://doi.org/10.1021/acsomega.6b00086.

Li, Q., Shen, R., Treger, J. S., Wanderling, S. S., Milewski, W., Siwowska, K., et al. (2015). Resting state of the human proton channel dimer in a lipid bilayer. *Proceedings of the National Academy of Sciences of the United States of America, 112*, E5926–E5935. https://doi.org/10.1073/pnas.1515043112.

Liang, Q., Anderson, W. D., Jones, S. T., Souza, C. S., Hosoume, J. M., Treptow, W., et al. (2015). Positive allosteric modulation of Kv channels by sevoflurane: Insights into the structural basis of inhaled anesthetic action. *PLOS One, 10*, e0143363. https://doi.org/10.1371/journal.pone.0143363.

Limongelli, V., Bonomi, M., & Parrinello, M. (2013). Funnel metadynamics as accurate binding free-energy method. *Proceedings of the National Academy of Sciences of the United States of America, 110*, 6358–6363. https://doi.org/10.1073/pnas.1303186110.

Lindahl, E., & Sansom, M. (2008). Membrane proteins: Molecular dynamics simulations. *Current Opinion in Structural Biology, 18*, 425–431. https://doi.org/10.1016/j.sbi.2008.02.003.

Linder, T., de Groot, B. L., & Stary-Weinzinger, A. (2013). Probing the energy landscape of activation gating of the bacterial potassium channel KcsA. *PLoS Computational Biology, 9*, e1003058. https://doi.org/10.1371/journal.pcbi.1003058.

Lioudyno, M. I., Birch, A. M., Tanaka, B. S., Sokolov, Y., Goldin, A. L., Chandy, K. G., et al. (2013). Shaker-related potassium channels in the central medial nucleus of the thalamus are important molecular targets for arousal suppression by volatile general anesthetics. *The Journal of Neuroscience, 33*, 16310–16322. https://doi.org/10.1523/JNEUROSCI.0344-13.2013.

Liu, L. T., Haddadian, E. J., Willenbring, D., Xu, Y., & Tang, P. (2010). Higher susceptibility to halothane modulation in open- than in closed-channel α4β2 nAChR revealed by molecular dynamics simulations. *The Journal of Physical Chemistry. B, 114*, 626–632. https://doi.org/10.1021/jp908944e.

Liu, P., Huang, X., Zhou, R., & Berne, B. J. (2005). Observation of a dewetting transition in the collapse of the melittin tetramer. *Nature, 437*, 159–162. https://doi.org/10.1038/nature03926.

Liu, L. T., Willenbring, D., Xu, Y., & Tang, P. (2009). General anesthetic binding to neuronal α4β2 nicotinic acetylcholine receptor and its effects on global dynamics. *The Journal of Physical Chemistry. B*, *113*, 12581–12589. https://doi.org/10.1021/jp9039513.
Lorber, D. M., & Shoichet, B. K. (2008). Flexible ligand docking using conformational ensembles. *Protein Science*, 7, 938–950. https://doi.org/10.1002/pro.5560070411.
Mattos, C., & Ringe, D. (1996). Locating and characterizing binding sites on proteins. *Nature Biotechnology*, *14*, 595–599. https://doi.org/10.1038/nbt0596-595.
McCammon, J. A., Gelin, B. R., & Karplus, M. (1977). Dynamics of folded proteins. *Nature*, *267*, 585–590. https://doi.org/10.1038/267585a0.
Medovoy, D., Perozo, E., & Roux, B. (2016). Multi-ion free energy landscapes underscore the microscopic mechanism of ion selectivity in the KcsA channel. *Biochimica et Biophysica Acta (BBA)—Biomembranes*, *1858*, 1722–1732. https://doi.org/10.1016/j.bbamem.2016.02.019.
Merz, K. M., & Kollman, P. A. (1989). Free energy perturbation simulations of the inhibition of thermolysin: Prediction of the free energy of binding of a new inhibitor. *Journal of the American Chemical Society*, *111*, 5649–5658. https://doi.org/10.1021/ja00197a022.
Miller, C. (1992). Ionic channels of excitable membranes. Second edition. *Cell*, *69*, 579. https://doi.org/10.1016/0092-8674(92)90220-7.
Miller, K. W. (2002). The nature of sites of general anaesthetic action. *British Journal of Anaesthesia*, *89*, 17–31. https://doi.org/10.1093/bja/aef167.
Mittal, J., & Hummer, G. (2010). Interfacial thermodynamics of confined water near molecularly rough surfaces. *Faraday Discussions*, *146*, 341–352. discussion 367–393, 395–401.
Mondal, J., Morrone, J. A., & Berne, B. J. (2013). How hydrophobic drying forces impact the kinetics of molecular recognition. *Proceedings of the National Academy of Sciences of the United States of America*, *110*, 13277–13282. https://doi.org/10.1073/pnas.1312529110.
Morris, G. M., Huey, R., Lindstrom, W., Sanner, M. F., Belew, R. K., Goodsell, D. S., et al. (2009). AutoDock4 and AutoDockTools4: Automated docking with selective receptor flexibility. *Journal of Computational Chemistry*, *30*, 2785–2791. https://doi.org/10.1002/jcc.21256.
Mukherjee, S., Pantelopulos, G. A., & Voelz, V. A. (2016). Markov models of the apo-MDM2 lid region reveal diffuse yet two-state binding dynamics and receptor poses for computational docking. *Scientific Reports*, *6* https://doi.org/10.1038/srep31631.
Noe, F., Schutte, C., Vanden-Eijnden, E., Reich, L., & Weikl, T. R. (2009). Constructing the equilibrium ensemble of folding pathways from short off-equilibrium simulations. *Proceedings of the National Academy of Sciences of the United States of America*, *106*, 19011–19016. https://doi.org/10.1073/pnas.0905466106.
Olsen, R. W., Li, G.-D., Wallner, M., Trudell, J. R., Bertaccini, E. J., Lindahl, E., et al. (2014). Structural models of ligand-gated ion channels: Sites of action for anesthetics and ethanol. *Alcoholism, Clinical and Experimental Research*, *38*, 595–603. https://doi.org/10.1111/acer.12283.
Ostmeyer, J., Chakrapani, S., Pan, A. C., Perozo, E., & Roux, B. (2013). Recovery from slow inactivation in K+ channels is controlled by water molecules. *Nature*, *501*, 121–124. https://doi.org/10.1038/nature12395.
OuYang, W., & Hemmings, H. C. (2007). Isoform-selective effects of isoflurane on voltage-gated Na+ channels. *Anesthesiology*, *107*, 91–98. https://doi.org/10.1097/01.anes.0000268390.28362.4a.
Ouyang, W., Herold, K. F., & Hemmings, H. C. (2009). Comparative effects of halogenated inhaled anesthetics on voltage-gated Na+ channel function. *Anesthesiology*, *110*, 582–590. https://doi.org/10.1097/ALN.0b013e318197941e.
Ouyang, W., Jih, T.-Y., Zhang, T.-T., Correa, A. M., & Hemmings, H. C. (2007). Isoflurane inhibits NaChBac, a prokaryotic voltage-gated sodium channel. *The Journal of*

Pharmacology and Experimental Therapeutics, *322*, 1076–1083. https://doi.org/10.1124/jpet.107.122929.

Palovcak, E., Delemotte, L., Klein, M. L., & Carnevale, V. (2014). Evolutionary imprint of activation: The design principles of VSDs. *The Journal of General Physiology*, *143*, 145–156. https://doi.org/10.1085/jgp.201311103.

Pande, V. S., Beauchamp, K., & Bowman, G. R. (2010). Everything you wanted to know about Markov state models but were afraid to ask. *Methods*, *52*, 99–105. https://doi.org/10.1016/j.ymeth.2010.06.002.

Payandeh, J., Scheuer, T., Zheng, N., & Catterall, W. A. (2011). The crystal structure of a voltage-gated sodium channel. *Nature*, *475*, 353–358. https://doi.org/10.1038/nature10238.

Plattner, N., & Noé, F. (2015). Protein conformational plasticity and complex ligand-binding kinetics explored by atomistic simulations and Markov models. *Nature Communications*, *6*, 7653. https://doi.org/10.1038/ncomms8653.

Pratt, L. R., & Pohorille, A. (2002). Hydrophobic effects and modeling of biophysical aqueous solution interfaces. *Chemical Reviews*, *102*, 2671–2692. https://doi.org/10.1021/cr000692.

Raju, S. G., Barber, A. F., LeBard, D. N., Klein, M. L., & Carnevale, V. (2013). Exploring volatile general anesthetic binding to a closed membrane-bound bacterial voltage-gated sodium channel via computation. *PLoS Computational Biology*, *9*, e1003090. https://doi.org/10.1371/journal.pcbi.1003090.

Rasaiah, J. C., Garde, S., & Hummer, G. (2008). Water in nonpolar confinement: From nanotubes to proteins and beyond. *Annual Review of Physical Chemistry*, *59*, 713–740. https://doi.org/10.1146/annurev.physchem.59.032607.093815.

Rastelli, G., Rio, A. D., Degliesposti, G., & Sgobba, M. (2010). Fast and accurate predictions of binding free energies using MM-PBSA and MM-GBSA. *Journal of Computational Chemistry*, *31*, 797–810. https://doi.org/10.1002/jcc.21372.

Ravindranath, P. A., Forli, S., Goodsell, D. S., Olson, A. J., & Sanner, M. F. (2015). AutoDockFR: Advances in protein-ligand docking with explicitly specified binding site flexibility. *PLoS Computational Biology*, *11*, e1004586. https://doi.org/10.1371/journal.pcbi.1004586.

Ravindranath, P. A., & Sanner, M. F. (2016). AutoSite: An automated approach for pseudo-ligands prediction—From ligand-binding sites identification to predicting key ligand atoms. *Bioinformatics*, *32*, 3142–3149. https://doi.org/10.1093/bioinformatics/btw367.

Roux, B., Allen, T., Bernche, S., & Im, W. (2004). Theoretical and computational models of biological ion channels. *Quarterly Reviews of Biophysics*, *37*, 15–103. https://doi.org/10.1017/S0033583504003968.

Rush, T. S., Grant, J. A., Mosyak, L., & Nicholls, A. (2005). A shape-based 3-D scaffold hopping method and its application to a bacterial protein-protein interaction. *Journal of Medicinal Chemistry*, *48*, 1489–1495. https://doi.org/10.1021/jm040163o.

Salari, R., Murlidaran, S., & Brannigan, G. (2014). Pentameric ligand-gated ion channels: Insights from computation. *Molecular Simulation*, *40*, 821–829. https://doi.org/10.1080/08927022.2014.896462.

Schütte, C., & Sarich, M. (2015). A critical appraisal of Markov state models. *The European Physical Journal Special Topics*, *224*, 2445–2462. https://doi.org/10.1140/epjst/e2015-02421-0.

Shen, R., Han, W., Fiorin, G., Islam, S. M., Schulten, K., & Roux, B. (2015). Structural refinement of proteins by restrained molecular dynamics simulations with non-interacting molecular fragments. *PLoS Computational Biology*, *11*, e1004368. https://doi.org/10.1371/journal.pcbi.1004368.

Sherman, W., Day, T., Jacobson, M. P., Friesner, R. A., & Farid, R. (2006). Novel procedure for modeling ligand/receptor induced fit effects. *Journal of Medicinal Chemistry*, *49*, 534–553. https://doi.org/10.1021/jm050540c.

Shukla, D., Hernández, C. X., Weber, J. K., & Pande, V. S. (2015). Markov state models provide insights into dynamic modulation of protein function. *Accounts of Chemical Research*, *48*, 414–422. https://doi.org/10.1021/ar5002999.

Stock, L., Souza, C., & Treptow, W. (2013). Structural basis for activation of voltage-gated cation channels. *Biochemistry (Mosc)*, *52*, 1501–1513. https://doi.org/10.1021/bi3013017.

Takeshita, K., Sakata, S., Yamashita, E., Fujiwara, Y., Kawanabe, A., Kurokawa, T., et al. (2014). X-ray crystal structure of voltage-gated proton channel. *Nature Structural & Molecular Biology*, *21*, 352–357. https://doi.org/10.1038/nsmb.2783.

Tarek, M., & Delemotte, L. (2013). Omega currents in voltage-gated ion channels: What can we learn from uncovering the voltage-sensing mechanism using MD simulations? *Accounts of Chemical Research*, *46*, 2755–2762. https://doi.org/10.1021/ar300290u.

Tiwari-Woodruff, S. K., Lin, M. A., Schulteis, C. T., & Papazian, D. M. (2000). Voltage-dependent structural interactions in the *Shaker* K^+ channel. *The Journal of General Physiology*, *115*, 123–138. https://doi.org/10.1085/jgp.115.2.123.

Totrov, M., & Abagyan, R. (2008). Flexible ligand docking to multiple receptor conformations: A practical alternative. *Current Opinion in Structural Biology*, *18*, 178–184. https://doi.org/10.1016/j.sbi.2008.01.004.

Treptow, W., & Klein, M. L. (2012). Computer simulations of voltage-gated cation channels. *Journal of Physical Chemistry Letters*, *3*, 1017–1023. https://doi.org/10.1021/jz300089g.

Tsai, C.-J., & Nussinov, R. (2014). A unified view of "how allostery works" *PLoS Computational Biology*, *10*, e1003394. https://doi.org/10.1371/journal.pcbi.1003394.

Uchime, O., Herrera, R., Reiter, K., Kotova, S., Shimp, R. L., Miura, K., et al. (2012). Analysis of the conformation and function of the Plasmodium falciparum merozoite proteins MTRAP and PTRAMP. *Eukaryotic Cell*, *11*, 615–625. https://doi.org/10.1128/EC.00039-12.

Vaitheeswaran, S., Yin, H., Rasaiah, J. C., & Hummer, G. (2004). Water clusters in nonpolar cavities. *Proceedings of the National Academy of Sciences of the United States of America*, *101*, 17002–17005. https://doi.org/10.1073/pnas.0407968101.

van Keulen, S. C., Gianti, E., Carnevale, V., Klein, M. L., Rothlisberger, U., & Delemotte, L. (2016). Does proton conduction in the voltage-gated H+ channel hHv1 involve Grotthuss-like hopping via acidic residues? *The Journal of Physical Chemistry. B*, *121*, 3340–3351. https://doi.org/10.1021/acs.jpcb.6b08339.

Vargas, E., Yarov-Yarovoy, V., Khalili-Araghi, F., Catterall, W. A., Klein, M. L., Tarek, M., et al. (2012). An emerging consensus on voltage-dependent gating from computational modeling and molecular dynamics simulations. *The Journal of General Physiology*, *140*, 587–594. https://doi.org/10.1085/jgp.201210873.

Vemparala, S., Domene, C., & Klein, M. L. (2008). Interaction of anesthetics with open and closed conformations of a potassium channel studied via molecular dynamics and normal mode analysis. *Biophysical Journal*, *94*, 4260–4269. https://doi.org/10.1529/biophysj.107.119958.

Vemparala, S., Domene, C., & Klein, M. L. (2010). Computational studies on the interactions of inhalational anesthetics with proteins. *Accounts of Chemical Research*, *43*, 103–110. https://doi.org/10.1021/ar900149j.

Voelz, V. A., Elman, B., Razavi, A. M., & Zhou, G. (2014). Surprisal metrics for quantifying perturbed conformational dynamics in Markov state models. *Journal of Chemical Theory and Computation*, *10*, 5716–5728. https://doi.org/10.1021/ct500827g.

Voigt, T. B., Heusser, S., Klement, G., Pouya, I., Mola, A. R., Ruel, T. M. D., et al. (2015). Physicochemical determinants of alcohol binding in a model ligand-gated ion channel. *Alcohol and Alcoholism*, *50*. https://doi.org/10.1093/alcalc/agv080.54.

Votapka, L., & Amaro, R. E. (2013). Multistructural hot spot characterization with FTProd. *Bioinformatics*, *29*, 393–394. https://doi.org/10.1093/bioinformatics/bts689.

Waghe, A., Rasaiah, J. C., & Hummer, G. (2012). Entropy of single-file water in (6,6) carbon nanotubes. *The Journal of Chemical Physics*, *137*, 044709. https://doi.org/10.1063/1.4737842.

Wang, S., Peretich, K., Zhao, Y., Liang, G., Meng, Q., & Wei, H. (2009). Anesthesia-induced neurodegeneration in fetal rat brains. *Pediatric Research*, *66*, 435–440. https://doi.org/10.1203/PDR.0b013e3181b3381b.

Wang, Y., Roose, B. W., Palovcak, E. J., Carnevale, V., & Dmochowski, I. J. (2016). A genetically encoded β-lactamase reporter for ultrasensitive ^{129}Xe NMR in mammalian cells. *Angewandte Chemie, International Edition*, *55*, 8984–8987. https://doi.org/10.1002/anie.201604055.

Wang, L., Wu, Y., Deng, Y., Kim, B., Pierce, L., Krilov, G., et al. (2015). Accurate and reliable prediction of relative ligand binding potency in prospective drug discovery by way of a modern free-energy calculation protocol and force field. *Journal of the American Chemical Society*, *137*, 2695–2703. https://doi.org/10.1021/ja512751q.

Wei, H., Liang, G., Yang, H., Wang, Q., Hawkins, B., Madesh, M., et al. (2008). The common inhalational anesthetic isoflurane induces apoptosis via activation of inositol 1,4,5-trisphosphate receptors. *Anesthesiology*, *108*, 251–260. https://doi.org/10.1097/01.anes.0000299435.59242.0e.

Weingarth, M., van der Cruijsen, E. A. W., Ostmeyer, J., Lievestro, S., Roux, B., & Baldus, M. (2014). Quantitative analysis of the water occupancy around the selectivity filter of a K^+ channel in different gating modes. *Journal of the American Chemical Society*, *136*, 2000–2007. https://doi.org/10.1021/ja411450y.

Willard, A. P., & Chandler, D. (2008). The role of solvent fluctuations in hydrophobic assembly. *The Journal of Physical Chemistry. B*, *112*, 6187–6192. https://doi.org/10.1021/jp077186+.

Willenbring, D., Liu, L. T., Mowrey, D., Xu, Y., & Tang, P. (2011). Isoflurane alters the structure and dynamics of GLIC. *Biophysical Journal*, *101*, 1905–1912. https://doi.org/10.1016/j.bpj.2011.09.026.

Woll, K. A., Dailey, W. P., Brannigan, G., & Eckenhoff, R. G. (2016). Shedding light on anesthetic mechanisms: application of photoaffinity ligands. *Anesthesia and Analgesia*, *123*, 1253–1262. https://doi.org/10.1213/ANE.0000000000001365.

Woll, K. A., Peng, W., Liang, Q., Zhi, L., Jacobs, J. A., Maciunas, L., et al. (2017). Photoaffinity ligand for the inhalational anesthetic sevoflurane allows mechanistic insight into potassium channel modulation. *ACS Chemical Biology*, *12*, 1353–1362. https://doi.org/10.1021/acschembio.7b00222.

Woll, K. A., Skinner, K. A., Gianti, E., Bhanu, N. V., Garcia, B. A., Carnevale, V., et al. (2017). Sites contributing to TRPA1 activation by the anesthetic propofol identified by photoaffinity labeling. *Biophysical Journal*, *113*, 2168–2172. https://doi.org/10.1016/j.bpj.2017.08.040.

Wong, C. F., & McCammon, J. A. (1986). Dynamics and design of enzymes and inhibitors. *Journal of the American Chemical Society*, *108*, 3830–3832. https://doi.org/10.1021/ja00273a048.

Yarov-Yarovoy, V., DeCaen, P. G., Westenbroek, R. E., Pan, C.-Y., Scheuer, T., Baker, D., et al. (2012). Structural basis for gating charge movement in the voltage sensor of a sodium channel. *Proceedings of the National Academy of Sciences of the United States of America*, *109*, E93–E102. https://doi.org/10.1073/pnas.1118434109.

Yu, F. H., & Catterall, W. A. (2004). The VGL-chanome: A protein superfamily specialized for electrical signaling and ionic homeostasis. *Science Signaling*, *2004*, re15. https://doi.org/10.1126/stke.2532004re15.

Zauhar, R. J., Gianti, E., & Welsh, W. J. (2013). Fragment-based shape signatures: A new tool for virtual screening and drug discovery. *Journal of Computer-Aided Molecular Design*, *27*, 1009–1036. https://doi.org/10.1007/s10822-013-9698-7.

Zauhar, R. J., Moyna, G., Tian, L., Li, Z., & Welsh, W. J. (2003). Shape signatures: A new approach to computer-aided ligand- and receptor-based drug design. *Journal of Medicinal Chemistry*, *46*, 5674–5690. https://doi.org/10.1021/jm030242k.

Zhang, B., Tian, M., Zhen, Y., Yue, Y., Sherman, J., Zheng, H., et al. (2012). The effects of isoflurane and desflurane on cognitive function in humans. *Anesth Analg*, *114*, 410–415. https://doi.org/10.1213/ANE.0b013e31823b2602.

Zhang, J., Qu, X., Covarrubias, M., & Germann, M. W. (2013). Insight into the modulation of Shaw2 Kv channels by general anesthetics: Structural and functional studies of S4–S5 linker and S6 C-terminal peptides in micelles by NMR. *Biochimica et Biophysica Acta (BBA)—Biomembranes*, *1828*, 595–601. https://doi.org/10.1016/j.bbamem.2012.09.025.

Zhou, D. W., Mowrey, D. D., Tang, P., & Xu, Y. (2015). Percolation model of sensory transmission and loss of consciousness under general anesthesia. *Physical Review Letters*, *115*(10), 108103 (Epub 2015 Sep 4. PubMed PMID: 26382705; PubMed Central PMCID:PMC4656020).

Zwanzig, R. W. (1954). High-temperature equation of state by a perturbation method. I. Nonpolar gases. *The Journal of Chemical Physics*, *22*, 1420–1426. https://doi.org/10.1063/1.1740409.

CHAPTER THREE

Molecular Mechanics Parameterization of Anesthetic Molecules

Thomas T. Joseph*,[1], Jérôme Hénin†
*Perelman School of Medicine, University of Pennsylvania, Philadelphia, PA, United States
†Laboratoire de Biochimie Théorique, Institut de Biologie Physico-Chimique, CNRS, Paris, France
[1]Corresponding author: e-mail address: thomas.joseph@uphs.upenn.edu

Contents

Abstract

Anesthetic drug molecules are being increasingly studied through the use of computational methods such as molecular dynamics (MD). Molecular mechanics force fields require the investigator to supply parameters for the force field equation, which are not available for novel molecules. Careful selection of these parameters is critical for simulations to produce meaningful results. Therefore, this chapter presents a state-of-the-art method for determining these parameters by comparison to quantum mechanics calculations and experimental quantities. Ketamine is used as an example to demonstrate the process.

Highlights

- Anesthetic molecules are often studied using computational methods.
- A method for rigorously choosing molecular mechanics parameters is given.
- This can enhance the accuracy of computational studies of anesthetics.

Methods in Enzymology, Volume 602
ISSN 0076-6879
https://doi.org/10.1016/bs.mie.2018.01.003

1. INTRODUCTION

1.1 Molecular Mechanics

Molecular dynamics (MD) simulations are often employed in the study of molecular systems relevant to anesthetic function. Essentially, an MD simulation considers atoms to be soft spheres with point charges and velocities, and propagates Newton's second law of motion over time, generating a *trajectory* of atomic paths and energies. This simplification of reality is called *molecular mechanics* (MM)—as opposed to quantum mechanics (QM). MM is designed to produce a reasonable approximation to the experimental behavior of a system. Various thermodynamic properties of MM systems can be derived using statistical mechanics, which is a subfield of physics that formalizes the relationship between the microscopic behavior of a collection of particles and macroscopic properties. Within the MM framework, the microscopic behavior of these particles is governed by a relatively simple potential energy function, an example of which is shown here:

$$\begin{aligned} V = & \sum_{\text{bonds}} k_b (b - b_0)^2 + \sum_{\text{angles}} k_\theta (\theta - \theta_0)^2 + \sum_{\text{Urey-Bradley}} k_u (u - u_0)^2 \\ & + \sum_{\text{dihedrals}} k_\phi [1 + \cos(n\phi - \delta)] + \sum_{\text{impropers}} (k_\omega - \omega_0)^2 + \sum_{\text{residues}} V_{\text{CMAP}}(\Phi, \Psi) \\ & + \sum_{\text{nonbonded}} \epsilon \left[\left(\frac{R_{ij}^{\min}}{r_{ij}} \right)^{12} - \left(\frac{R_{ij}^{\min}}{r_{ij}} \right)^{6} \right] + \frac{q_i q_j}{\epsilon r_{ij}} \end{aligned}$$

This is the CHARMM force field equation (Brooks et al., 2009). While this equation looks long and complicated, it is actually conceptually quite simple. Different types of interactions are deemed to be separable and additive, and are expressed with simple functional forms. Many excellent treatments of the theory behind MM exist which justify these approximations (Cramer, 2004; Leach, 2001); therefore, we will limit the scope of this chapter to practical aspects of preparing ligand parameters, with a special emphasis on anesthetics. We will also limit our scope to "all-atom" force fields, which represent each atom with one discrete particle. By contrast, "united-atom" or "coarse-grained" force fields may use a single coarse-grained particle to represent more than one atom, in the process sacrificing precision for computational efficiency.

Many parameters (k_b, k_u, etc.), which are specific to the particular atoms being considered, must be chosen. This has been done already for most atoms in biomacromolecular systems, and these parameters have been extensively validated. But it may not have been done for a particular molecule of interest. Here, we describe our methodology for choosing these parameters, a process called "parameterization." What may be the optimal parameterization methodology is a matter of debate, but we hope to provide a good starting point. This chapter is written with the CHARMM all-atom force field (Brooks et al., 2009), NAMD molecular dynamics software (Phillips et al., 2005), and Gaussian QM software (Gaussian, Inc., Wallingford, CT, USA) in mind, but the concepts are applicable to others. Ketamine, a drug in common use as an anesthetic, is presented as an example.

1.2 Bonded vs Nonbonded Terms

Bonded terms restrict the motion of atoms that are connected by bonds and encompass bonds, angles, Urey–Bradley angles, dihedral angles ("dihedrals" for short), and improper dihedral angles (impropers). Bonded terms define bonds and angles between atoms as harmonic oscillators, with an equilibrium value and force constant. This represents the fact that stretching or squeezing a bond, or distorting an angle, is energetically unfavorable. An example of this would be the equilibrium O–H bond length of 0.9572 Å and H–O–H angle of 104.5 degrees in water, which are explicitly made part of the widely used TIP3P water model (Jorgensen, Chandrasekhar, Madura, Impey, & Klein, 1983). A crucial property of harmonic bonds is that they are nondissociative: energy grows rapidly as a function of bond length, so that the bond may not break. Thus, the bonded chemical structure remains fixed in such MM simulations, which may therefore not describe chemical reactions. Those require more complex and computationally expensive strategies, such as "reactive" force fields or QM simulations.

Dihedral terms allow for refining conformational preferences around rotatable bonds that are not accurately modeled by other parts of the MM equation. Improper dihedral terms are typically used to enforce the planarity of sets of four atoms.

Nonbonded terms in the equation describe van der Waals and Coulombic (electrostatic) interactions. Van der Waals interactions are usually modeled with the 6–12 Lennard-Jones potential (Jones, 1924), which includes an attractive term describing dispersion forces between atoms, and a very short-distance core repulsion, which combine to yield an equilibrium distance. The Lennard-Jones equation is computationally efficient: the dispersion energy behaves as the inverse 6th power of the distance and the repulsion term is defined as the inverse 12th power. These can be easily calculated as the products of second and third powers. So even though more accurate potentials exist, the Lennard-Jones potential is preferred in MM calculations.

As noted earlier, each atom is assigned a *partial charge*, which is related to the overall molecular charge distribution and does not need to be (and rarely is) an integer, as it represents a notional partitioning of the overall charge density among the particles in the simulation. Note that this approximation of a continuous, fluctuating charge density as a constellation of static point charges is a very real limitation of MM. These charges are used with the Coulomb equation to determine the electrostatic interaction energy, and hence forces. In biomacromolecules at physiological temperature, electrostatic interactions normally contribute the bulk of the potential energy. Be sure to consider the pK_a of the ligand being parameterized, which normally is available experimentally. If it is likely to exist in a protonated form at physiological pH, you will need to study the protonated form. Finally, consider whether environment effects such as a hydrophobic or charged cavity might create enough of a pK_a shift to change the dominant species.

2. OVERVIEW OF PARAMETERIZATION METHOD

The overall workflow suggested here is a progressive refinement of an MM energy surface by fitting to a QM energy surface, along the way choosing force field parameters that optimize the fit. The resulting parameters can be used to compute experimental quantities such as partition coefficients, as well as solvation and binding energies, for validation. A detailed approach for CHARMM all-atom models is outlined in Mackerell et al. (1998).

The QM level of theory and basis set should be chosen to capture the behavior of the ligand in question, and normally should be in accordance with the development philosophy of the force field employed. In the case of CHARMM, this is something like B3LYP/6-31G**. A treatment of QM methods is outside the scope of this chapter. Suggested introductory reading includes (Cramer, 2004; Leach, 2001).

It is imperative to start from an energy-minimized structure when generating any type of QM energy surface for fitting. Such a structure represents a stationary point on the potential energy surface and is easily obtained using the geometry optimization function of a QM software, such as with the *Opt* keyword in Gaussian. If this is not done, the derived parameters will erroneously stabilize what would be an experimentally unstable structure, and in particular the Hessian matrix (discussed later) will include spurious derivatives reflecting restorative forces to a stationary point.

A potential energy surface is generated by varying some parameter along a degree of freedom of interest and calculating a single-point energy at each of those points. Here, each of these points represents a new ligand configuration. For example, one might vary a particular dihedral angle involving a bulky group that would be expected to have high-energy interactions in some positions: this is depicted in Fig. 1.

At each point, a constrained QM geometry optimization (keeping the selected coordinate constant and optimizing the rest) and single-point energy calculation are done. At each of the same points, a short MM minimization of the structure is undertaken by analogy to the QM geometry optimization, for practical reasons, a strong harmonic restraint is often used instead of a geometric constraint. Then, a single-point MM energy is computed, in gas phase, without periodic boundary conditions. Because MM potentials do not have a physically significant energy reference, the energy zero points will not match, but the energy surfaces can be aligned with zero energy being defined as the original optimized molecular configuration.

It is most important to meticulously fit the low-energy regions, because the lowest energy states are the most probable states. In a Boltzmann distribution, the probability $P(x)$ of a state x is related to its energy $E(x)$ by

$$P(x) \sim \exp[-E(x)/RT]$$

Here, R represents the molar gas constant and T represents the temperature (see Fig. 2 for a plot of this equation).

A high-energy peak is extremely unlikely to be accessed in MD simulation because these simulations are normally done at near physiological temperatures. For example, if QM calculations identify a peak with height 50 kcal/mol, it is acceptable for the MM energy surface to peak at, say, 45 or 55 kcal/mol, because the probability of the states with those energies is incredibly small. On the other hand, if QM calculations identify a peak with

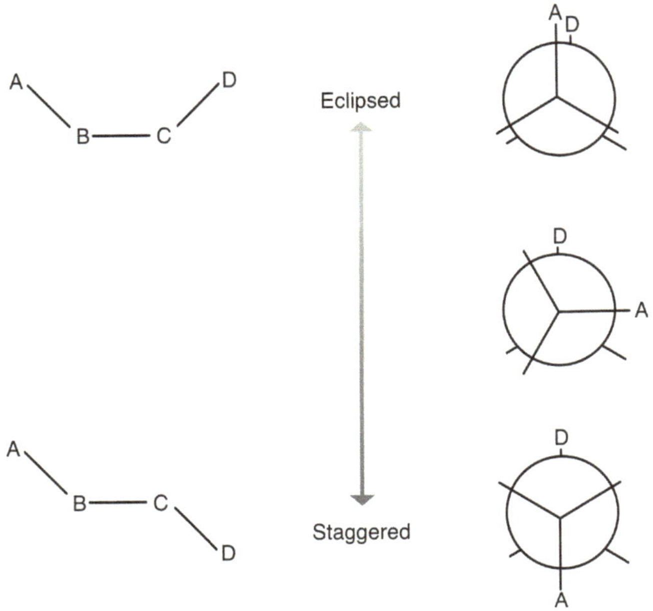

Fig. 1 Notional diagram of butane. The four carbon atoms are labeled A, B, C, and D, and hydrogens are omitted for clarity. Again ignoring hydrogens, bond terms are A–B and C–D, which would be equivalent, and B–C. Angle terms are A–B–C and B–C–D, also equivalent, and the dihedral term is A–B–C–D. The eclipsed version of the dihedral is at the *top* and staggered at the *bottom*. Newman projections are at the *right*. The eclipsed conformation is higher energy than the staggered conformation, because carbons A and D are further apart in the staggered conformation.

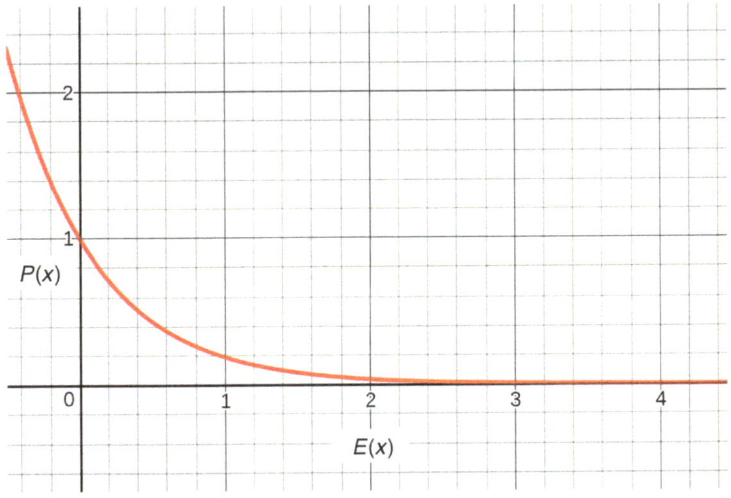

Fig. 2 Probability of a state as a function of energy. Note that the probability drops as $\exp(-E)$, which means that the probability of high energy states is quite low.

height 1 kcal/mol, the MM energy surface should reproduce this to yield realistic transitions across that barrier.

Some of these steps can be automated with software such as Force Field Toolkit (FFTK) (Mayne, Saam, Schulten, Tajkhorshid, & Gumbart, 2013) which is included with recent versions of VMD (Humphrey, Dalke, & Schulten, 1996). Indeed, we describe its use as such below. Keep in mind though that while it is a time saver, it is imperative to double-check its output at every stage, e.g., by visualizing the results of QM geometry optimizations or conformational scans.

3. METHODS FOR CHOOSING PARAMETERS

3.1 Implementation Details

Here, we discuss briefly the implementation details of working with the CHARMM force field. The interested reader is advised to peruse the CHARMM force field website and documentation for more information (http://mackerell.umaryland.edu/charmm_ff.shtml), particularly including the details of the relevant file formats. These include the Residue Topology File (.rtf) and parameter file (.prm). These are both text files that may be combined into a single "stream" (.str) file; this capability is an artifact of its original use with the CHARMM molecular dynamics software. The .rtf file describes how atoms are connected in each type of molecule. Amino acids, nucleic acids, and various lipids are defined, as well as water, ions, and others. Atom types, masses, and charges are specified. Atom charges are molecule-specific, and the same atom type may occur multiple times in a single residue with different charges. The specific MM force field parameters used in the simulation are then a function of the atom types and how they are connected. Those parameters—bond, angle, dihedral terms, as well as Lennard-Jones parameters—reside in the .prm file.

Each atom in an MM system has a *type* and a *name*. An atom type identifies the correct parameters to use for that atom. Chemically similar atoms may be assigned the same atom type if energetically appropriate. For example, the C2 and C3 atoms in phenol are both assigned the CG2R61 atom type, even though one is ortho and the other is meta from the hydroxyl group. On the other hand, an aliphatic carbon behaves differently from an aromatic carbon, and these require separate parameters: among other differences, a C–C aromatic bond has a different equilibrium length and force constant than a C–C alkane bond. The C2 and C3 atoms in cyclohexanone are assigned the CG321 atom type.

We assume that atom types and their force field terms can be reused across different molecules; this assumption is called *transferability*. Atom names, on the other hand, are just names used as identifiers, and are unique to atoms within the ligand or residue. When possible these follow common conventions for ease of reading.

Before constructing .rtf and .prm files of your own, it will be extremely helpful to peruse these files as included in the CHARMM force field in conjunction with the file format documentation to understand how parameters for a given molecule are retrieved and plugged into the force field equation. The suggested approach here involves starting with the CGenFF extension of CHARMM, which is contained in the top_all36_cgenff.rtf and par_all36_cgenff.prm files. For the example provided earlier, see the parameters including CG2R61 and CG321 atom types, respectively, as well as PHEN and CHON residues.

3.2 Automatic Parameterization Software

Some force fields include molecular fragments that have already been parameterized, and all that may be required is to stitch these fragments together to produce a complete molecule. This is an ideal starting point because these fragments have usually been extensively validated for use with the other parameters in the force field. Various software is available to automate this task, such as ParamChem and SwissParam for CHARMM-compatible parameters, and Antechamber for AMBER-compatible parameters. However, these services may not always provide an adequate set of parameters. This can occur when:

- There is a potential energy term that cannot be varied independently (e.g., two dihedrals with interacting groups) and further correction terms are required (propofol hydroxyl group, see Section 3.5);
- Not all dihedral parameters are present or accurate (dexmedetomidine);
- A novel functional group is present (diazirines);
- Atoms joining functional groups are not yet parameterized (ketamine).

Some software, such as ParamChem, gives a penalty score that provides an idea how close the match is. If the software does not think it made a good match, you should validate and possibly tweak the parameters, by using the generated parameters as a starting point for the procedures described later. Even if the parameters themselves are completely wrong, this type of software automates the creation of skeletal .rtf and .prm files that can be subsequently modified.

3.3 Creating a Residue Topology With Atom Types and Partial Charges

The first step is to create the topology of the molecule by choosing atom types and charges and defining the bonds among them yourself, or using automatic parameterization software to provide a starting point. A reasonable starting point for assigning atom types and partial charges is an existing molecule in a given force field. For example, CHARMM includes CGenFF (Vanommeslaeghe & MacKerell, 2012; Vanommeslaeghe, Raman, & MacKerell, 2012), which contains parameters for many common functional groups that can be assembled into a complete molecule.

In most cases, atom types may be taken from groups containing the same elements in similar chemical environments; the associated Lennard-Jones parameters can then be reasonably expected to transfer. If this process does not give good results and refinement of Lennard-Jones parameters is necessary, a higher level of care and expertise might be necessary, because Lennard-Jones parameters are generally underdetermined and reliable data (usually pure condensed phase properties) to fit them are not always available. Therefore, this falls outside the scope of this piece.

Once this is done, refinement of partial charges can begin. Many partial charges in the CHARMM force field were refined by fitting their interaction energy with water using the TIP3P model from MM calculations against those from QM calculations (Mackerell et al., 1998) and are therefore a useful approach for anesthetics (Hénin, Brannigan, Dailey, Eckenhoff, & Klein, 2010).

Here is our suggested approach to assigning partial charges:

1. Find the most similar functional groups available in the force field to those comprising your molecule. For each atom, consider valence, bond order, and whether the atom is part of a ring. Using these as a guide, write the topology of your ligand. You must define all bonds between atoms. Or, use a parameterization server as described earlier to generate an initial guess, which will also generate a valid topology file. Some of the parameters chosen may be wrong, but you will refine them. In certain cases, for very similar molecules to compounds that are already parameterized, the parameters generated by such a server may be adequate as they are, and the rest of the steps can be avoided.

2. Ensure that the sum of the charges is correct. A neutral ligand should of course have a total charge of zero. In any event, the ligand cannot have a noninteger charge for an accurate simulation. NAMD will produce a warning but will not stop execution for a noninteger charge. Make an educated guess as to how to tweak the charges by examining other similar molecules if available.
3. Determine which atoms have the most questionable charges. This is to some extent a judgment call as you do not know a priori how reasonable the assigned partial charges are. Of note, charges with a magnitude greater than 1 are usually too large. Generally, functional groups appearing in your molecule should be similar to those already in the CHARMM or CGenFF residue library.
4. For each of these atoms, run a QM geometry optimization, with a water molecule in a hydrogen bonding configuration with that atom. Repeat with the water in multiple initial positions in an arc around the atom. In Gaussian, this type of "arc scan" can be automated by specifying the water part of the system as a *Z*-matrix and varying the appropriate angle using the *Opt=ModRedundant* keyword. The end-point interaction energy from each of these calculations is printed with units of hartrees in the output.
5. Evaluate the same ligand–water configurations as an MD single-point interaction energy calculation after a short minimization. Compare this energy profile with the QM energy profile and tweak charges as necessary to achieve a closer fit. Repeat as necessary.

While tools such as FFTK can automate the derivation of partial charges, unsupervised charge fitting calculations can produce bizarre values that are at odds with standard values in the force field. For example, in most six-membered aromatic rings with the CHARMM force field, the carbons have charge −0.115 and the hydrogens +0.115, but FFTK will not necessarily end up at these values.

Ketamine example. For simplicity, we will discuss deprotonated ketamine, even though at physiologic pH the protonated form dominates. We consider here the S-(+) enantiomer because that is more clinically potent. Ketamine consists of cyclohexanone and chlorobenzene functional groups joined together, with a methylamine group bonded to the cyclohexanone (see Fig. 3). The former two functional groups are already available in CGenFF as the CHON and CHLB residues, respectively, and so they were copied directly. The methylamine was built by analogy with the dimethylamine (DMAM) residue. The charges did not add up to zero so the charges near

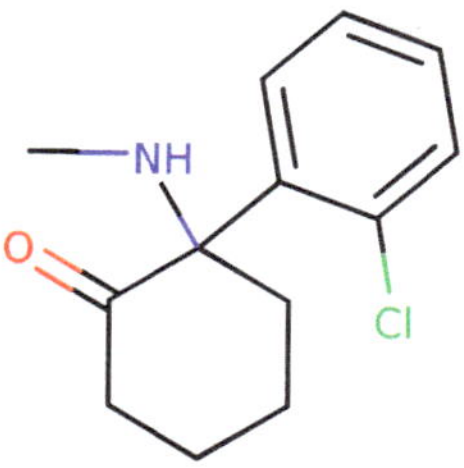

Fig. 3 Diagram of ketamine. Note the constituent functional groups: cyclohexanone, chlorobenzene, and methylamine.

the interfaces between the groups were adjusted by hand to produce a reasonable starting point. To refine the charges, arc QM water scans on the nitrogen and chlorine were done as described earlier.

3.4 Bonds and Angles

Recall that in MM bonds are modeled as harmonic springs. Given sufficient force separating the two-bonded atoms, a real atomic bond will break, and the electron density will redistribute between the two pieces. As molecular dynamics is not intended to sample very high energy states, it does not attempt to model covalent bonds breaking, and so the bond potential equation can be kept simple for computational efficiency. Thus in the harmonic spring approximation, as the atoms are separated, the restoring force increases linearly and energy required increases quadratically. At great distances this is wildly inaccurate, but since we are only concerned with relatively low-energy configurations close to equilibrium, this is acceptable.

Nominally, bond and angle terms should reproduce vibrational spectra. However, these may not be readily available for comparison for your molecule of choice. Therefore, we adopt the approach taken by the FFTK (Mayne et al., 2013), which automates this calculation through the use of the Hessian matrix, which contains partial second derivatives of the energy with respect to each atom coordinate, and can be calculated ab initio from QM.

1. Make a list of the atom types in your ligand and make note of all bond or angle terms already present that include any of these atom types. Make note of any bonded atoms in your ligand that do not have bond or angle terms. These are the terms that will need to be predicted or refined.
2. Using as a starting point the geometry-optimized structure already determined using QM optimization, calculate the Hessian matrix at the same

level of theory and with the same basis set used for the geometry optimization. In Gaussian this is done using the *Freq* keyword.

3. Using the Hessian matrix, you can approximate the local shape of the potential energy surface by small perturbations of bonds and angles, and use the Hessian force constants to calculate the resulting potential energy. The MM bond and angle force constants can then be fitted to reproduce these local energy surfaces. This step is automated by the FFTK.

Urey–Bradley terms allow modifications of the potential energy as a function of the distance between the first and third atoms in an angle: effectively, it adds a bond between these atoms. These are poorly transferable and are only useful for reproducing subtle aspects of vibrational spectra. Therefore, they are not required for pharmacological applications, and will not be further discussed here.

Ketamine example. Many of the bond and angle terms exist in CGenFF, so these are used directly. However, bond and angle terms involving the atoms joining the three main functional groups in ketamine did not exist in CGenFF. Therefore, the Hessian was calculated and FFTK was used to automate the process of fitting these bond and angle terms.

3.5 Dihedrals

Dihedral terms give a potential energy as a function of dihedral angle. They are important when other terms are not sufficient to properly reproduce the potential energy surface.

Consider butane. The four carbon atoms—let's call them A, B, C, and D—can be twisted, with the midpoint defined as a plane perpendicular to the B–C bond. Importantly, the potential energy surface of this rotation is periodic. This lends itself to the functional form of the dihedral potential energy expression. Dihedral terms are expressed with a force constant multiplied by a cosine which provides periodicity and phase. As with the other types of potential energy terms, dihedral terms are additive: it is often the case that multiple dihedral terms with different coefficients will be needed for the same four atoms in order to properly reproduce the potential energy surface as the atoms are rotated—but not always. When several cosine terms with different periods are added, the dihedral potential is effectively defined as a truncated Fourier series. One should strive to use the fewest possible dihedral terms to avoid adding spurious, underdetermined parameters.

1. Choose the dihedrals that are worth refining, or any that are required but not already included in the force field. Energetically important dihedrals describe those rotatable bonds that may produce transitions between low-energy conformers.
2. Run a dihedral torsion scan at the appropriate QM level. In Gaussian, this can be done with the *Mod=OptRedundant* keyword. We tend to use an angle increment of 15 degrees, for a total of 24 points.
3. Using your parameter set in progress, run an MM dihedral torsion scan at the same angles. To minimize the effect of small differences between MM and QM minima, at each step a short MM minimization is necessary prior to the single-point energy calculation.
4. Modify or add dihedral terms as necessary to improve the fit of the MM surface to the QM surface, and repeat the MM torsion scan. Remember that the QM and MM zero energies will be different, so they must be aligned by adding a constant.

In some cases, the energy profile of a given dihedral depends on another dihedral angle. This can happen when two bulky or charged groups, otherwise separated, can be brought near each other by torsional rotation. Such dihedrals must be optimized self-consistently by adjusting each of them in turn and iterating. In some cases, however, the nature of the coupling makes it impossible to capture using isolated dihedral potentials. An example is the hydroxyl group of propofol, where the hydrogen interacts with the isopropyl groups depending on their relative rotations (LeBard, Hénin, Eckenhoff, Klein, & Brannigan, 2012). In such cases, the CHARMM force field allows an explicit correction to be applied in the form of a potential energy term that couples both dihedrals. Initially developed for protein backbone corrections (Best et al., 2012), such “CMAP” correction terms have proven useful in ligand parameterization as well.

“Impropers” are a form of dihedral term that is used to maintain either planarity or a given chirality in groups of four atoms. They are not necessary for most molecules, and so they will not be discussed further here.

Ketamine example. Since N, O, and Cl atoms are in relatively close proximity, dihedrals involving the amine group as well as the atoms connecting the two rings were evaluated at 15-degree intervals by conducting QM geometry optimizations at each point. Dihedral terms were modified to closely reproduce the fit of MM to QM energy surface, with a particular focus on the low-energy regions. The amine methyl was thought likely to interact with the chlorobenzene, so torsions of these groups with respect to the central carbon were scanned together. Fitting MM to QM energies produced a table of CMAP correction terms.

3.6 Comparison With Experiment

There is no universally accepted method of validating new parameters. However, since the ultimate objective of molecular simulation is to predict real life, one reasonable approach is to use the parameters in an MD simulation to estimate some experimental result to show concordance of the simulation with reality. A dilemma may occur when a limited amount of precise experimental data must be divided between a training set and a test set for the model. It is then tempting to use most of the information in the training set to improve parameter quality, thus leaving little reference data for tests. Extensive use of QM calculations for parameterization brings the benefit of requiring little in the way of experimental observables, possibly paving the way for more extensive a posteriori tests.

A common and relatively simple experimental quantity is the octanol–water partition coefficient. Octanol and water are poorly miscible and will separate into layers when mixed. A third added chemical will partition itself between the two, reaching an equilibrium distribution. The ratio of the concentrations is an equilibrium constant, which can be related to a free energy change by the relation $\Delta G = -RT\ln K$, where R is the gas constant, T is the temperature, and K is the equilibrium constant. This describes the thermodynamic cost of transferring the chemical from one layer to another on a molar basis. In practice, the base-10 logarithm of the equilibrium constant is reported, and this is referred to as "log P" or "log K_{ow}." Note that predicted log P values are often reported as "cLogP" values, which are merely the result of a proprietary computation based on an empirical model (BioByte Inc., Claremont, CA) rather than experiment.

To compute log P of a molecule, one could simulate the process of moving one molecule from octanol to water or vice versa and determine the free energy cost of the geometrical translation, e.g., using the adaptive biasing force method (Darve & Pohorille, 2001) as implemented in NAMD and LAMMPS molecular simulation software (Hénin, Fiorin, Chipot, & Klein, 2010). Another method is to use free energy perturbation (FEP) to calculate the free energy of decoupling the molecule from octanol and recoupling it to water. Because free energy is a state function—free energy changes do not depend on the path—the sum is equal to the energy of moving the molecule from octanol to water. Our preferred approach is to simulate decoupling of the molecule from each solvent and subtracting the decoupling free energies to yield a value for ΔG. The log P can then be calculated according to

$$\log P = \Delta G / 2.303RT$$

The constant 2.303 is used for conversion between natural and base-10 logarithms.

Dissociation constants of the ligand with a known protein can be determined by experiment, and in some cases these can be compared in simulation. For example, many diazirine photoaffinity ligands bind to horse spleen apoferritin (HSAF) in a presumed binding site with known affinity (Oakley et al., 2012; Vedula et al., 2009). A standard free energy of binding can be computed using alchemical FEP. This is itself a substantial undertaking and convergence may be an issue, particularly when coupling/decoupling charged compounds. For HSAF, one needs only to simulate one dimer out of the 24-mer because the presumed binding site is located at a monomer interface. The details of implementing such a calculation are out of the scope of this review, but the interested reader is encouraged to review other literature such as Gilson and Zhou (2007).

4. SUMMARY

We described a state-of-the-art method for deriving MM parameters for small anesthetic molecules by comparison to QM energies as a gold standard, as well as certain experimental quantities. The parameters are compatible with the CHARMM force fields for proteins and lipids, allowing for simulations of drug–receptor interactions in complex environments, as well as binding affinity predictions. Ketamine was presented as an example molecule.

ACKNOWLEDGMENTS

This work was supported by the National Institutes of Health [Grant number 5T32GM112596-03] and by the French National Research Agency through Grant ANR-11-LABX-0011-01 (LABEX DYNAMO).

REFERENCES

Best, R. B., Zhu, X., Shim, J., Lopes, P. E. M., Mittal, J., Feig, M., et al. (2012). Optimization of the additive CHARMM all-atom protein force field targeting improved sampling of the backbone ϕ, ψ and side-chain $\chi 1$ and $\chi 2$ dihedral angles. *Journal of Chemical Theory and Computation*, *8*, 3257–3273.

Brooks, B. R., Brooks, C. L., Mackerell, A. D., Nilsson, L., Petrella, R. J., Roux, B., et al. (2009). CHARMM: The biomolecular simulation program. *Journal of Computational Chemistry*, *30*, 1545–1614.

Cramer, C. J. (2004). *Essentials of computational chemistry: Theories and models*. John Wiley & Sons, Ltd.

Darve, E., & Pohorille, A. (2001). Calculating free energies using average force. *The Journal of Chemical Physics*, *115*, 9169–9183.

Gilson, M. K., & Zhou, H.-X. (2007). Calculation of protein-ligand binding affinities. *Annual Review of Biophysics and Biomolecular Structure*, *36*, 21–42.

Hénin, J., Brannigan, G., Dailey, W. P., Eckenhoff, R., & Klein, M. L. (2010). An atomistic model for simulations of the general anesthetic isoflurane. *The Journal of Physical Chemistry. B*, *114*, 604–612.

Hénin, J., Fiorin, G., Chipot, C., & Klein, M. L. (2010). Exploring multidimensional free energy landscapes using time-dependent biases on collective variables. *Journal of Chemical Theory and Computation*, *6*, 35–47.

Humphrey, W., Dalke, A., & Schulten, K. (1996). VMD: Visual molecular dynamics. *Journal of Molecular Graphics*, *14*(33–38), 27–28.

Jones, J. E. (1924). On the determination of molecular fields. II. From the equation of state of a gas. *Proceedings of the Royal Society of London A: Mathematical, Physical and Engineering Sciences*, *106*, 463–477.

Jorgensen, W. L., Chandrasekhar, J., Madura, J. D., Impey, R. W., & Klein, M. L. (1983). Comparison of simple potential functions for simulating liquid water. *The Journal of Chemical Physics*, *79*, 926–935.

Leach, A. (2001). *Molecular modelling: Principles and applications*. Pearson.

LeBard, D. N., Hénin, J., Eckenhoff, R. G., Klein, M. L., & Brannigan, G. (2012). General anesthetics predicted to block the GLIC pore with micromolar affinity. *PLoS Computational Biology*, *8*, e1002532.

Mackerell, A., Bashford, D., Bellott, M., Dunbrack, R., Evanseck, J., Field, M., et al. (1998). All-atom empirical potential for molecular modeling and dynamics studies of proteins. *The Journal of Physical Chemistry. B*, *102*, 3586–3616.

Mayne, C. G., Saam, J., Schulten, K., Tajkhorshid, E., & Gumbart, J. C. (2013). Rapid parameterization of small molecules using the Force Field Toolkit. *Journal of Computational Chemistry*, *34*, 2757–2770.

Oakley, S., Vedula, L. S., Bu, W., Meng, Q. C., Xi, J., Liu, R., et al. (2012). Recognition of anesthetic barbiturates by a protein binding site: A high resolution structural analysis. *PLoS One*, *7*, e32070.

Phillips, J. C., Braun, R., Wang, W., Gumbart, J., Tajkhorshid, E., Villa, E., et al. (2005). Scalable molecular dynamics with NAMD. *Journal of Computational Chemistry*, *26*, 1781–1802.

Vanommeslaeghe, K., & MacKerell, A. D. (2012). Automation of the CHARMM General Force Field (CGenFF) I: Bond perception and atom typing. *Journal of Chemical Information and Modeling*, *52*, 3144–3154.

Vanommeslaeghe, K., Raman, E. P., & MacKerell, A. D. (2012). Automation of the CHARMM General Force Field (CGenFF) II: Assignment of bonded parameters and partial atomic charges. *Journal of Chemical Information and Modeling*, *52*, 3155–3168.

Vedula, L. S., Brannigan, G., Economou, N. J., Xi, J., Hall, M. A., Liu, R., et al. (2009). A unitary anesthetic binding site at high resolution. *The Journal of Biological Chemistry*, *284*, 24176–24184.

CHAPTER FOUR

Insights Into Receptor-Based Anesthetic Pharmacophores and Anesthetic–Protein Interactions

Victoria S. Fahrenbach*,†, Edward J. Bertaccini*,†,1
*Stanford University School of Medicine, Stanford, CA, United States
†Palo Alto VA Health Care System, Palo Alto, CA, United States
[1]Corresponding author: e-mail address: edwardb@stanford.edu

Contents

Abstract

General anesthetics are thought to allosterically bind and potentiate the inhibitory currents of the $GABA_A$ receptor through drug-specific binding sites. The physiologically relevant isoform of the $GABA_A$ receptor is a transmembrane ligand-gated ion channel consisting of five subunits (γ-α-β-α-β linkage) symmetrically arranged around a central chloride-conducting pore. Although the exact molecular structure of this heteropentameric $GABA_A$ receptor remains unknown, molecular modeling has allowed significant advancements in understanding anesthetic binding and action. Using the open-channel conformations of the homologous glycine and glutamate-gated chloride receptors as templates, a homology model of the $GABA_A$ receptor was constructed using the Discovery Studio computational chemistry software suite. Consensus structural alignment of the homology templates allowed for the construction of a three-dimensional heteropentameric $GABA_A$ receptor model with (γ_2-β_3-α_1-β_3-α_1) subunit linkage. An anesthetic binding site was identified within the transmembrane

Methods in Enzymology, Volume 602
ISSN 0076-6879
https://doi.org/10.1016/bs.mie.2018.01.004

α/β intersubunit space by the convergence of three residues shown to be essential for anesthetic activity in previous studies with mutant mice (β_3-N265, β_3-M286, α_1-L232). Propofol derivatives docked into this binding site showed log-linear correlation with experimentally derived $GABA_A$ receptor potentiation (EC_{50}) values, suggesting this binding site may be important for receptor activation. The receptor-based pharmacophore was analyzed with surface maps displaying the predominant anesthetic–protein interactions, revealing an amphiphilic binding cavity incorporating the three residues involved in anesthetic modulation. Quantum mechanics calculations of the bonding patterns found in complementary high-resolution receptor systems further elucidated the complex nature of anesthetic–protein interactions.

1. INTRODUCTION

General anesthetics like propofol and etomidate are thought to mediate their actions within the CNS by potentiating the inhibitory currents through transmembrane ligand-gated ion channels (LGICs) via drug-specific binding sites. The heteropentameric gamma-aminobutyric acid type A receptor ($GABA_AR$) in particular has been shown to be potentiated by the presence of general anesthetics (Jurd et al., 2003). This LGIC is a member of the Cys-loop superfamily and consists of five subunits (γ-α-β-α-β linkage) symmetrically arranged around a central ion-conducting pore. The extracellular domain is characterized by a large component of beta sheet secondary structure and contains the binding site for the native ligand (GABA), which is believed to convert the channel from a closed to open conformation upon binding, allowing chloride ion flux through the central pore with subsequent hyperpolarization of the neuron (Spurny et al., 2012). The transmembrane domain of each subunit is composed largely of four-helix bundles. It is within this transmembrane region that anesthetics and alcohols are thought to bind and convey the majority of their channel modulation, with mutation studies revealing specific residues essential for anesthetic activity (β_3-N265, β_3-M286, α_1-L232) (Bertaccini, Wallner, Trudell, & Lindahl, 2010; Bertaccini, Yoluk, Lindahl, & Trudell, 2013). By identifying anesthetic binding cavities within this region, novel drugs can be designed to bind specifically within these sites, with the hypothesis that ideal pharmacodynamics can be achieved by optimizing ligand–receptor interactions.

Since the LGICs are transmembrane proteins, their isolation and purification have proven difficult. High-resolution crystal structures of the heteropentameric $GABA_AR$ in the open state, which anesthetics are thought to stabilize, do not exist (Spurny et al., 2012). However, significant progress has been made toward understanding the structure of LGICs and their interactions with anesthetics via methods of molecular modeling, in which theories of molecular mechanics and quantum chemistry are used in conjunction with advanced computational science to simulate the behavior of molecules at the atomic level. The initial method of constructing and studying receptors of unknown structure lies in the techniques of homology modeling.

Homology modeling is the method by which the amino acid sequence of a protein of unknown structure is aligned and threaded over that of a closely related amino acid sequence from a homologous protein with known three-dimensional structure, such that the structural coordinates of the known protein can be transferred to those of the unknown. While homology modeling involves a great deal of computational theory, it is also very dependent on experimentally described coordinates of "template" proteins with high-sequence homology to the desired protein (Bertaccini et al., 2010; Murail, Wallner, Trudell, Bertaccini, & Lindahl, 2011). In recent years, several templates with significant homology to the LGICs have been determined via cryo-electron microscopy, X-ray crystallography, and nuclear magnetic resonance (NMR) spectroscopy and have made model construction more robust. Open-channel structures of the glutamate-gated chloride (GluCl) and glycine (GlyR) receptors in particular were recently published (Du, Lu, Wu, Cheng, & Gouaux, 2015; Hibbs & Gouaux, 2011), demonstrating the conformational state to which anesthetics bind and stabilize. As members of the Cys-loop superfamily of LGICs, these structures can be used as homologous templates for the construction of a heteropentameric $GABA_AR$ model. With a validated homology model, receptor-binding sites can be determined and characterized for drug discovery purposes using specialized modeling software. Small molecules are docked in silico into binding pockets and scored based on their binding affinity, which is a function of the relative free energy change of ligand binding. Docking protocols determine molecules with the highest binding affinities, as reflected by the lowest (most stable) bound energy states (Bertaccini et al., 2013).

Such exploration and characterization of potential anesthetic binding sites within $GABA_AR$ are necessary to better understand anesthetic ligand interactions and receptor modulation. While this approach allows for high-throughput in silico ligand screening to discover novel anesthetic compounds, quantum mechanics can offer a detailed analysis of the specific anesthetic–protein interactions occurring within receptor-binding cavities.

More refined analyses can be performed on higher resolution, experimentally derived structures that are not as prone to the variability one might find in a homology model. By utilizing in silico computational chemistry to examine currently available high-resolution crystal structures of bound anesthetics to a variety of proteins, additional insights into the mechanisms of anesthesia have also been gained.

2. HOMOLOGY MODELING

In order to construct an in silico three-dimensional model, the amino acid sequence of the target (model) receptor must be aligned to homologous receptors with known biologically relevant protein conformations. Homologous receptors are identified via their sequence similarity to the target protein and serve as structural "templates" for the modeling process. As noted later, careful selection of appropriate receptor isoforms is critical for the success and reliability of model construction.

2.1 Creating a Homology Model of the $GABA_AR$

The amino acid sequences for the alpha-1 (α_1), beta-3 (β_3), and gamma-2 (γ_2) subunits of the human $GABA_AR$ ($hGABA_AR$) were obtained from the National Center for Biotechnology Information (NCBI) (https://www.ncbi.nlm.nih.gov/). Subunit isoforms were selected based on the specific sequence characteristics listed in the FASTA and GenPept reports. For instance, the gamma-2 isoform with the shortest sequence (gam2iso2pre) allowed for easier model construction due to the truncated intracellular TM3–4 loop.

By inputting the mature subunit sequences of the $hGABA_AR$, homologous template receptors with known three-dimensional structures were identified based on sequence similarity via redundant Blast and RCSB search functions. In Discovery Studio (Dassault Systèmes BIOVIA, Discovery Studio, Software version 4.1, San Diego: Dassault Systèmes), a Blast search was performed by importing the $hGABA_AR$ subunit sequences and applying the "Search Sequence Similarity" tool with default NCBI server settings. Similarly, a sequence alignment search for homologous protein receptors was performed online via the Research Collaboratory for Structural Bioinformatics, Protein Data Bank website (RCSB.org). Both search protocols yielded similar results, and the α_1-GlyR (PDB ID 3JAF = *alpha-1 Glycine Receptor (ECM)—glycine and ivermectin* bound) and GluCl receptor (PDB

ID 3RIF = *C. elegans GluCl with Fab, ivermectin, glutamate*) were selected as homology templates based on their high level of sequence similarity to $hGABA_AR$. These receptors, both members of the Cys-loop receptor superfamily, are structurally similar with five subunits symmetrically arranged around a central ion pore (Wolters & Bickelhaupt, 2012). The specific 3JAF GlyR and 3RIF GluCl receptor files were selected due to the presence of both orthosteric and allosteric binding sites in addition to their open or partially open state. Although the ligands were removed (along with all other heteroatoms) in preparation for model construction, the agonist-induced open conformations of these receptor templates were utilized to build the $hGABA_AR$ model. Recent evidence suggests that for the LGICs, which mediate a hyperpolarizing chloride ion flux, the effects of anesthetics on these channels are to bind to and stabilize the open state of the receptor (Forman & Miller, 2011; Sauguet, Shahsavar, & Delarue, 2015). This becomes critically relevant given that most of the homologous templates to the $hGABA_AR$ sequence in question represent closed, partially closed, or desensitized states.

In preparation for the modeling process, the 3JAF GlyR and 3RIF GluCl PDB structures were cleaned and modified in Discovery Studio, beginning with the deletion of ligands from their binding sites. For the GluCl receptor (3RIF), the extra side antibody subunits involved in the precipitation process and attached to subunits A–E were also removed so that only the conserved core remained for modeling, consistent with the GlyR 3JAF and heteropentameric $hGABA_AR$ structures, as evidenced in the literature (Wolters & Bickelhaupt, 2012).

Variants of the Align123 protocol from Discovery Studio were used to align multiple protein sequences based on the ClustalW progressive pairwise alignment algorithm and BLOSUM 30 pairwise alignment scoring matrix. A template sequence alignment profile was created by superimposing the 3JAF and 3RIF PDB files using secondary structure prediction to align the main chain atoms (see Fig. 1). The sequence for each $hGABA_AR$ subunit was then individually aligned to the corresponding subunit of the template sequence alignment profile, generating five sequence alignment threads (one for each subunit).

The aligned sequence profiles were manually modified to facilitate model construction (Bertaccini et al., 2013). Using the linked secondary structure of the template sequence profile as a guide, the $hGABA_AR$ TM3–4 intracellular loop was replaced with seven glycine residues. Gaps within the 3JAF/3RIF template sequence were deleted along with the

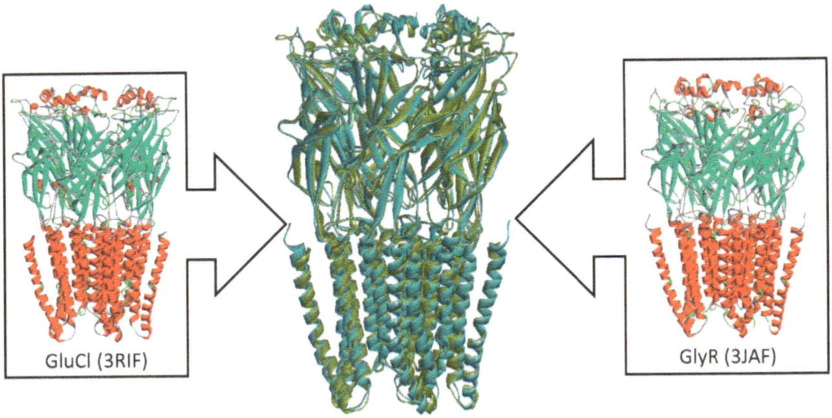

Fig. 1 Individual and overlapped solid ribbon models of the templates chosen for homology modeling. The glutamate-gated chloride channel (GluCl, 3RIF, *upper left*) and the alpha-1 glycine homomer (GlyR, 3JAF, *upper right*) showing transmembrane alpha helices (*red*) and extracellular beta sheets (*light blue*). These were overlapped using the Align and Superimpose Structures Algorithm within the Discovery Studio 2016 Suite. The *middle figure* shows the high degree of three-dimensional structural similarity between 3RIF (*green*) and 3JAF (*blue*).

corresponding $hGABA_AR$ residues. These modifications simplified subunit regions that lacked corresponding template elements without sacrificing regions involved in ligand binding. The modified $hGABA_AR$ subunit sequence profiles were then connected to create one continuous sequence thread representative of the complete heteropentameric $hGABA_AR$. The subunits were linked in γ_2-β_3-α_1-β_3-α_1 arrangement as is suggested in the clockwise orientation work done by Trudell (2002). This continuous sequence thread was then manually realigned to the template sequence profile based on the individual subunit alignment "blueprints" that were initially created.

The three-dimensional $hGABA_AR$ homology model was then created by crossreferencing the GABA, 3JAF, and 3RIF sequence alignment profile to the "Linked Template Structures" (3JAF, 3RIF). To construct a three-dimensional model of the $hGABA_AR$ (Bertaccini et al., 2013), the Modeler algorithm assigned coordinates to the aligned $hGABA_AR$ residues using the GluCl/GlyR structural alignment profile as a three-dimensional blueprint. Loop construction and side-chain refinement were also performed using the Modeler algorithm. A CHARMm-based receptor minimization with fixed backbone constraints optimized the overall three-dimensional $hGABA_AR$ structure shown in Fig. 2A.

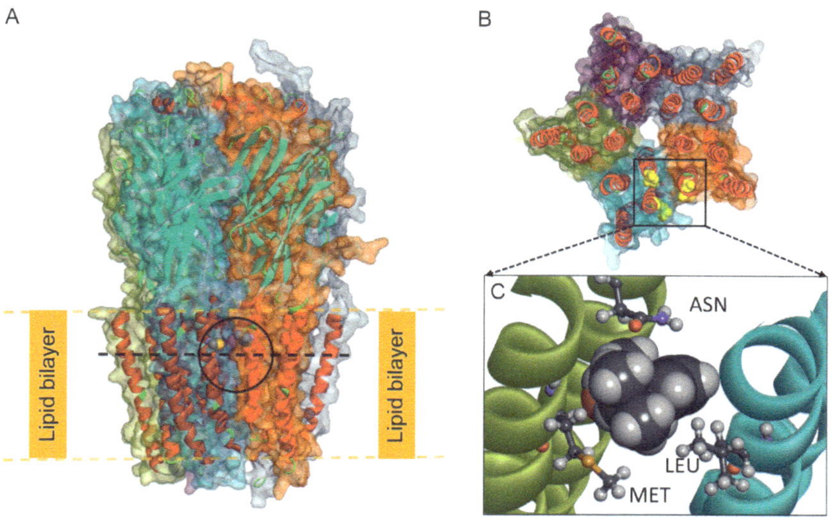

Fig. 2 (A) Homology model of the $hGABA_AR$ with an anesthetic binding pocket (ABP) located within the outer third of the transmembrane domain (*circled region*). The surface of each subunit is individually *colored* for distinction, yet transparent to reveal the underlying secondary structures. (B) Cross section of the transmembrane domain composed of a pentamer of four-helix bundle subunits (*red ribbons*) at the level of the *black dashed line* in (A). The ABP is in the *rectangular region*. (C) Magnified intersubunit anesthetic binding site with propofol in space-filling rendering. Note the surrounding three key amino acids that when mutated are known to modulate anesthetic potentiation of $GABA_AR$. They are β_3-ASN265, β_3-MET286, and α_1-LEU232.

2.2 Creating an Anesthesia Binding Site Within the $hGABA_AR$ Model

After obtaining an energetically minimized, optimized homology model of the heteropentameric $hGABA_AR$, it was necessary to create a ligand-binding cavity in order to help verify the reliability of the model since optimization in vacuo allowed occlusion of the binding site due to side-chain rotations. This binding cavity would also be used in future studies involving in silico high-throughput ligand screening for drug discovery purposes.

Based on previous work that revealed a putative anesthetic binding site within the transmembrane intersubunit space (Bertaccini et al., 2013), the α/β subunits were extracted from the $hGABA_AR$ homology model while maintaining the alignment of the intersubunit cleft, and displayed in three-dimensional space via Discovery Studio. Computational requirements were significantly reduced by including only these two subunits in modeling calculations going forward.

Three residues shown to be essential for anesthetic activity in previous studies with mutant mice (β_3-N265, β_3-M286, α_1-L232) (Bertaccini et al., 2010, 2013) were mapped within the α/β intersubunit space, converging on a putative anesthetic binding pocket. A propofol molecule was then manually docked utilizing three-dimensional technology to avoid steric hindrance and ensure proper orientation within the cavity. An in situ ligand minimization using CHARMm was performed on the propofol molecule to optimize its position within the receptor-binding pocket. The Flexible Docking algorithm surveyed the conformational space in relation to varied propofol ligand poses to achieve an energetically optimized binding cavity. With a spherical binding domain of 10 Å, this algorithm applied flexible protein side chains while docking the flexible propofol ligand via the ChiFlex algorithm to achieve an energetically optimized induced fit binding cavity conformation (see Fig. 2). The rigid CDOCKER algorithm was then applied to this cavity configuration to dock a series of propofol derivatives. The calculated binding affinities were correlated with known $GABA_AR$ EC_{50} values to test model reliability.

2.3 Characterizing the Receptor-Based Pharmacophore of the $hGABA_AR$ Model

Until the exact structure of the heteropentameric $hGABA_AR$ is solved and published, successful modeling (and understanding) of the $GABA_A$ receptor is wholly dependent upon the availability of biologically relevant, high-resolution structures of homologous proteins. Bertaccini and Trudell's original $GABA_AR$ homology models published in 2013 (Bertaccini et al., 2013) was based on GluCl and the prokaryotic pH-sensing channel known as GLIC (*Gloeobacter violaceus* ion channel: GLIC, PDB ID 3EAM). In their original models, the $GABA_AR$ subunits were modeled from the GluCl subunits with the GLIC receptor serving as a guide for some of the assemblies of the heteropentameric ion channel and intersubunit spaces.

The updated $hGABA_AR$ homology model described here expands on previous work by including the glycine receptor as a structural template in addition to the GluCl receptor. The publication of the cryo-electron microscopy-derived α_1-glycine receptor by the Gouaux lab in 2015 offered an opportunity to revise the $GABA_AR$ model using high-resolution structural data from this highly homologous receptor (Du et al., 2015). The glycine/ivermectin-bound structural variant of the glycine receptor (PDB 3JAF) is thought to represent at least a partially open state as opposed to the closed or desensitized state of GLIC. This open conformational state is believed to be integral for the gating mechanism of Cys-loop receptor ion channels as noted earlier. Also for this reason, the 3JAF GlyR was included as a homology template instead of the recently published crystal structure of the human $GABA_AR$ β_3 homopentamer, as the latter is thought to exhibit a closed-channel conformation (Miller & Aricesc, 2014).

The updated heteropentameric $hGABA_AR$ model constructed from the GluCl and GlyR homology templates showed similar overall three-dimensional structure to the original model: (γ_2-β_3-α_1-β_3-α_1) linked subunits arranged symmetrically around a central ion-conducting pore, with an extracellular domain comprised of beta sheets and an alpha-helical transmembrane ligand-binding domain (secondary features common to all members of the Cys-loop superfamily of LGICs) (Wolters & Bickelhaupt, 2012). As seen previously, the putative anesthetic binding cavity was molded within the tetrameric alpha-helical bundles that comprise the transmembrane domain of each subunit. This site was identified by the convergence of the three essential residues (β_3-N265, β_3-M286, α_1-L232) shown to modulate anesthetic activity within the α/β intersubunit space in previous studies with knockout mice (Bertaccini et al., 2013).

The Receptor–Ligand Interaction tool in Discovery Studio was used to map the various interaction surfaces of the propofol molecules docked within the putative $hGABA_AR$ anesthetic binding site. The rigid CDOCKER algorithm resulted in 10 equivalent propofol poses with less than 1.5 kcal/mol binding energy difference (−29.9667 to −28.6123 kcal/mol). Hydrophobic interactions between the propofol ligand and binding cavity were predominant and included π–π stacking, alkyl, and π-alkyl bonding. β_3-M286 was shown to contribute to this hydrophobic stabilization, while the other two residues shown to be essential for anesthetic activity, α_1-L232 and β_3-N265, displayed van der Waals interactions between the isopropyl groups of the docked propofol molecule (diisopropylphenol) in several of the ligand poses. These pharmacophore findings could partially explain the obliterated response to propofol seen in rats with the N265M mutation (Forman, 2011; Jonsson Fagerlund, Sjodin, Krupp, & Dabrowski, 2010). Additional interactions of significance included a carbon–hydrogen bond between α_1-P233 and the hydroxyl group of the propofol molecule and strong hydrophobic π-π stacking with β_3-F289. As illustrated in Fig. 3, the $hGABA_AR$ receptor-based pharmacophore is best described as amphiphilic in nature, with both hydrophilic and hydrophobic stabilization forces.

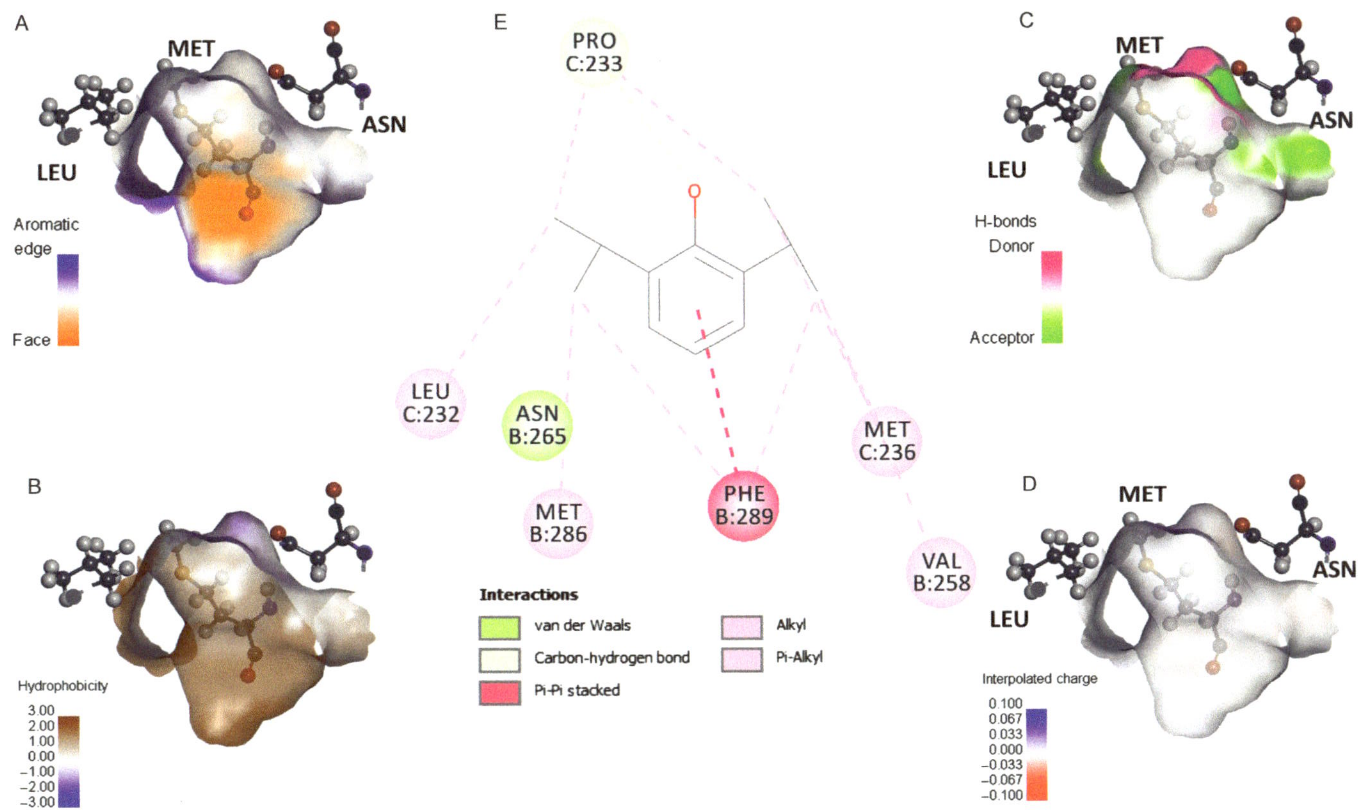

Fig. 3 Cutaway views to expose intersubunit anesthetic binding site interactions surface. Each is *color coded*, revealing propensity for aromatic (A), hydrophobic (B), hydrogen bonding (C), and polar charge (D). The *middle diagram* (E) is 2D schematic from one higher scoring pose that shows individual types of interactions from points on propofol to specific amino acids. Note the prevalence of hydrophobic interactions from the aromatic Phe, as well as the weak hydrogen bond potential arising from a hydrogen shared between a carbon on Pro and the oxygen on Propofol.

2.4 Validating the $hGABA_AR$ Model and Anesthetic Binding Site

In order to verify the $hGABA_AR$ binding site, etomidate and propofol derivatives with known $hGABA_AR$ agonist activity were docked into the rigid binding cavity and scored via the CDOCKER algorithm for correlation with experimentally derived $GABA_AR$ potentiation EC_{50} values. The rigid CDOCKER algorithm was utilized to dock a series of propofol derivatives into an energetically optimized conformation of the essential residue convergence site binding cavity. This protocol allowed for flexible ligand poses, while the receptor-binding cavity remained rigid. The series of docked propofol derivatives included etomidate, propofol (2,6-diisopropylphenol), 2,6-dimethylphenol, phenol, 2,6-diethylphenol, 2-isopropylphenol, and 2,6-disecbutylphenol. The *cis* and *trans* forms of the nonimmobilizer compound 1,2-dichlorohexafluorocyclobutane (F6), which lack anesthetic activity, were included in these docking trials as negative controls. Affinity scores of propofol derivatives docked into the α/β intersubunit binding cavity showed log-linear correlation with experimentally derived $GABA_AR$ potentiation EC_{50} values ($R^2 = 0.919$) as shown in Fig. 4. The nonimmobilizer F6 molecule displayed unfavorable docking energies, indicating a lack of affinity for the binding site in accordance with its clinical profile. These results validate the three-dimensional structure of the $hGABA_AR$ model and suggest a binding site that may be important for receptor activation.

Affinity scores of only the propofol derivatives (not etomidate) docked into the α/β intersubunit binding cavity showed even greater log-linear

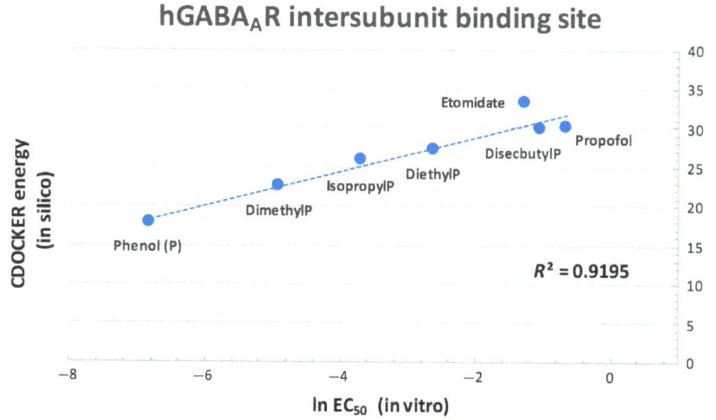

Fig. 4 Log-linear correlation of in silico docking scores for propofol derivatives and etomidate within the α/β intersubunit binding cleft with experimental $GABA_AR$ potentiation EC_{50} values. P = Phenol.

correlation with experimentally derived $GABA_AR$ potentiation EC_{50} values ($R^2=0.9804$), providing further validation for this putative anesthesia binding site. As compared to equivalent docking trials from the original $GABA_AR$ model by Bertaccini and Trudell (Bertaccini et al., 2013), these results point to the creation of a more robust model with stronger correlation between in silico and experimental data ($R^2=0.919$ for original model vs $R^2=0.9804$ for updated $hGABA_AR$ model described here). In order to further verify the relevance of this site for anesthetic binding, (R)-etomidate was docked alongside the series of propofol derivatives. (R)-Etomidate, a potent anesthetic that acts as an allosteric agonist at the $GABA_AR$, is believed to bind within the transmembrane α/β (β_3 specifically) intersubunit space (Forman, 2011). Its anesthetic activity was diminished in the previously described mutation studies characterizing the three essential residues. Although increased steric hindrance was likely encountered while docking this larger molecule within the rigid binding cavity, the fact that significant log-linear correlation was still seen when etomidate was included in the docking trials ($R^2=0.9195$, see Fig. 4) provides further evidence for the structural validity of this revised $hGABA_AR$ homology model, and supports the hypothesis that the essential residue convergence site may be involved in anesthetic binding and receptor activation.

3. RECEPTOR-BASED PHARMACOPHORE BASED ON ANESTHETICS COMPLEXED WITH NON-LGIC PROTEINS

3.1 The Problem With Deriving the Ligand-Based Anesthetic Pharmacophore

Anesthetics range in size from simple atomic xenon to more complicated, extended, or aromatic, often highly halogenated hydrocarbons. While their chemical complexities vary, there are a variety of similarities based on individual physicochemical characteristics that have been pursued as a means of establishing a ligand-based pharmacophore. The work of Sewell and Sear (Sewell & Halsey, 1997; Sewell & Sear, 2006) has pursued this to great lengths and found that among a structurally similar series of compounds, comparative molecular field analyses (CoMFA) can be used to derive a pharmacophore specific to the series studied as well as physiologic endpoint. However, it seems that this analysis breaks down when attempting to apply it across the wide range of all anesthetics. Therefore, more recent work has focused on the few available known complexes of actual anesthetics bound to a variety of crystallizable proteins, in an effort to look at specific ligand–protein interactions.

3.2 The Available Systems for Examining Anesthetic–Protein Complexes

Given that anesthetic molecules in general are not very large, it is reasonable to suggest that their overall binding motifs may have commonalities across several types of complexes. There have been many crystallographic studies of anesthetic–protein complexes that can be examined. These include bromoform-luciferase (Franks, Jenkins, Conti, Lieb, & Brick, 1998), halothane-albumin (Bhattacharya, Curry, & Franks, 2000), halothane-cholesterol oxidase (Bertaccini, Trudell, & Franks, 2007; Vrielink, Lloyd, & Blow, 1991), and dichloroethane-dehalogenase (Verschueren, Seljee, Rozenboom, Kalk, & Dijkstra, 1993). More recently, horse spleen apoferritin (HSAF) has also been crystallized with halothane, isoflurane, and a series of propofol-like derivatives (Liu, Loll, & Eckenhoff, 2005; Vedula et al., 2009). Other less well-defined crystallographic studies have included complexes of dichloroethane-insulin (Gursky, Fontano, Bhyravbhatla, & Caspar, 1994), halothane-adenylate kinase (Sachsenheimer, Pai, Schulz, & Schirmer, 1977), and halothane complexed with rhodopsin (Ishizawa, Sharp, Liebman, & Eckenhoff, 2000). Synthetic four-helix bundles complexed with anesthetics (Johansson, Scharf, Davies, Reddy, & Eckenhoff, 2000) have been examined spectroscopically. Additionally, various azi-alcohol derivatives have been used to characterize anesthetic binding sites within protein-kinase C (Das et al., 2004), nAChR (Ziebell, Nirthanan, Husain, Miller, & Cohen, 2004), GABA receptors (Chiara et al., 2012; Jayakar et al., 2014; Savechenkov et al., 2017), the Kv1 potassium channel (Bu et al., 2018) and the TRPA1 channel (Woll et al., 2017).

While it is clear from the Meyer–Overton correlation that hydrophobic character must play a significant role in anesthetic potencies, it is also important to note that Koblin and Eger (Koblin et al., 1994, 1995) have robustly demonstrated several classes of compounds that deviate from the Meyer–Overton rule. These include a series of compounds whose lipid solubilities would predict reasonable anesthetic potency but whose actual anesthetic effect is negligible. Within this latter series, it was even noted that anesthetic potency tended to correlate with hydrophilicity and not as much with lipophilicity. Thus, we have employed methods of in silico computational chemistry in order to reveal anesthetic interactions within what turns out to be the amphiphilic and forcefully polarizing anesthetic binding pocket of these complexes.

3.3 The Use of In Silico Computational Chemistry to Reveal the Complex Nature of Anesthetic–Protein Interactions

While it would be ideal to study every anesthetic–protein interaction at the highest level of computational chemistry theory, such computations would outstrip current supercomputer capabilities. Further, given the relatively limited range of the nonbonded interactions between an anesthetic ligand and the moieties with which it interacts in a given binding site, this is probably also unnecessary.

Initially using the capabilities within the Hyperchem Software Suite (Hypercube, Inc., Gainesville, FL), we studied the interactions of several anesthetics within their binding sites for the aforementioned systems: bromoform-luciferase, halothane-albumin, halothane-cholesterol oxidase, and dichloroethane-dehalogenase (Bertaccini et al., 2007). These systems were chosen due to the existence of reasonably high-resolution coordinate structures. Computational efficiency was enhanced by using two key principles. The first was to limit the size of the system over which calculations would be performed by carving out a specific portion of the protein within a given range of the ligand, often around 10–12 Å. Further efficiency was introduced by using the method of combined quantum mechanics (semiempirical AM1 level)-molecular mechanics (AMBER) calculations. Systems had hydrogens added and were hydrated with a TIP3P solvation algorithm, and underwent a series of energy optimizations. Calculations of partial charges for both the in situ and in vacuo versions of the ligand showed not only obvious alterations in conformational geometry, but also definite polarizations of partial atomic charges induced by the presence of the amphiphilic electric field within the binding site. Additional interactions included weak hydrogen bonds, other weak polar interactions as well as the expected van der Waals interactions explaining hydrophobicity.

Increases in computing power have subsequently allowed similar calculations at a much higher level of quantum mechanics rigor and across a more expansive set of anesthetic–ligand complexes (Qiu, Lin, & Bertaccini, 2015). Specifically, the models that could then be studied were based on the X-ray crystal structures determined for the complexes of firefly luciferase with bromoform (PDB 1BA3), human serum albumin with halothane (PDB 1E7B), haloalkane dehalogenase with dichloroethane (PDB 2DHC), and HSAF with the inhalational anesthetics halothane (PDB 1XZ1), isoflurane (PDB 1XZ3), and a homologous series of injectable anesthetic propofol-like compounds such as propofol (PDB 3F33), 2,6-diethylphenol (PDB 3F34),

2-isopropylphenol (PDB 3F36), 2,6-dimethylphenol (PDB 3F37), and phenol (PDB 3F39), as well as the injectable anesthetics pentobarbital (PDB 3RAV) and thiopental (PDB 3RD0). A far more rigorous QMMM calculation was performed that utilized a three-layer ONIOM ("our own n-layered integrated molecular orbital and molecular mechanics" from Gaussian, Inc., Wallingford, CT) partitioning scheme method combined with electronic embedding (EE). The highest level of computation involved the anesthetic ligand alone and was computed using density functional theory with the functional and basis set M06-2X/6-31 + G*. The intermediate layer included residues up to 4 Å away from the ligand, so as to treat any possible hydrogen bonds with at least a semiempirical quantum mechanics level using the PM6 method. The remainder of each protein complex was treated with AMBER molecular mechanics. These robust calculations not only reinforced the polar and nonpolar interactions of anesthetics with proteins but also further delineated the existence of not only weak hydrogen bonding patterns, but a newer and unique form of interaction known as a halogen bond (see Fig. 5).

Further elucidation of detailed anesthetic–protein interactions has been gleaned from the differential study of diisopropyl phenol (propofol) and its inactive congener fropofol (1-fluro-2,6-diisopropylbenzene) in three separate systems: human serum albumin (HSA), the pH-gated ion channel from *Gloeobacter violaceus* (GLIC), and HSAF (Qiu et al., 2017). Each system has a well-characterized propofol-protein crystallograph which could have its propofol hydroxyl changed in silico to a fluorine for subsequent analysis. Each system was studied at the similarly robust level noted above (three-layer ONIOM M06-2X/6-31 + G*:PM6:AMBER). The ONIOM calculations, subsequent atoms in molecules analyses and electrostatic potential analyses all demonstrate the significance of hydrogen bond, halogen bond, and hydrophobic interactions in promoting proper molecular recognition and binding. The binding interaction energies of propofol in each complex were generally more favorable than those for fropofol and could explain their differential anesthetic potencies. However, even though the hydrogen-bonding effect of the hydroxyl moiety was prominent in propofol, the substitution of the 1-hydroxyl by a fluorine atom did not prevent fropofol from binding to the protein via the aforementioned unique halogen bonding interaction. This further emphasizes the point that multiple specific interactions, rather than just hydrogen or halogen bonds, must be incorporated into any explanation of this complex binding motif.

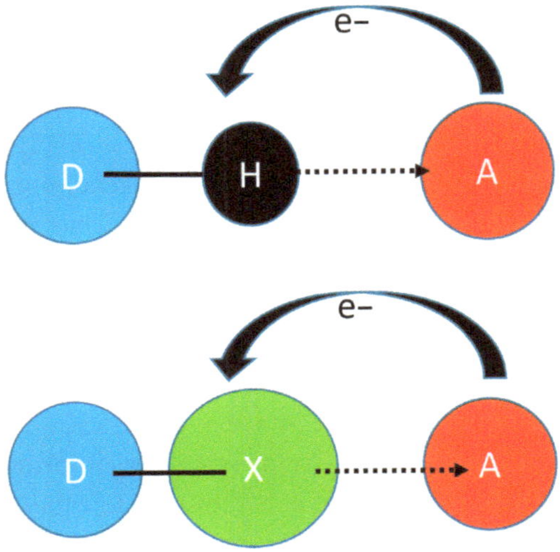

Fig. 5 Figure illustrating and contrasting halogen bonds (*top*) and hydrogen bonds (*bottom*). "D" is donor of either halogen or hydrogen, *arrow* indicates direction of nuclear donation, "A" indicates acceptor of nuclear donation. Each is further detailed with the direction of actual electron flow. While the nuclear donation "suspends" the donated species in between donor (D) and acceptor (A), the additional electron sharing (shown by *curved arrow*) occurs in such a way as to satisfy the incomplete outer molecular orbital of each donated species. In the case of hydrogen (H), the outer S orbital only has one electron, needing one more to be complete. In the case of a halogen atom (X), the outer p orbital has seven electrons, also needing only one more to be complete. This latter phenomenon gives these bonds their status as partially covalent, between simple nonbonded polar and truly bonded covalent interactions. Donor is typically a carbon atom, and the acceptor is often an oxygen, nitrogen, or sulfur atom in both cases. The acceptor can also be the π cloud of an aromatic ring.

4. SUMMARY AND CONCLUSIONS

Consensus structural alignment of GlyR and GluCl homologous templates allowed for the construction of a three-dimensional human heteropentameric $GABA_AR$ model with (γ_2-β_3-α_1-β_3-α_1) subunit linkage. The three residues shown to be essential for anesthetic activity in previous studies with mutant mice (β_3-N265, β_3-M286, α_1-L232) were mapped within this model, with all three essential residues converging together within the transmembrane α/β intersubunit space, suggestive of a putative anesthetic binding cavity that may be important for receptor activation. The three-dimensional structure of the model binding cavity was validated by the log-linear correlation found between in silico docking scores and experimental $GABA_AR$ potentiation for a series of propofol derivatives and etomidate.

Closer inspection of the receptor–ligand interactions between the docked propofol molecules and putative anesthetic binding pocket revealed an amphiphilic cavity with weak polar interactions (believed to account for reversible anesthetic action) surrounding a hydrophobic core (consistent with the Meyer–Overton theory of anesthetic activity). As predicted, the convergence site residues proved to be major contributors to the receptor–ligand binding scheme, with α_1-L232 and β_3-N265 consistently engaging in anesthetic van der Waals interactions and β_3-types of anesthetic–protein interactions were highlighted by an in-depth quantum mechanics analysis of high-resolution images of anesthetics bound to a variety of other proteins whose structures were experimentally derived at very high resolution.

While these results reinforce the validity and utility of the updated $hGABA_AR$ homology model and the amphiphilic nature of the anesthetic binding site, further exploration of anesthetic binding is necessary to better understand receptor modulation and will allow for high-throughput in silico screening of novel anesthetic compounds.

REFERENCES

Bertaccini, E. J., Trudell, J. R., & Franks, N. P. (2007). The common chemical motifs within anesthetic binding sites. *Anesthesia and Analgesia*, *104*(2), 318–324.

Bertaccini, E. J., Wallner, B., Trudell, J. R., & Lindahl, E. (2010). Modeling anesthetic binding sites within the glycine alpha one receptor based on prokaryotic ion channel templates: The problem with TM4. *Journal of Chemical Information and Modeling*, *50*(12), 2248–2255.

Bertaccini, E. J., Yoluk, O., Lindahl, E. R., & Trudell, J. R. (2013). Assessment of homology templates and an anesthetic binding site within the gamma-aminobutyric acid receptor. *Anesthesiology*, *119*(5), 1087–1095.

Bhattacharya, A. A., Curry, S., & Franks, N. P. (2000). Binding of the general anesthetics propofol and halothane to human serum albumin. High resolution crystal structures. *The Journal of Biological Chemistry*, *275*(49), 38731–38738.

Bu, W., Liang, Q., Zhi, L., Maciunas, L., Loll, P. J., Eckenhoff, R. G., et al. (2018). Sites and functional consequence of alkylphenol anesthetic binding to Kv1.2 channels. *Molecular Neurobiology*, *55*(2), 1692.

Chiara, D. C., Dostalova, Z., Jayakar, S. S., Zhou, X., Miller, K. W., & Cohen, J. B. (2012). Mapping general anesthetic binding site(s) in human alpha1beta3 gamma-aminobutyric acid type A receptors with [(3)H]TDBzl-etomidate, a photoreactive etomidate analogue. *Biochemistry*, *51*(4), 836–847.

Das, J., Addona, G. H., Sandberg, W. S., Husain, S. S., Stehle, T., & Miller, K. W. (2004). Identification of a general anesthetic binding site in the diacylglycerol-binding domain of protein kinase Cdelta. *The Journal of Biological Chemistry*, *279*(36), 37964–37972.

Du, J., Lu, W., Wu, S., Cheng, Y., & Gouaux, E. (2015). Glycine receptor mechanism elucidated by electron cryo-microscopy. *Nature*, *526*(7572), 224–229.

Forman, S. A. (2011). Clinical and molecular pharmacology of etomidate. *Anesthesiology*, *114*(3), 695–707.

Forman, S. A., & Miller, K. W. (2011). Anesthetic sites and allosteric mechanisms of action on Cys-loop ligand-gated ion channels. *Canadian Journal of Anaesthesia*, *58*(2), 191–205.

Franks, N. P., Jenkins, A., Conti, E., Lieb, W. R., & Brick, P. (1998). Structural basis for the inhibition of firefly luciferase by a general anesthetic. *Biophysical Journal*, *75*(5), 2205–2211.

Gursky, O., Fontano, E., Bhyravbhatla, B., & Caspar, D. L. (1994). Stereospecific dihaloalkane binding in a pH-sensitive cavity in cubic insulin crystals. *Proceedings of the National Academy of Sciences of the United States of America*, *91*, 12388–12392.

Hibbs, R. E., & Gouaux, E. (2011). Principles of activation and permeation in an anion-selective Cys-loop receptor. *Nature*, *474*(7349), 54–60.

Ishizawa, Y., Sharp, R., Liebman, P. A., & Eckenhoff, R. G. (2000). Halothane binding to a G protein-coupled receptor in retinal membranes by photoaffinity labeling. *Biochemistry*, *39*(29), 8497–8502.

Jayakar, S. S., Zhou, X., Chiara, D. C., Dostalova, Z., Savechenkov, P. Y., Bruzik, K. S., et al. (2014). Multiple propofol-binding sites in a gamma-aminobutyric acid type A receptor ($GABA_AR$) identified using a photoreactive propofol analog. *The Journal of Biological Chemistry*, *289*(40), 27456–27468.

Johansson, J. S., Scharf, D., Davies, L. A., Reddy, K. S., & Eckenhoff, R. G. (2000). A designed four-alpha-helix bundle that binds the volatile general anesthetic halothane with high affinity. *Biophysical Journal*, *78*(2), 982–993.

Jonsson Fagerlund, M., Sjodin, J., Krupp, J., & Dabrowski, M. A. (2010). Reduced effect of propofol at human {alpha}1{beta}2(N289M){gamma}2 and {alpha}2{beta}3(N290M){gamma}2 mutant GABA(A) receptors. *British Journal of Anaesthesia*, *104*(4), 472–481.

Jurd, R., Arras, M., Lambert, S., Drexler, B., Siegwart, R., Crestani, F., et al. (2003). General anesthetic actions in vivo strongly attenuated by a point mutation in the GABA(A) receptor beta3 subunit. *The FASEB Journal*, *17*(2), 250–252.

Koblin, D. D., Chortkoff, B. S., Laster, M. J., Eger, E. I., 2nd, Halsey, M. J., & Ionescu, P. (1994). Polyhalogenated and perfluorinated compounds that disobey the Meyer–Overton hypothesis. *Anesthesia and Analgesia*, *79*(6), 1043–1048.

Koblin, D. D., Chortkoff, B. S., Laster, M. J., Eger, E. I., Halsey, M. J., & Ionescu, P. (1995). Nonanesthetic polyhalogenated alkanes and deviations from the Meyer–Overton hypothesis. *Progress in Anesthesiology, Molecular Mechanisms of Anesthesia*, *3*, 451–456.

Liu, R., Loll, P. J., & Eckenhoff, R. G. (2005). Structural basis for high-affinity volatile anesthetic binding in a natural 4-helix bundle protein. *The FASEB Journal*, *19*(6), 567–576.

Miller, P. S., & Aricescu, A. R. (2014). Crystal structure of a human GABA receptor. *Nature*, *512*, 270–275.

Murail, S., Wallner, B., Trudell, J. R., Bertaccini, E., & Lindahl, E. (2011). Microsecond simulations indicate that ethanol binds between subunits and could stabilize an open-state model of a glycine receptor. *Biophysical Journal*, *100*(7), 1642–1650.

Qiu, L., Lin, J., & Bertaccini, E. J. (2015). Insights into the nature of anesthetic-protein interactions: An ONIOM study. *The Journal of Physical Chemistry. B*, *119*(40), 12771–12782.

Qiu, L., Lin, J., Liu, Q., Wang, S., Lv, G., Li, K., et al. (2017). The role of the hydroxyl group in propofol-protein target recognition: Insights from ONIOM studies. *The Journal of Physical Chemistry. B*, *121*(24), 5883–5896.

Sachsenheimer, W., Pai, E. F., Schulz, G. E., & Schirmer, R. H. (1977). Halothane binds in the adenine-specific niche of crystalline adenylate kinase. *FEBS Letters*, *79*(2), 310–312.

Sauguet, L., Shahsavar, A., & Delarue, M. (2015). Crystallographic studies of pharmacological sites in pentameric ligand-gated ion channels. *Biochimica et Biophysica Acta*, *1850*(3), 511–523.

Savechenkov, P. Y., Chiara, D. C., Desai, R., Stern, A. T., Zhou, X., Ziemba, A. M., et al. (2017). Synthesis and pharmacological evaluation of neurosteroid photoaffinity ligands. *European Journal of Medicinal Chemistry, 136*, 334–347.

Sewell, J. C., & Halsey, M. J. (1997). Shape similarity indices are the best predictors of substituted fluoroethane and ether anaesthesia. *European Journal of Medicinal Chemistry, 32*, 731–737.

Sewell, J. C., & Sear, J. W. (2006). Determinants of volatile general anesthetic potency: A preliminary three-dimensional pharmacophore for halogenated anesthetics. *Anesthesia and Analgesia, 102*(3), 764–771. https://doi.org/10.1213/01.ane.0000195421.46107.d0.

Spurny, R., Ramerstorfer, J., Price, K., Brams, M., Ernst, M., Nury, H., et al. (2012). Pentameric ligand-gated ion channel ELIC is activated by GABA and modulated by benzodiazepines. *Proceedings of the National Academy of Sciences of the United States of America, 109*(44), E3028–E3034.

Trudell, J. (2002). Unique assignment of inter-subunit association in GABA(A) alpha 1 beta 3 gamma 2 receptors determined by molecular modeling. *Biochimica et Biophysica Acta, 1565*(1), 91–96.

Vedula, L. S., Brannigan, G., Economou, N. J., Xi, J., Hall, M. A., Liu, R., et al. (2009). A unitary anesthetic binding site at high resolution. *The Journal of Biological Chemistry, 284*(36), 24176–24184.

Verschueren, K. H. G., Seljee, F., Rozenboom, H. J., Kalk, K. H., & Dijkstra, B. W. (1993). Crystallographic analysis of the catalytic mechanism of haloalkane dehalogenase. *Nature, 363*, 693–698.

Vrielink, A., Lloyd, L. F., & Blow, D. M. (1991). Crystal structure of cholesterol oxidase from Brevibacterium sterolicum refined at 1.8 Å resolution. *Journal of Molecular Biology, 219*, 533–554.

Woll, K. A., Skinner, K. A., Gianti, E., Bhanu, N. V., Garcia, B. A., Carnevale, V., et al. (2017). Sites contributing to TRPA1 activation by the anesthetic propofol identified by photoaffinity labeling. *Biophysical Journal, 113*(10), 2168–2172.

Wolters, L. P., & Bickelhaupt, F. M. (2012). Halogen bonding versus hydrogen bonding: A molecular orbital perspective. *Chemistry Open, 1*(2), 96–105.

Ziebell, M. R., Nirthanan, S., Husain, S. S., Miller, K. W., & Cohen, J. B. (2004). Identification of binding sites in the nicotinic acetylcholine receptor for [3H] azietomidate, a photoactivatable general anesthetic. *The Journal of Biological Chemistry, 279*(17), 17640–17649.

CHAPTER FIVE

Understanding Anesthetic Mechanisms: Analysis of the Complex Kinetics of Ligand-Gated Ion Channels

Robert S. Cantor[1]
Dartmouth College, Hanover, NH, United States
[1]Corresponding author: e-mail address: robert.s.cantor@dartmouth.edu

Contents

Abstract

Anesthetics modulate the response of ligand-gated ion channels to their neurotransmitter agonists, in a way that is consistent with clinical anesthesia: inhibition of synaptic transmission, by activation of inhibitory receptors and/or inhibition of excitatory receptors. Electrophysiological results for receptors such as $GABA_AR$ indicate that this modulation can be remarkably kinetically complex, characterized by concentration-dependent changes in the extent and (multiple) time scales of desensitization and deactivation. The full range of these features cannot be reproduced by a kinetic model in which anesthetic acts only by binding to putative protein sites, without having multiple sites with varying affinities, as well as many additional conformational states beyond the canonical set of three (resting, open, and desensitized). So, we discuss the implementation of a kinetic approach that incorporates only these three states, but accounts for effects of adsorption of anesthetic and agonist to the membrane in which the receptor is embedded, which modulates the conformational free energy landscape of the protein. As a result, the rate constants of conformational transitions become time dependent (non-Markovian), requiring nonstandard methods of kinetic analysis that can readily be implemented using available computational software.

Methods in Enzymology, Volume 602
ISSN 0076-6879
https://doi.org/10.1016/bs.mie.2018.02.002

1. INTRODUCTION

The clinical effects of many general anesthetics are largely mirrored by their effects on postsynaptic ligand-gated ion channels (LGICs): potentiation of anion flux through inhibitory LGICs and/or inhibition of cation flux through excitatory LGICs. Although simply stated in terms of total ion flux, the details of the time dependence of ion currents, the integral of which gives the flux, are far more complex. Current traces obtained from fast-perfusion electrophysiological studies of receptors such as $GABA_AR$ reveal a remarkably complex dependence of the features of those traces (e.g., the extent and multiple time constants of desensitization and of deactivation) on agonist and anesthetic concentration. The ability to reproduce and predict these features thus provides an essential test of any mechanistic, molecular understanding of anesthesia: how the details of receptor desensitization and deactivation depend on concentrations, the length of the agonist pulse, whether anesthetic is coapplied with agonist or continuously present, etc. To that end, it is necessary to generate physically reasonable kinetic models of receptor activity and examine the extent to which they can reproduce and predict experimental results.

Kinetic models include transitions among a set of conformational states of the receptor, each of which can bind agonist. The minimal set of three conformations comprises a resting state (R), ion-conducting open state (O), and a desensitized state (D). Models using this set can reproduce many of the *overall* features of desensitization and deactivation, but not the details, which may include multiple time scales for each, with a dependence on agonist and anesthetic concentration that continues well beyond binding saturation. One approach has been to consider models with multiple additional conformational states (Burkat, Yang, & Gingrich, 2001; Celentano & Wong, 1994; Haas & MacDonald, 1999; Jones & Westbrook, 1995; Lema & Auerbach, 2006). By adjusting the kinetic parameters, it is possible to reproduce some of the electrophysiological results for a given receptor with regard to its activation by neurotransmitter. Such an approach has not yet successfully been applied to the effect of anesthetics, which adds a far greater level of complexity.

Such models implicitly assume that all transitions are Markovian, i.e., the rate constant of each remains constant. However, if the environment of the receptor changes during the process then the rate constant can change as well. (A common example in chemical reaction kinetics is the effect of a temperature change during the reaction.) More precisely, the receptor exists along a conformational free energy landscape (FEL) with a generalized reaction coordinate that connects conformational states (free energy minima) separated pairwise by transition state maxima. Adsorption of aqueous solutes (anesthetics and/or agonist) to the membrane bilayer in which the receptor is embedded—an essential part of the environment of the receptor—alters the receptor's FEL. In general, the sensitivity to a given solute (how rapidly the receptor free energy varies with increasing solute membrane adsorption) will vary with location on the FEL. So, for example, if adsorption of a given amount of anesthetic to the bilayer changes the free energy of the R state to a different degree than it changes that of the transition state that separates the R and O states, then the free energy *barrier* for the opening conformational transition ($R \rightarrow O$) will change. As discussed later, the transition rate constant is an extremely sensitive function of the free energy barrier. In an electrophysiological experiment, membrane adsorption does not occur instantaneously upon application of a pulse of agonist or anesthetic, nor does desorption occur instantaneously upon washout. During the adsorption and desorption processes, the FEL thus changes gradually with time, so the rate constants of conformational transitions become time dependent. It has been shown (Lee, Albershardt, & Cantor, 2015) that even with only the canonical set of three conformational states $\{R, O, D\}$, the full complexity of the response of $GABA_A$ receptors to their agonists, modulation by anesthetics, and the activation of $GABA_AR$ by superclinical concentrations of anesthetic in the absence of agonist, can all be reproduced in the context of a model that incorporates effects of adsorption of agonist and anesthetic to the membrane.

2. METHODS

The goal is to obtain a best fit of the model to sets of electrophysiological data, i.e., to multiple traces taken under widely different conditions: pulse length, the full range of agonist, and anesthetic concentrations, whether anesthetic is coapplied with agonist or continuously present, etc. In principle, each trace is in the form of current vs time: $I(t)$. If (as is largely supported by single-channel results) it is assumed that the conductance of an open channel is constant, independent of agonist ligation, then the current will be proportional to the total fraction of receptors in the open state. The fraction of protein in conformational state X with *i* bound agonist molecules is denoted as $f_{A_iX}(t)$, so $f_{A_iO}(t)$ represents the fraction of receptors in the open state with i agonist molecules bound. Let the total number of binding sites be n (for example, $n = 2$ for $GABA_AR$, and the sites are equivalent). Then the total open fraction is given by

$$f_O(t) = f_{A_0O}(t) + f_{A_1O}(t) + \ldots + f_{A_nO}(t) \tag{1}$$

If it is possible to estimate the maximal current of each experimental trace, I_{max} (i.e., with all channels conducting), then for purposes of adjusting the parameters of the model to obtain a best fit to the data, the observed currents can be reexpressed as the total open fraction, $f_O(t) = I(t)/I_{max}$.

The details of the representation of the model as a set of coupled first-order differential equations have been described in detail elsewhere (Lee et al., 2015). In summary, there are six possible conformational transitions among the three conformational states: $\{R \rightleftarrows O;\ O \rightleftarrows D;\ R \rightleftarrows D\}$ as depicted in Fig. 1. It is the rate constants of these transitions that may vary with solute adsorption, and thus with time. The binding and unbinding of agonist to receptor in each of its three states obey standard Markovian kinetics. If, as is the case for $GABA_AR$, there are two equivalent agonist-binding sites on the receptor, then there are three ligation states (0, 1, or 2 agonists).

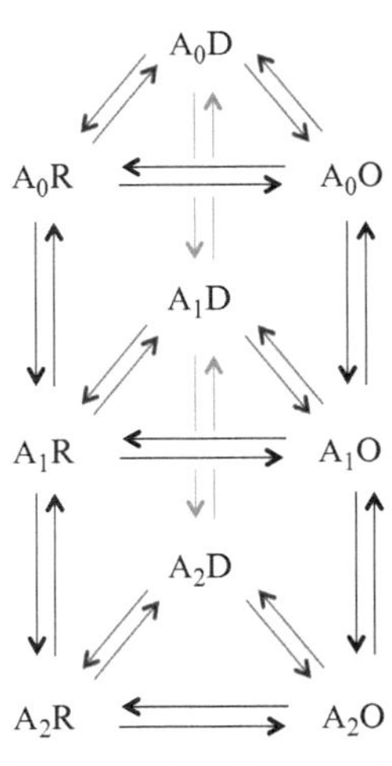

Fig. 1 General kinetic scheme, incorporating three conformational states {R, O, D}; each has two equivalent agonist-binding sites. *Vertical arrows* indicate agonist binding and unbinding; the *remaining arrows* indicate conformational transitions at fixed agonist ligation.

The rate of change of $f_{\mathrm{A_iX}}(t)$, for any one of the three states $\mathrm{X}=\{\mathrm{R, O, D}\}$, where Y and Z represent the other two states, is given by

$$\begin{aligned}
df_{\mathrm{A_0X}}/dt &= k_{\mathrm{A_0Y \to A_0X}} f_{\mathrm{A_0Y}} - k_{\mathrm{A_0X \to A_0Y}} f_{\mathrm{A_0X}} + k_{\mathrm{A_0Z \to A_0X}} f_{\mathrm{A_0Z}} \\
&\quad - k_{\mathrm{A_0X \to A_0Z}} f_{\mathrm{A_0X}} - k_{\mathrm{b,X}} c_{\mathrm{ag}} f_{\mathrm{A_0X}} + k_{\mathrm{u,X}} f_{\mathrm{A_1X}} \\
df_{\mathrm{A_1X}}/dt &= k_{\mathrm{A_1Y \to A_1X}} f_{\mathrm{A_1Y}} - k_{\mathrm{A_1X \to A_1Y}} f_{\mathrm{A_1X}} + k_{\mathrm{A_1Z \to A_1X}} f_{\mathrm{A_1Z}} \\
&\quad - k_{\mathrm{A_1X \to A_1Z}} f_{\mathrm{A_1X}} + k_{\mathrm{b,X}} c_{\mathrm{ag}} f_{\mathrm{A_0X}} - k_{\mathrm{u,X}} f_{\mathrm{A_1X}} - k_{\mathrm{b,X}} c_{\mathrm{ag}} f_{\mathrm{A_1X}} \\
&\quad + k_{\mathrm{u,X}} f_{\mathrm{A_2X}} \\
df_{\mathrm{A_2X}}/dt &= k_{\mathrm{A_2Y \to A_2X}} f_{\mathrm{A_2Y}} - k_{\mathrm{A_2X \to A_2Y}} f_{\mathrm{A_2X}} + k_{\mathrm{A_2Z \to A_2X}} f_{\mathrm{A_2Z}} \\
&\quad - k_{\mathrm{A_2X \to A_2Z}} f_{\mathrm{A_2X}} + k_{\mathrm{b,X}} c_{\mathrm{ag}} f_{\mathrm{A_1X}} - k_{\mathrm{u,X}} f_{\mathrm{A_2X}}
\end{aligned} \tag{2}$$

with ligand binding and unbinding rate constants $k_{\mathrm{b,X}}$ and $k_{\mathrm{u,X}}$, respectively, aqueous agonist concentration c_{ag}, and transition rate constants in the form $k_{\mathrm{A_iX \to A_iY}}$. The fractions necessarily add to 1, i.e.,

$$\sum_{i=0}^{2} (f_{\mathrm{A_iR}} + f_{\mathrm{A_iO}} + f_{\mathrm{A_iD}}) = 1 \tag{3}$$

so the distribution is defined by any eight of these nine fractions. Taking the time derivative of Eq. (3) gives

$$\sum_{i=0}^{2} \left(\frac{df_{\mathrm{A_iR}}}{dt} + \frac{df_{\mathrm{A_iO}}}{dt} + \frac{df_{\mathrm{A_iD}}}{dt} \right) = 0 \tag{4}$$

That this adds to zero is confirmed by summing the right hand side of all nine equations defined in Eq. (2). Thus only eight of the nine equations are independent. Note that the kinetics of conformational transitions and agonist binding/unbinding have been modeled using standard forms for the rate equations, expressed in terms of the fractions $f_{\mathrm{A_iX}}$.

As discussed earlier, adsorption of solutes (agonist or anesthetic) onto the lipid bilayer of the postsynaptic membrane alters the physical properties of the bilayer. Changes in these properties can affect the protein free energy, to a degree that will vary with position on the FEL. Let x represents a generalized conformational coordinate and $G^\circ(x)$ the standard FEL of the protein, where the standard state (°) refers to the landscape in the absence of any adsorbed solutes. Let θ_s represents the fractional bilayer adsorption of solute s (agonist or anesthetic), i.e., the fraction of its maximal adsorption. The free energy $G(x)$ varies with sensitivity $\lambda_s(x)$ that depends on both location on the conformational landscape (x) and the identity of the solute (and receptor). In the first approximation, the change will be proportional to θ_s, i.e.,

$$G(x) = G^\circ(x) + \theta_s \lambda_s(x) \tag{5}$$

(Note that this represents the first term in a Taylor expansion in θ_s; the quality of the approximation will decrease with increasing adsorption.) Consider a conformational transition $A_iX \rightarrow A_iY$. Let $A_iXY^{\ddagger}$ represent the activated complex, at which the free energy reaches a maximum along the trajectory $A_iX \rightarrow A_iY$. To good approximation, the transition rate constant $k_{A_iX \rightarrow A_iY}$ depends exponentially on $\Delta G^{\ddagger}$, the free energy barrier between the initial state ($x = A_iX$) and the activated complex ($x = A_iXY^{\ddagger}$), with

$$k_{A_iX \rightarrow AiY} \propto \exp\left\{-\Delta G^{\ddagger}/RT\right\} \tag{6}$$

and

$$\Delta G^{\ddagger}/RT = \left[G\left(A_iXY^{\ddagger}\right) - G(A_iX)\right]/RT = \Delta G^{o\ddagger}/RT + \theta_s\alpha_{s,X\rightarrow Y} \tag{7}$$

where the dimensionless parameter $\alpha_{s,X\rightarrow Y} = [\lambda_s(XY^{\ddagger}) - \lambda_s(X)]/RT$ accounts for the difference in sensitivities (to adsorption of solute s) of the protein in the activated complex and in its initial state. Since the sensitivity presumably depends on structural properties of the protein in the transmembrane domain, which (at given location on the FEL) is not affected by agonist binding, it is reasonable to assume that the sensitivities are independent of ligation state. If $\alpha_{s,X\rightarrow Y} \neq 0$, then membrane adsorption/desorption of solute will cause the transition rate constant k to vary with θ_s:

$$k_{A_iX \rightarrow A_iY} = k^{\circ}{}_{A_iX \rightarrow A_iY} \exp\left[-\theta_s\alpha_{s,X\rightarrow Y}\right] \tag{8}$$

where k° represents the standard rate constant, i.e., at $\theta_s = 0$. If both agonist and anesthetic are present and if it is assumed that they adsorb independently, then Eq. (8) becomes

$$k_{A_iX \rightarrow A_iY} = k^{\circ}{}_{A_iX \rightarrow A_iY} \exp\left[-\left(\theta_{an}\alpha_{an,X\rightarrow Y} + \theta_{ag}\alpha_{ag,X\rightarrow Y}\right)\right] \tag{9}$$

Because θ_s is time dependent (adsorption and desorption of solutes do not occur instantaneously), and because of the exponential dependence, the rate constants of conformational transitions can be strongly time dependent. The time dependence of adsorption, $\theta_s(t)$, can be approximated using a simple Langmuir model of adsorption/desorption kinetics, which is given by the solution to the first-order differential rate equation

$$d\theta_s/dt = k_{s,on}c_s(1 - \theta_s) - k_{s,off}\theta_s \tag{10}$$

where $k_{s,on}$ and $k_{s,off}$ represent the rate constants of bilayer adsorption and desorption, respectively, of solute s (agonist or anesthetic). At this level of approximation, if agonist or anesthetic is applied in pulses (constant c_s during each pulse) then the expression for $\theta_s(t)$ can be expressed analytically as decaying exponential functions of time, both during and after the agonist pulse.

The parameters of the model comprise four groups: (1) the standard rate constants of the conformational transitions (i.e., without adsorbed solutes), $k_{A_iX \rightarrow A_iY}$; (2) rate constants of binding and unbinding to the three conformational states ($k_{b,X}$, $k_{u,X}$); (3) rate constants of adsorption and desorption of solute s (=anesthetic or agonist) to the bilayer ($k_{s,on}$, $k_{s,off}$); and (4) the (dimensionless) sensitivities of the transition rate constants to solute adsorption, $\alpha_{s,X \rightarrow Y}$. Of these, the first two groups are independent of any bilayer adsorption effects; in contrast, the third is independent of the protein and depends only on solute–bilayer interactions. Only the fourth depends both on the receptor and on anesthetic or agonist adsorption. Constraints arising from thermodynamic cycles reduce the number of independent parameters considerably, as discussed in detail elsewhere (Lee et al., 2015).

As described earlier, for a given experimental protocol (agonist and anesthetic concentrations, pulse length, etc.) the time-dependent distribution of protein states is described by the fractions $f^j_{A_iX}(t)$, with X = {R, O, D} and i = {0, 1, 2}, where j labels the experimental protocol used for a given trace. For a given set of parameter values, the $f^j_{A_iX}(t)$ for each protocol j are determined from numerical solutions of the coupled differential equations given in Eq. (2). In the software package Mathematica (Wolfram Research Inc., 2015), this is implemented with the function "NDSolve." However, what is needed is to adjust the parameters so as to determine the set of values that best fit the resulting predictions of $f^j_O(t)$, to the tabulated experimental current traces j. This is also possible within Mathematica, using the "FindFit" function, which automates the fitting process. In short, in its simplest form, FindFit simultaneously minimizes the sums of squares of differences (a least-squares fit) between predicted and experimental $f^j_O(t)$, for each trace j.

The user supplies an initial set of values of the parameters. The choice of these values is an important part of the process, since if the parameters result in predicted $f^j_O(t)$ that are vastly different from the experimental traces, FindFit may be unable to converge on a solution. It is thus useful to check, using NDSolve, that the initial guesses for the parameters result in a predicted set of $f^j_O(t)$ that are reasonably similar to each of the experimental traces.

FindFit starts by using NDSolve to predict the $f^j_O(t)$ for each experiment j, at the initial set of parameter values, and determines the quality of the fit, using least-squares (or some other normalization condition that can optionally be defined by the user). It then changes very slightly the value of each parameter, one at a time, in each case using NDSolve to predict the resulting (slightly different) set of $f^j_O(t)$, recalculates the fit to each trace, and redetermines the overall fit quality. In this manner, it determines the gradient of the overall quality of the fit with respect to variation in each parameter. It then extrapolates to a new parameter set that (hopefully) results in better fits to the data, and recalculates $f^j_O(t)$ using NDSolve with the new parameter set. This process is reiterated, in each cycle adjusting the parameter values until it reaches convergence, at which point the difference between predicted and experimental traces is minimized. (The user can select among various algorithms for this process, although the default approach usually works well.) Note that while this process entails a best fit to experimental data, it is not a fit to an *analytical* function, but to the result of a numerical solution to a set of differential equations: This is a particularly powerful feature of Mathematica.

Although FindFit may converge on a solution, that only indicates a *local* minimum in the fit quality. Given the complexity of the system, it is difficult, if not impossible to explore sufficiently the full parameter space to ensure that the calculated solution represents a *global* minimum. One important check is the parameter values themselves, since many have been measured independently (e.g., binding and unbinding rate constants, some of the standard conformational transition rate constants, and the corresponding equilibrium constants). Perhaps the most demanding test is to perform the fit using a large number of experimental traces obtained under widely different experimental conditions, particularly the range of concentrations of anesthetic and agonist.

Because each experimental trace typically contains thousands of data points, calculating best fits using all the points would be prohibitively slow, so it is necessary to reduce significantly the number of data points per trace. Such data thinning can be achieved by smoothing the experimental data in various ways. Note that the reduction in the density of data points (number of points per unit time) need not be done uniformly; this can be useful if there are regions of the traces where current is changing rapidly (typically at the beginning of the pulse and at the beginning of washout) during which the density of points can remain higher.

3. DISCUSSION

The current measured in an electrophysiological experiment is proportional to the total fraction of protein channel in an open conformation, $f_O(t)$. Such data thus provide very incomplete information about the time development of the receptor state distribution. For example, flow between the resting and desensitized states is not observed, since both states are non-conducting. Because the numerical solution of the kinetic equations (Eq. 2) provides the complete protein state distribution $\{f_{A_iX}(t)\}$ as well as the time derivatives of each fraction, it also describes the flow of probability among the states, as a function of time. It thus reveals the predominant kinetic pathways followed by the receptor under various agonist/anesthetic conditions, and the origins of many of the characteristics of the traces. For example, it can show under what conditions the desensitized state is accessed from the resting state (the "branched" pathway) or from the open state (the "linear" pathway). As another example, it also indicates the pathways followed from the diliganded desensitized state A_2D—the dominant state after activation by a long pulse of agonist at saturating concentration—back to the unliganded resting state A_0R upon washout at the end of the pulse.

Animation software is available for download that dynamically illustrates the time-dependent flow of probability among protein states in the kinetic scheme. This file, labeled "KineticSchemeAnimation.nb," requires either Mathematica or PlayerPro™ (Wolfram) to be run. The parameters have been fixed at the values obtained for $GABA_AR$ and isoflurane (Lee et al., 2015). It runs interactively, allowing the user to vary the three experimental parameters: c_{ag} (agonist concentration), c_{an} (anesthetic concentration), and τ (pulse duration). The user can select whether the anesthetic is coapplied with agonist during the pulse, or is continuously present.

That experimental traces give incomplete information on the state distribution can easily lead to misinterpretation; it is particularly relevant at low agonist concentrations. An example is given in Fig. 2. The predicted response of the receptor to a pulse of GABA at concentration $c_{ag}=0.05\ K_{d,R}$ (i.e., at 5% of the equilibrium dissociation constant of agonist binding to the resting state) is shown, using parameter values obtained for $GABA_AR$ and isoflurane (Lee et al., 2015). As has been observed experimentally, the current is predicted to rise to a constant plateau value (Fig. 2A). It might then be expected that the entire state distribution has reached equilibrium, but that is not the case; as shown in Fig. 2B and C, the predicted fraction in the desensitized state continues to increase, with compensating decrease of the fraction in the resting state. In this case, the system takes an order of magnitude longer to reach equilibrium (and maximal desensitization) than to reach the steady-state current. It points to the danger of associating increased occupancy of the desensitized conformational state, with functional desensitization (decrease in current from its initial peak during the agonist pulse). In general, the fact that the time scale of

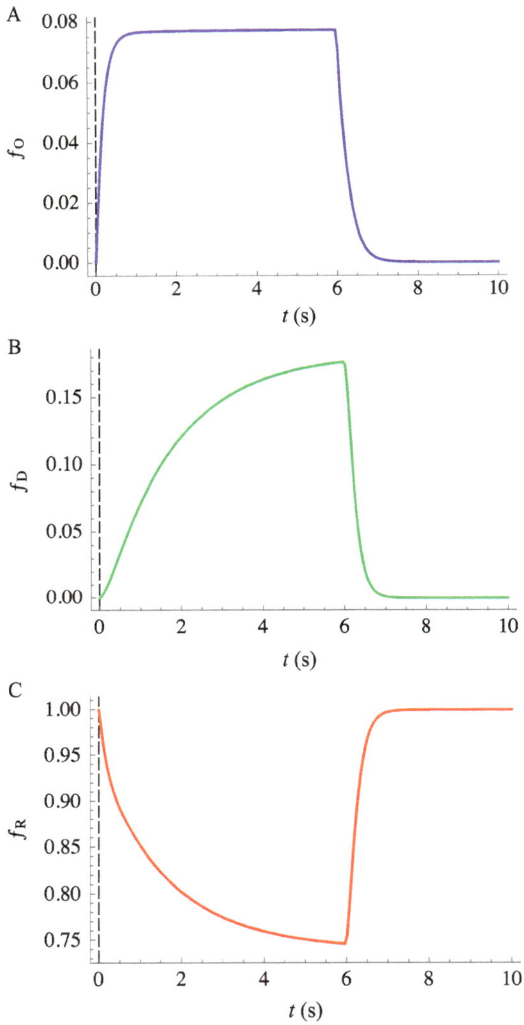

Fig. 2 Predictions of kinetic model for the response of $GABA_AR$ to a long (6 s) pulse of GABA at subsaturating concentration $= 0.05\ K_{d,R}$, coapplied with 0.1 m*M* isoflurane. (A) Total fraction of receptor in the open state, $f_O = f_{A_0O} + f_{A_1O} + f_{A_2O}$; (B) total fraction of receptor in the desensitized state, $f_D = f_{A_0D} + f_{A_1D} + f_{A_2D}$; (C) total fraction of receptor in the resting state, $f_R = f_{A_0R} + f_{A_1R} + f_{A_2R}$. Note that although the open fraction (and thus the observed current) reaches a steady-state value within 1 s, the system has not yet equilibrated; the net flow of probability from the resting to the desensitized state continues for a far longer time.

recovery from desensitization can change with increasing pulse duration, even for pulses much longer than the time needed to reach the steady-state current, is closely related to the fact that the receptor's state distribution has not yet reached equilibrium, even after the current has attained a constant value.

Examples of the extent to which this model can reproduce the complex features of electrophysiological traces are presented elsewhere (Lee et al., 2015), for the $GABA_A$ receptor and the anesthetic isoflurane. (These can all be explored using the animation software that is available for download, as described earlier.) As described therein, comprehensive electrophysiological data are available for $GABA_AR$, over a broad range of concentrations of agonist and various inhalation anesthetics, including extremely high agonist concentrations (orders of magnitude above the binding affinity to the resting state of the receptor), and anesthetic concentrations that are orders of magnitude above clinical concentrations (Adelsberger, Wilde, Franke, & Dudel, 1998; Banks & Pearce, 1999; Hapfelmeier, Haseneder, Eder, et al., 2001; Hapfelmeier, Haseneder, Kochs, Beyerle, & Zieglgänsberger, 2001; Hapfelmeier, Schneck, & Kochs, 2001; Hapfelmeier, Zieglgänsberger, Haseneder, Schneck, & Kochs, 2000; Haseneder, Rammes, Zieglgänsberger, Kochs, & Hapfelmeier, 2002; Krampfl, Bufler, Lepier, Dudel, & Adelsberger, 2000; Mercik, Zamowska, Mandat, & Mozrzymas, 2002; Neumahr et al., 2000). This includes changes in the peak current and rates of desensitization, and the development of multiple time scales of desensitization with increasing agonist concentration, the relative amounts and time constants of which can be strongly concentration dependent. It also reproduces the effects of increasing coapplied anesthetic for a long pulse of agonist at subsaturating concentrations: the initial increase (and subsequent decrease) in peak current and rise in the rate of desensitization, and the development of a rebound (tail) current on washout. Importantly, it predicts the very unusual features observed in traces of anesthetic-induced currents in the absence of agonist: with increasing anesthetic, the rise and subsequent fall of the steady-state current, and the monotonic increase in the magnitude of the rebound current observed upon washout of anesthetic. Also, and perhaps of greatest direct relevance to clinical anesthesia, it predicts how traces induced by a very short agonist pulse are affected by increasing concentrations c_{an} of continuously present anesthetic (before, during, and after the agonist pulse): the slowing of deactivation, resulting in a significant increase in ion flux at clinical anesthetic concentrations. Representative predicted traces are shown in Fig. 3, along with a plot of $Q(c_{an})/Q(0)$, the total ion flux $Q(c_{an})$ relative to its value in the absence of anesthetic $Q(0)$, as a function of anesthetic concentration c_{an}. This behavior has been observed experimentally by various groups (Benkwitz, Banks, & Pearce, 2004; Hapfelmeier, Haseneder, Eder, et al., 2001). Note that at sufficiently high c_{an}, the peak current is predicted to drop, as is also observed.

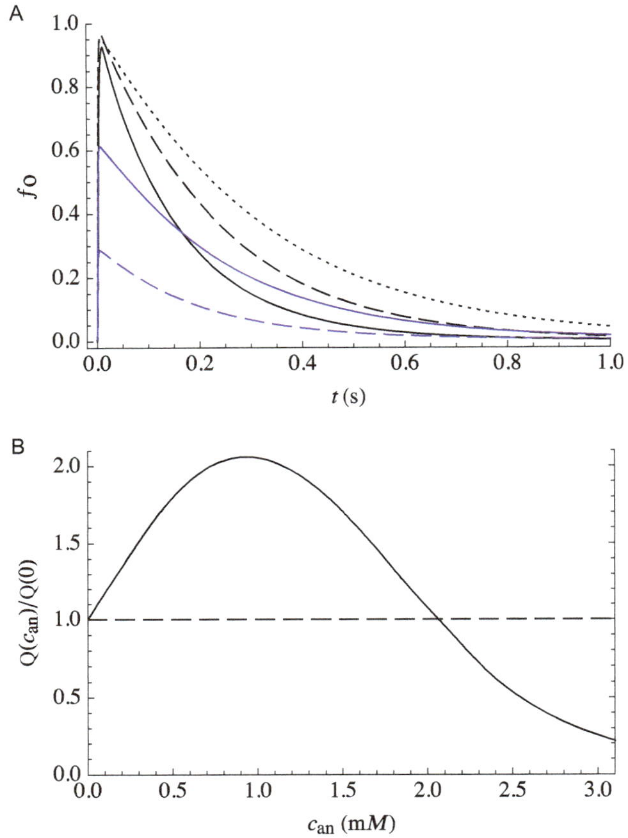

Fig. 3 (A) Response to a short pulse (10 ms) of agonist ($c_{ag} = 10\ K_{d,R}$), with isoflurane continuously present (before, during, and after the agonist pulse), as a function of anesthetic concentration: $c_{an} = 0$ (*black*), 0.3 m*M* (*black, dashed*), 0.9 m*M* (*black, dotted*), 2.0 m*M* (*blue*), and 2.7 m*M* (*blue, dashed*). Note the increase in the time scale of deactivation with increasing c_{an}. (B) Dependence of total ion flux $Q(c_{an})$ on isoflurane concentration c_{an}, relative to Q(0), its value in the absence of anesthetic. Note that it rises through the clinical range to a maximum; it decreases thereafter due to the predicted (and observed) decreases in peak current at very high c_{an}.

4. CONCLUSIONS

There is extensive evidence that postsynaptic LGICs play an essential role in clinical anesthesia. Any proposed molecular mechanism of anesthesia must be able to account for its modulation of the complex kinetic response of receptors to their agonists, under a broad range of experimental conditions. Conventional kinetic models, in which agonist and anesthetic only interact with the receptor through binding to well-defined protein sites, are unable to do so (at least for $GABA_AR$), unless the number of conformation states is increased well beyond the canonical set of three, even without accounting for effects of anesthetics. In this chapter, the implementation of an alternative model has been described. The receptor is assumed only to have three states, but the model accounts for modulation of the receptor's conformational FEL arising from absorption of anesthetic and agonist to the membrane in which the receptor is embedded. Nonuniform changes to the FEL alter the rate constants of the conformational transitions, and since adsorption/desorption does not occur instantaneously, the rate constants become time dependent, i.e., the kinetics of transitions are not Markovian. The resulting nonlinear differential equations can be solved numerically, using the computational software package Mathematica (Wolfram Research Inc., 2015), for given values of the parameters (rate constants); it also allows adjustment of the parameters to provide a best fit to experimental data (current traces under a wide range of experimental conditions). This has been shown to be quite successful for $GABA_AR$, with the anesthetic isoflurane, since extensive electrophysiological data were available in which experimental parameters (concentrations, pulse duration, etc.) were systematically varied over a very broad range. It is hoped that such extensive electrophysiological data will become available for other LGICs, and that the methodology described here can then be applied to analyze those results and test the validity of the hypothesis on which the model is based.

ACKNOWLEDGMENT

I am extremely grateful to Lucyna Bojanowka for her wisdom and support, without whom this work would not have been possible.

REFERENCES

Adelsberger, H., Wilde, J., Franke, C., & Dudel, J. (1998). Multiple mechanisms of block by the anesthetic isoflurane of a γ-aminobutyric acid activated chloride channel in crayfish. *Journal of Comparative Physiology A*, *182*, 51–58.

Banks, M. I., & Pearce, R. A. (1999). Dual actions of volatile anesthetics on GABAA IPSCs. *Anesthesiology*, *90*, 120–134.

Benkwitz, C., Banks, M. I., & Pearce, R. A. (2004). Influence of GABAA receptor gamma2 splice variants on receptor kinetics and isoflurane modulation. *Anesthesiology*, *101*, 924–936.

Burkat, P. M., Yang, J., & Gingrich, K. J. (2001). Dominant gating governing transient GABA(A) receptor activity: A first latency and Po/o analysis. *The Journal of Neuroscience*, *21*, 7026–7036.

Celentano, J. J., & Wong, R. K. (1994). Multiphasic desensitization of the GABAA receptor in outside-out patches. *Biophysical Journal, 66*, 1039–1050.

Haas, K. F., & MacDonald, R. L. (1999). GABAA receptor subunit gamma2 and delta subtypes confer unique kinetic properties on recombinant GABAA receptor currents in mouse fibroblasts. *The Journal of Physiology, 514*(Pt. 1), 27–45.

Hapfelmeier, G., Haseneder, R., Eder, M., Adelsberger, H., Kochs, E., Rammes, G., et al. (2001). Isoflurane slows inactivation kinetics of rat recombinant α1β2γ2L GABAA receptors: Enhancement of GABAergic transmission despite an open-channel block. *Neuroscience Letters, 307*, 97–100.

Hapfelmeier, G., Haseneder, R., Kochs, E., Beyerle, M., & Zieglgänsberger, W. (2001). Coadministered nitrous oxide enhances the effect of isoflurane on GABAergic transmission by an increase in open-channel block. *The Journal of Pharmacology and Experimental Therapeutics, 298*, 201–208.

Hapfelmeier, G., Schneck, H., & Kochs, E. (2001). Sevoflurane potentiates and blocks GABA-induced currents through recombinant α1β2γ2 GABAA receptors: Implications for an enhanced GABAergic transmission. *European Journal of Anaesthesiology, 18*, 377–383.

Hapfelmeier, G., Zieglgänsberger, W., Haseneder, R., Schneck, H., & Kochs, E. (2000). Nitrous oxide and xenon increase the efficacy of GABA at recombinant mammalian GABA(A) receptors. *Anesthesia and Analgesia, 91*, 1542–1549.

Haseneder, R., Rammes, G., Zieglgänsberger, W., Kochs, E., & Hapfelmeier, G. (2002). GABAA receptor activation and open-channel block by volatile anaesthetics: A new principle of receptor modulation? *European Journal of Pharmacology, 451*, 43–50.

Jones, M. V., & Westbrook, G. L. (1995). Desensitized states prolong GABAA channel responses to brief agonist pulses. *Neuron, 15*, 181–191.

Krampfl, K., Bufler, J., Lepier, A., Dudel, J., & Adelsberger, H. (2000). Desensitization characteristics of rat recombinant GABAA receptors consisting of α1β2γ2S and α1β2 subunits expressed in HEK293 cells. *Neuroscience Letters, 278*, 21–24.

Lee, D. K., Albershardt, D. A., & Cantor, R. S. (2015). Exploring the mechanism of general anesthesia: Kinetic analysis of GABAA receptor electrophysiology. *Biophysical Journal, 108*, 1081–1093.

Lema, G. M. C., & Auerbach, A. (2006). Modes and models of GABAA receptor gating. *The Journal of Physiology, 572*, 183–200.

Mercik, K., Zamowska, E. D., Mandat, M., & Mozrzymas, J. W. (2002). Saturation and self-inhibition of rat hippocampal GABAA receptors at high GABA concentrations. *The European Journal of Neuroscience, 16*, 2253–2259.

Neumahr, S., Hapfelmeier, G., Scheller, M., Schneck, H., Franke, C., & Kochs, E. (2000). Dual action of isoflurane on the γ-aminobutyric acid (GABA)-mediated currents through recombinant α1β2γ2L-GABAA-receptor channels. *Anesthesia and Analgesia, 90*, 1184–1190.

Wolfram Research, Inc.. (2015). *Mathematica, Version 10.2.* Champaign, IL: Wolfram Research Inc.

SECTION II

Genetics and Model Organisms

CHAPTER SIX

General Genetic Strategies

Louise M. Steele*[,1], Philip G. Morgan†[,1]
*Kent State University, Salem, OH, United States
†University of Washington and Seattle Children's Research Institute, Seattle, WA, United States
[1]Corresponding author: e-mail address: philip.morgan@seattlechildrens.org

Contents

Abstract

It is difficult to study the genetics and molecular mechanisms of anesthesia in humans. Fortunately, the genetic approaches in model organisms can, and have, led to profound insights as to the targets of anesthetics. In turn, the organization of these putative targets into meaningful pathways has begun to elucidate the mechanisms of action of these agents. However, it is important to first appreciate the strengths, and limitations, of genetic approaches to understand the anesthetic action. Here we compare the commonly used genetic model organisms, various anesthetic endpoints, and different modes of genetic screens. Coupled with the more specific data presented in subsequent chapters, this chapter places those results in a framework with which to analyze the discoveries across organisms and eventually extend the resulting models to humans.

Methods in Enzymology, Volume 602
ISSN 0076-6879
https://doi.org/10.1016/bs.mie.2018.01.005

1. INTRODUCTION

It is increasingly evident that congruent behavioral responses to volatile anesthetics (VAs) exist across the animal kingdom. As a result, genetic studies in model organisms occupy a special position in attempts to understand molecular control of VA action. Responses to a neuroactive drug must depend on genetic information regardless of whether the anesthetic target is a protein, lipid, carbohydrate, or something else. We are aided by the explosion of knowledge during the past 40 years concerning both the molecular nature and the manipulation of genetics (i.e., molecular genetics). It is now often possible to identify a specific change in the DNA of an organism that gives rise to a particular change in behavior. The DNA sequence is used to predict the amino acid sequence of the resulting protein. The amino acid sequences of a massive number of proteins with known functions have been identified making it feasible to determine the function of new sequences by comparison. This makes possible the following general experimental approach (Fig. 1).

2. ORGANISMS

Several researchers have undertaken forward genetic screens to identify the site of action of VAs. Studies in mice (Baker, Melchior, & Deitrich, 1980; Koblin & Deady, 1981; McCrae, Gallaher, Winter, & Firestone, 1993; Quintana, Morgan, Kruse, Palmiter, & Sedensky, 2012; Zimin et al., 2016), fruit flies (Gamo et al., 2003; Krishnan & Nash, 1990), and nematodes (Kayser, Morgan, Hoppel, & Sedensky, 2001; Morgan & Cascorbi, 1985; Sedensky & Meneely, 1987; van Swinderen et al., 1999) have identified genetic changes that affect responses to VAs. Several of these genetic changes alter behavior in some VAs differently than others (Campbell & Nash, 1994; Morgan, Sedensky, & Meneely, 1990). The reader is referred to these chapters for the details of individual genetic screens as well as the initial results. In addition, the references listed in this chapter for each technique are those that describe methods useful for the reader.

Model organisms have individual strengths and weaknesses. In a forward screen, mutations are induced and the resulting offspring are observed for alterations in response to the anesthetic, with no presupposition as to a putative anesthetic target (Jorgensen & Mango, 2002). It is helpful to start with isogenic (genetically identical) organisms to determine the "normal" or

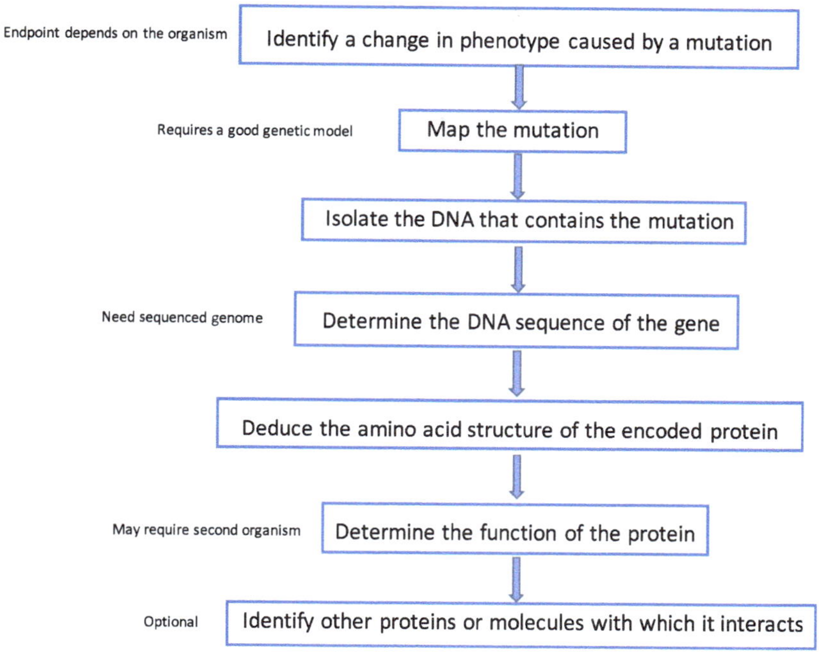

Fig. 1 A general flow of experimentation for genetic studies in anesthetic experiments. First the researcher needs to pick a reasonable endpoint in a useful organism. Once an animal with an altered response is identified, one needs to map the mutation and isolate the altered gene or sequence. Since essentially all major organisms have sequenced genomes at this time, these data allow one to identify the altered gene, deduce the resulting protein and, often, the proteins which interact with the initial protein. It is important also to determine the extent of the gene function throughout phylogeny as a measure of the probable importance in humans.

"wild-type" response. Having an isogenic organism (or at least an inbred strain) simplifies identification of the single change causing the new phenotype (Brenner, 1974). We first describe some of the organisms which have been used for molecular genetic studies of VA action with advantages and disadvantages (Table 1).

2.1 Yeast (*Saccharomyces cerevisiae*)

Yeast are an extremely powerful genetic model and were among the first to have a sequenced genome. Libraries of yeast have been generated with knockout mutations in about 6000 genes of the estimated 6200 total genes (Giaever & Nislow, 2014). Such libraries can be screened rapidly for changes in response to a drug. In addition, libraries exist to screen for protein–protein

Table 1 Advantages and Disadvantages of Major Model Systems for Genetic Studies

	Forward Screen	Targeted Screen	Other Advantages	Disadvantages	Genome
Yeast	+++	+++	Very short generation time (2 h) Extremely powerful genetics Metabolism/organelles conserved Complete knockout libraries Two hybrid screens	Lack Nervous System No organ systems	~6000 genes 46% orthologous human
Nematode	+++	++	Superb genetics Short generation time (3 days) Many mutants isolated RNAi effective Knockout libraries Nervous system defined (302 neurons) Cell lineage defined Isogenic wild type Self-fertilizing hermaphrodites Transparent	Neurons small Difficult electrophysiology Cuticle an obstacle Limited behaviors Limited anatomy	~20,000 genes 35%–40% orthologous human
Fruit fly	+++	+++ p-Elements	Short generation time (11 days) Many mutants isolated siRNA Knockout libraries Nervous system More behaviors than Ce	Neurons small	~17,000 genes 61% orthologous human

Zebrafish	++	++ (Morpholinos)	Morpholinos Talens Zinc finger nucleases	Water	~26,000 genes 69% orthologous human PMCID: PMC3703927
Mouse		++++ Overwhelming advantage	Mammalian behavior Complex CNS Larger neurons	Long generation time	~23,000 genes ~95% orthologous human

The column for targeted screens refers to prior studies. With the advent of CRISPR/Cas9 technology, targeted screens are available in each of these organisms.

interactions in yeast (two-hybrid screen), allowing extension of any identified protein to interacting members of pathways (Silva, Freitas, Felgueiras, & Fardilha, 2015). The primary disadvantage of yeast is that they do not have a nervous system. Although excellent for studying anesthetic effects on gene expression, development, homeostatic pathways, and metabolism (Keil, Wolfe, Reiner, Peterson, & Riley, 1996; Wolfe, Hester, & Keil, 1998), they are not very useful for studying determinants of neuronal function.

2.2 Nematodes

The nematode, *Caenorhabditis elegans*, is an animal with the simplest nervous system that has been used to study genetic control of VA sensitivity (Morgan et al., 1990; Sedensky & Meneely, 1987; van Swinderen et al., 1999). The hermaphrodite has a short generation time (3 days), is relatively cheap and easy to grow on agar plates, and has a plethora of identified mutations in genes affecting the nervous system (Brenner, 1974). It is particularly useful in studies using forward screens or in the use of interference RNA (RNAi) (see later) (Fire et al., 1998; Montgomery, Xu, & Fire, 1998). The anatomy of *C. elegans* can be described at the level of individual cells; the cell lineages are completely known and a complete wiring diagram of the 302 neuron nervous system has been constructed (White, Southgate, Thomson, & Brenner, 1986). Finally, the organism is transparent, which allows the use of fluorescent markers to determine expression patterns of important genes and also use of optogenetics to affect anesthetic sensitivity (Singaram et al., 2011).

Limitations in the nematode include (1) difficulty in performing electrophysiology since neurons are quite small; (2) determining what anesthetic endpoints are optimal; and (3) limited, if any, effects of parenteral anesthetics. A common concern about results in *C. elegans* has been whether the findings would be applicable to mammals; however, results have been remarkably translatable into more complex organisms (Herring, Xie, Marks, & Fox, 2009; Humphrey et al., 2007; Speca et al., 2010; Zalucki et al., 2015; Zimin et al., 2016). Nematodes will be discussed in detail in Chapter "Approaches to anesthetic mechanisms: The *C. elegans* model" by Steele and Sedensky.

2.3 Flies

The fruit fly, *Drosophila melanogaster*, is similar to *C. elegans* in that it is a powerful genetic tool for forward screens (St Johnston, 2002). The fly also has a short generation time (about 11 days at room temperature), is relatively cheap to grow and has a long history of elegant genetic studies that relies on mating to maintain/establish strains. The fly's sequenced genome has an enormous list of mutations; techniques exist to knockout mutations in any desired gene. It possesses an elaborate nervous system, organs like eyes and heart and lungs, and multiple morphologic characteristics. As a result, it has been extensively used to study VA action (Campbell & Nash, 1994; Krishnan & Nash, 1990). Limitations in the fly are similar to those in the nematode. However, targets identified in the fly are orthologous to targets identified in the nematode and have now been extended in the mouse (Humphrey et al., 2007). Flies will be discussed in detail in Chapter "Using *Drosophila* to understand general anesthesia: From synapses to behavior" by Karunanithi, Troup, and van Swinderen.

2.4 Zebrafish

Zebrafish are rapidly gaining traction as a genetic model. They share many of the advantages of nematodes and flies (short generation time, easy genetics, and sequenced genome). As a genetic model, they now rival the fly and the nematode with ability to generate targeted mutations using zinc fingered nucleases, transcription activator-like effector nucleases (TALENs) in addition to the well-known use of morphilinos (Howe et al., 2013; Yoganantharjah & Gibert, 2017). However, their aqueous environment has limited use in the study of volatile agents. Zebrafish will be discussed in detail in Chapter "Zebrafish: A pharmacogenetic model for anesthesia" by Bedell et al.

2.5 Mouse

The mouse presents a significantly different set of advantages and disadvantages than do flies and nematodes. A more complicated nervous system allows a more apparent extension into human genetics. Neurons are bigger and electrophysiology is manageable in mice. However, the generation time of the mouse is much longer than the nematode or fly and the number of offspring is much less. As a result, it is extremely difficult to do large forward screens in the mouse. However, using homologous recombination and Cre recombinase, the ability to knockout specific genes is tremendously developed in the mouse and has been its primary advantage in studying CNS function (Palmiter, 2011). In addition, at the time of this writing, given new capabilities using a technique known as CRISPR/Cas9 technology, researchers are planning to generate lines of mice containing mutations in every known gene (Joung et al., 2017). While this will be a tremendous resource, testing every strain of mice for responses to a drug (for example) will still be prohibitive both in time and in expense. The mouse will be discussed in detail in Chapter "The mouse as a model organism for assessing anesthetic sensitivity" by Wasilczuk, Maier, and Kelz.

2.6 Human

Humans present a completely different genetic problem than any of the model systems. On the one hand, classical genetic experiments are completely impossible. However, studies across the human population present a variation in the genetic range simply not available in model systems in the laboratory. Thus, one can use human responses to drugs to identify possible gene targets by comparing large numbers of patients and comparing the large datasets. The results must then be confirmed using one or more of the model systems to isolate specific genes or gene groups.

3. ENDPOINTS

The choice of anesthetic endpoints is crucial in all model organisms. The problem is that we are studying an endpoint (loss of consciousness or response to pain) that is difficult to quantify not only in model systems but also in humans. As a result, any behavioral endpoint must be interpreted with caution. Proof of validity for extension into humans always awaits final studies in mammals and humans. Each of these organisms and their endpoints will be discussed in Chapters "Approaches to anesthetic mechanisms: The *C. elegans* model" by Steele and Sedensky; "Using *Drosophila* to understand general anesthesia: From synapses to behavior" by Karunanithi, Troup, and van Swinderen; "Zebrafish: A pharmacogenetic model for anesthesia" by Bedell et al.; and "The mouse as a model organism for assessing anesthetic sensitivity" by Wasilczuk, Maier, and Kelz. The short presentation here is to assist with comparison of the endpoints across the organisms (Table 2).

3.1 Yeast (*S. cerevisiae*)

Since yeast do not have a nervous system, behavioral endpoints were not used. Keil and colleagues used inhibition of cell division as the anesthetic endpoint (Keil et al., 1996). Follow-up studies showed that yeast sensitivity to VAs correlated with VA potency in mammals (Wolfe et al., 1998) and that protein translation was inhibited (Palmer, Shoemaker, Baptiste, Wolfe, &

Table 2 Comparison of Endpoints Used for Anesthetic Studies in Model Organisms

	Anesthetic	Endpoint	Concentration/Dose Anesthetic
Yeast	Volatile	Growth limitation	~6% (halothane) ~12% (isoflurane)
Nematode	Volatile	Uncoordination/mating/ chemotaxis	~0.5%–1% (halothane) ~1% (isoflurane)
	Volatile	Immobility	~3% (halothane) 6% (isoflurane)
Fruit fly	Volatile	Falling	~0.5%–1% (halothane) ~0.5% (isoflurane)
Zebrafish	Volatiles	Not done	NA
	Ketamine	Circular swimming	20 mg/L
Mouse	Volatiles	Loss of righting reflex (LORR) Response to tail pinch Multiple others	~0.8% (halothane) ~1.3% (halothane)
	Propofol	LORR	~70–140 mg/kg (IP)
	Ketamine	LORR	~40–70 mg/kg (IP)

Anesthetic concentrations are given for comparison. Other anesthetics have been studied and are in references in text.

Keil, 2005). While the results may ultimately shed light on the broad range of VA action (such as neurotoxicity), they were not felt to closely relate to the neuronal-based endpoints in higher organisms. The lack of a behavioral endpoint is felt to be the major limitation for yeast as a model system for studying anesthetics.

3.2 Nematodes (*C. elegans*)

In *C. elegans*, two types of endpoints were studied. One, reversible immobility, required higher concentrations of VAs than needed to anesthetize humans (Morgan et al., 1990; Morgan & Cascorbi, 1985). Therefore, several related endpoints were identified which required concentrations of anesthetic that more closely those used clinically in humans (Metz, Dasgupta, Liu, Hunt, & Crowder, 2007; van Swinderen et al., 1999). The relative endpoints and anesthetic concentrations in the nematode are shown in Fig. 2. Interestingly, both approaches predicted orthologous targets which

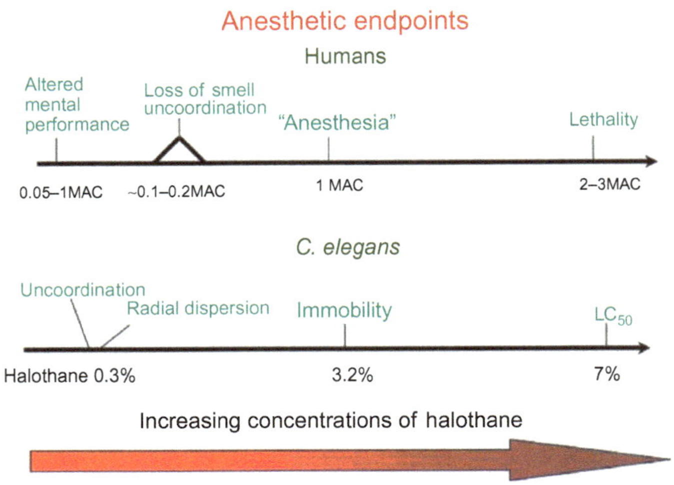

Fig. 2 Relative EC_{50}s for several endpoints in humans (*top*) and in *C. elegans* (*bottom*). Note the relatively higher anesthetic concentrations for the range of behavioral endpoints in the nematode compared to humans. It is not known which endpoints in the nematode correspond to specific endpoints in humans. However, the relative positions of different endpoints in humans can be compared those for endpoints in the nematode. Similar relationships can be drawn for other organisms. It is important to study targets which respond in the range of those causing behavioral changes in the organism studied.

controlled the tail clamp and loss of righting reflex responses in the mouse. Similar concentration-specific endpoints are likely present in all genetic model organisms.

3.3 Fruit Flies (*D. melanogaster*)

In fruit flies, a completely analogous approach to that in nematodes was taken by the Nash laboratory. Using an inebriometer, measuring loss of walking ability, the researchers identified mutations in two genes that were orthologous to genes identified in the nematode (Krishnan & Nash, 1990; Leibovitch, Campbell, Krishnan, & Nash, 1995). The collaboration between two laboratories demonstrated the value of comparing genetic results across organisms (Humphrey et al., 2007). Since those studies, similar genes in the mouse have also been shown to affect VA sensitivity (Speca et al., 2010). These studies show the power of model organisms, the importance of behavioral endpoints, and the crucial steps of extending the studies into more complicated organisms such as mammals.

3.4 Mice (*Mus musculus*)

Mice of course are much closer phylogenetically to humans than are any of the simpler model systems. The behaviors are correspondingly more familiar. Generally, loss of righting reflex (the reflex in mice to right themselves when placed on their backs) is used as a surrogate for loss of consciousness (Sonner et al., 2007). Failure to respond to a tail pinch is used to substitute for loss of movement in response to a painful stimulus in humans (Sonner et al., 2007; Sonner, Gong, & Eger, 2000). Other, more subtle, endpoints such as learning, anxiety, and memory have also been used in the mouse (Dai, Perouansky, & Pearce, 2012; Zarnowska et al., 2015).

4. STRATEGIES

4.1 Forward Screens

Any genetic study using a model organism must take into consideration the generation time, the size of the genome, and the similarity of the genome to that of humans. An overview of these details is presented in Table 3.

Classical forward genetics have been used widely in model organisms (such as yeast, nematodes, fruit flies, and zebrafish) to characterize an enormous array of developmental and physiologic pathways (Jorgensen & Mango, 2002; Joung et al., 2017; St Johnston, 2002; Yoganantharjah & Gibert, 2017). Forward genetics randomly mutagenizes DNA in a defined genetic background (as shown in Fig. 1). The offspring (first or second generation) is then screened for an observable change (phenotype) of interest (in this case, an altered response to anesthetics). The presumption is that

Table 3 Comparison of Developmental Times and Genomes in Model Organisms

	Generation Time	Genome Size	Genome Similarity to Human	Special Points
Yeast	~2 h	~6000 genes	~46%	Asexual reproduction
Nematode	~3 days	~19,000 genes	~35%–43%	Self-fertilizing hermaphrodites
Fruit fly	~10 days	~20,000 genes	~61%	
Zebrafish	~2–3 months	~26,000 genes	~70%	
Mouse	~2 months	~23,000 genes	~95%	

Special points refer to organism-specific characteristics affecting genetic studies only.

phenotype is the result of a single gene mutation. The change in DNA is then determined by genetic crosses to "map" the change relative to other known DNA markers. The markers can be other known mutations (i.e., mapped genes) or sequence markers (single-nucleotide polymorphisms, SNPs) already identified in the DNA. These techniques are well described and beyond the scope of this review; the reader is referred to Chapter "Approaches to anesthetic mechanisms: The *C. elegans* model" by Steele and Sedensky.

Using a forward genetic approach in a whole animal is appealing because no assumptions are made about the cellular components that should play a role in the mechanisms of anesthetic action. Forward screens in worms, flies, and now zebrafish can be of varying difficulty (Jorgensen & Mango, 2002; St Johnston, 2002). Screening for resistance to an anesthetic could simply look for survivors after exposure to a lethal dose of drug. Assaying for subtle changes in behavior, like uncoordination or mating behavior, can be more difficult but have the advantage of occurring at lower concentrations of anesthetic.

4.1.1 Mutagenesis

Mutagenesis in model organisms usually involves treating a control animal (often wild type but not always) with a mutagen to induce new random genetic changes. In general, young fertile animals are exposed to a chemical mutagen like ethylmethylsulfonate (EMS) designed to cause base pair changes in the DNA or irradiation designed to induce DNA damage. Many other techniques to alter the DNA randomly do exist and are not described here. Whatever mutagenesis is used, after inducing the DNA changes, one then screens the first generation offspring for dominant mutations, and the second for recessive changes. It is usually desirable to screen thousands of second generation offspring to survey the entire genome for changes. In the cases of nematodes and flies, researchers generally screen 20,000 or more offspring with each mutagenesis and perform the screen multiple times (Jorgensen & Mango, 2002; St Johnston, 2002). Since the generation time is short, the first round of screening often can be done in matter of weeks. However, the first screening for interesting mutations is only the beginning of the isolation. The animals of interest must be crossed (mated) to control animals multiple times to "isolate" the gene of interest from other newly induced mutations in the rest of the genome. Then the gene must be mapped and isolated from the DNA (cloned) to identify the gene and the form of the mutation. Thus, even in these simple model organisms, an approach with no preconceptions is time consuming.

4.1.2 RNAi

A simpler approach to forward genetics is now possible by use of RNAi (various forms in some organisms are called silencing RNA or siRNA) (Lee et al., 2004). RNAi consists of double-stranded RNA that is introduced into cells or animals in a variety of ways (Hammell & Hannon, 2012; Kamath & Ahringer, 2003; Kamath et al., 2003). Once the double-stranded RNA (RNAi) is introduced, an endogenous mechanism exists in most cells to digest and eliminate mRNA with stretches of sequence that match the introduced RNAi. Thus, to silence expression of gene A, one needs only to introduce RNAi with a stretch of sequence closely matching gene A. Use of RNAi to knockdown function of a gene is often employed when one has identified a putative ortholog in worms or flies of a gene already known to cause an interesting phenotype in other models or in humans. For this approach, RNAi can be used for targeted screens, similar to other targeted screens described later.

However, it is possible to use RNAi for a modified forward screen as well. In both nematodes and flies, RNAi libraries have been constructed that allow screening individual clones that cover essentially the entire genome (Kamath & Ahringer, 2003). As an example, in the nematode such libraries exist that contain clones representing about 20,000 different genes. By growing nematodes on individual plates, each seeded with bacteria containing one of these clones, it is possible to screen the entire genome (although it is tedious and requires 20,000 individual plates). There is a tremendous benefit once a clone is identified that causes the phenotype of interest. Since the gene is known which is present in each bacterial clone, no further work to genetically map the gene is necessary. Thus, there is a tremendous efficiency at the back end of the screen. With the advent of high-throughput techniques, the use of RNAi has largely replaced blind forward screens—but not completely. In addition, RNAi knockdown genes in a dose-dependent manner, which may uncover genes whose complete loss of function would be lethal. However, this dose-dependent knockdown may lead to one serious limitation; RNAi seldom causes complete loss of a gene, nor does it alter genes. Dominant and missense mutations are therefore missed by this approach and the true null phenotype may also not be apparent.

4.1.3 Recombinant Inbred Screens

Variation may exist between isolated populations of organisms in the response to a drug or exposure of interest. In those cases, individuals from the separate populations may be mated and the resulting offspring separated to isolate the variation in response (these are the recombinant inbred strains—RIS) (van Swinderen, Galifianakis, & Crowder, 1998). The mating may be thought of as mixing the chromosomes randomly between the groups and then recollecting the chromosomes or regions of chromosomes responsible for the difference by observing the offspring (van Swinderen et al., 1997). Those offspring with the original divergent responses are then inbred and the location(s) of the important gene(s) are mapped by noting the positions of SNPs which differ between the original populations. This inbreeding produced the first populations of mice sensitive or resistant to nitrous oxide. Repositories of mice and rats have now been generated that differ in only one chromosome from a parent population (congenic strains) limiting study of a changed behavior from the parent to a single chromosome in each strain (Nadeau, Singer, Matin, & Lander, 2000; Rapp, 2000). In addition, these repositories often have strains in which the differing region is limited to even smaller portions of a chromosome as an added advantage.

The advantages to this approach are that it exploits naturally occurring variation; one is not limited by the luck of whether a potentially important gene is hit by random mutagenesis. In addition, when a response results from protein–protein interactions, occasionally the RIS will show variation in excess of what is seen in the original populations (van Swinderen et al., 1997). However, while this is a very powerful technique, the primary limitation is that it often proves difficult to experimentally limit the change to one small region or gene. Other problems can occur, but they basically all stem from having a complicated array of genetic regions which can be quite difficult to sort.

4.2 Targeted Screens

Targeted genetics usually refers to studying individual genes which are suspected to be important because they do something potentially relevant. In many cases, a single gene has been implicated in one organism (e.g., nematode or fly) but would be useful to study in another organism (e.g., mouse). It can be useful then to mutate the specific orthologous gene of interest in the second organism. Conversely, genes identified in a complex organism can also be manipulated in a simpler organism to perform experiments not possible in the complex organism. The function of the genes studied in the simpler organisms may not even be known. Often, it is the sequence of those genes, which show striking homology to the genes in the original organism, which leads to their study. However, once their importance is established in the simpler model organism, all of the genetic power can be brought to bear to identify function, interaction, and physiologic pathways.

4.2.1 RNAi

RNAi, discussed earlier, can easily be used for targeted screens as well. For example, Falk et al. knocked down every highly conserved subunit of mitochondrial complex I of the electron transport chain once it was clear that the 49 kDa protein of complex I was important in controlling behavior in anesthetics (Falk, Kayser, Morgan, & Sedensky, 2006). The use of RNAi is relatively straightforward in worms, as one can simply grow worms on a lawn of *E. coli* expressing an individual clone, which is obtainable from one of several RNAi libraries. Worms ingest these RNAi expressing bacteria, which is sufficient to knockdown gene function, often within a matter of hours. Not all genes may be equally susceptible to knockdown, with genes expressed in neurons said to be the most difficult to suppress. This approach, however, is a powerful way to test a broad group of genes which may be of interest for any phenotype. A similar approach can be used in the larval forms of fruit flies and in zebrafish.

Another method for identification of gene products affecting the site of action of VAs is the study of gene interaction. If several gene products function together at a particular site, then mutations in a second or third gene may alter or compensate for mutations in an initial gene. The phenotypes of double mutants could provide valuable information about such interactions. Again, the use of RNAi in targeted screens is an excellent way to canvas such an array of genes.

4.2.2 Site-Directed Mutagenesis (SDM)

4.2.2.1 Homologous Recombination

A second example of targeted genetics is SDM which mutates a single specific gene. Most often performed in the mouse, SDM initially used homologous recombination to insert a desired mutation. Forman used SDM to specifically alter a region of the $GABA_A$ receptor, demonstrating the importance of that region in etomidate binding and activity on that receptor (Forman & Miller, 2011, 2016). While this approach carried great value for studying specific sites in a protein, the low efficiency of recombination and long time necessary to obtain homozygotes were serious limitations. In addition, these recombinations needed to be done in embryonic stem cell lines and implanted into mouse embryos, a technique not available at many institutions.

4.2.2.2 Cre Recombinase

For the past 2 decades, the Cre recombinase approach has been used to generate specific targeted mutations (Condie, 2016; Rossant & Nagy, 1995). The general approach is to insert a modified gene containing two specific DNA sequences (termed lox sites), in flanking a native gene in the mouse genome (the gene is said to be floxed—for *fl*anking *lox*). The lox sites can be recognized by an endonuclease, Cre recombinase, which removes the DNA at those positions. Generally, the lox containing sequences are inserted by homologous recombination, and the resulting animal is crossed to a mouse containing the Cre recombinase in its genome (generated earlier by homologous recombination and available from a mouse genetic repository). The offspring then contain the knockout of the gene and all that remains are to make the mutated gene homozygous. Originally, this approach had much the same advantages and disadvantages as homologous recombination described earlier. In fact, the initial step still involved the low-efficiency step. The advantage was that a Cre-containing mouse could be mated to any mouse containing the lox sites to generate the DNA removal or knockout.

4.2.2.3 Inducible Cre Recombinase

A tremendous extension to targeted genetics was developed by putting Cre recombinase under the control of specific promoters (Hnasko et al., 2006). These promoters were developmentally controlled which, in turn, allowed for the ability to knock a gene out in a specific tissue or cell line or at a specific time during development. The advantages of this approach are obvious; by mixing and matching the targeted floxed genes with strains containing Cre under different promoters, it is possible to test for important regions, cell types, or times of expression. The disadvantage is that this still requires the ability to generate homologous recombination lines in a mouse as a necessary step.

4.2.2.4 Optogenetics

The development of optogenetics presents a powerful mechanism to manipulate neuronal firing acutely using light as the trigger. A gene coding for a light sensitive protein (for example, channelrhodopsin or halorhodopsin) is inserted into cells of the nervous system. These proteins are capable of altering the transmembrane potential in response to a fiberoptic light stimulus causing localized activation or inhibition of nerve cells. If hyperpolarization is an important effect of VAs, an artificial depolarization of the important neurons (as caused by channelrhodopsin) of anesthetized animals should reverse anesthetic effects. On the contrary, artificial hyperpolarization (as caused by stimulation of halorhodopsin) of the cells in the presence of stimulating light accentuate anesthetic effects. This approach has been used to reverse VA-induced immobility in the nematode (Singaram et al., 2011) and has now been used to identify CNS regions important to the anesthetic response in the mouse (Taylor et al., 2016).

4.2.2.5 CRISPR/Cas9

A powerful technique to generate essentially any mutation in any gene in any organism is recently available. This technique is the CRISPR-Cas9 system. CRISPR, an abbreviation of clustered regularly interspaced short palindromic repeats, was identified as an immune function in bacteria for protection from viral infection (Friedland et al., 2013; Mali et al., 2013; Mali, Esvelt, & Church, 2013). Cas9 is the CRISPR-associated protein 9, an endonuclease which is responsible for cutting the target DNA. The use of the CRISPR system is probably the most exciting development in gene editing in recent years. For targeted mutagenesis studies, this is clearly the most powerful technique presently available.

The techniques using CRISPR are evolving so quickly that any in-depth description of them will likely be obsolete by the time this chapter is in press. The general approach is to construct a short, ~20 nucleotide RNA fragment (a CRISPR RNA or crRNA) which is complementary to the DNA target to be removed. The RNA fragment is accompanied (either fused or not fused) by a transactivating RNA; the combination of the two is termed a single guide (sg)RNA. The combination is introduced into the host cell with the Cas9 endonuclease. The crRNA binds (by standard Watson Crick base pairing) to the target sequence, and the Cas9 system excises the target sequence from the host DNA. If no template for repair is available, the DNA is repaired but is missing the targeted sequence. If a template for repair is supplied (generally containing a desired mutation), the host cell will use that template to make the repair. In this way, specific mutations can be introduced into the host gene. This is very superficial explanation of the mechanism and the reader is encouraged to read further on the topic. However, since this technique does not require any host machinery other than DNA repair, this approach is likely to completely alter our abilities to create targeted changes in genes of interest.

REFERENCES

Baker, R., Melchior, C., & Deitrich, R. (1980). The effect of halothane on mice selectively bred for differential sensitivity to alcohol. *Pharmacology, Biochemistry, and Behavior, 12*, 691–695.

Brenner, S. (1974). The genetics of Caenorhabditis elegans. *Genetics, 77*, 71–94.

Campbell, D. B., & Nash, H. A. (1994). Use of Drosophila mutants to distinguish among volatile general anesthetics. *Proceedings of the National Academy of Sciences of the United States of America, 91*, 2135–2139.

Condie, B. G. (2016). The untapped potential of the GENSAT mice—A valuable resource for developmental biology. *Genesis, 54*, 245–256.

Dai, S., Perouansky, M., & Pearce, R. A. (2012). Isoflurane enhances both fast and slow synaptic inhibition in the hippocampus at amnestic concentrations. *Anesthesiology, 116*, 816–823.

Falk, M. J., Kayser, E. B., Morgan, P. G., & Sedensky, M. M. (2006). Mitochondrial complex I function modulates volatile anesthetic sensitivity in C. elegans. *Current Biology, 16*, 1641–1645.

Fire, A., Xu, S., Montgomery, M. K., Kostas, S. A., Driver, S. E., et al. (1998). Potent and specific genetic interference by double-stranded RNA in Caenorhabditis elegans. *Nature, 391*, 806–811.

Forman, S. A., & Miller, K. W. (2011). Anesthetic sites and allosteric mechanisms of action on cys-loop ligand-gated ion channels. *Canadian Journal of Anaesthesia, 58*, 191–205.

Forman, S. A., & Miller, K. W. (2016). Mapping general anesthetic sites in heteromeric gamma-aminobutyric acid type A receptors reveals a potential for targeting receptor subtypes. *Anesthesia and Analgesia, 123*, 1263–1273.

Friedland, A. E., Tzur, Y. B., Esvelt, K. M., Colaiacovo, M. P., Church, G. M., et al. (2013). Heritable genome editing in C. elegans via a CRISPR-Cas9 system. *Nature Methods, 10*, 741–743.

Gamo, S., Tomida, J., Dodo, K., Keyakidani, D., Matakatsu, H., et al. (2003). Calreticulin mediates anesthetic sensitivity in Drosophila melanogaster. *Anesthesiology, 99*, 867–875.

Giaever, G., & Nislow, C. (2014). The yeast deletion collection: A decade of functional genomics. *Genetics, 197*, 451–465.

Hammell, C. M., & Hannon, G. J. (2012). Inducing RNAi in C. elegans by feeding with dsRNA-expressing E. coli. *Cold Spring Harbor Protocols, 2012*, 1263–1266.

Herring, B. E., Xie, Z., Marks, J., & Fox, A. P. (2009). Isoflurane inhibits the neurotransmitter release machinery. *Journal of Neurophysiology, 102*, 1265–1273.

Hnasko, T. S., Perez, F. A., Scouras, A. D., Stoll, E. A., Gale, S. D., et al. (2006). Cre recombinase-mediated restoration of nigrostriatal dopamine in dopamine-deficient mice reverses hypophagia and bradykinesia. *Proceedings of the National Academy of Sciences of the United States of America, 103*, 8858–8863.

Howe, K., Clark, M. D., Torroja, C. F., Torrance, J., Berthelot, C., et al. (2013). The zebrafish reference genome sequence and its relationship to the human genome. *Nature, 496*, 498–503.

Humphrey, J. A., Hamming, K. S., Thacker, C. M., Scott, R. L., Sedensky, M. M., et al. (2007). A putative cation channel and its novel regulator: Cross-species conservation of effects on general anesthesia. *Current Biology, 17*, 624–629.

Jorgensen, E. M., & Mango, S. E. (2002). The art and design of genetic screens: Caenorhabditis elegans. *Nature Reviews Genetics, 3*, 356–369.

Joung, J., Konermann, S., Gootenberg, J. S., Abudayyeh, O. O., Platt, R. J., et al. (2017). Genome-scale CRISPR-Cas9 knockout and transcriptional activation screening. *Nature Protocols, 12*, 828–863.

Kamath, R. S., & Ahringer, J. (2003). Genome-wide RNAi screening in Caenorhabditis elegans. *Methods, 30*, 313–321.

Kamath, R. S., Fraser, A. G., Dong, Y., Poulin, G., Durbin, R., et al. (2003). Systematic functional analysis of the Caenorhabditis elegans genome using RNAi. *Nature*, *421*, 231–237.
Kayser, E. B., Morgan, P. G., Hoppel, C. L., & Sedensky, M. M. (2001). Mitochondrial expression and function of GAS-1 in Caenorhabditis elegans. *The Journal of Biological Chemistry*, *276*, 20551–20558.
Keil, R. L., Wolfe, D., Reiner, T., Peterson, C. J., & Riley, J. L. (1996). Molecular genetic analysis of volatile-anesthetic action. *Molecular and Cellular Biology*, *16*, 3446–3453.
Koblin, D. D., & Deady, J. E. (1981). Anaesthetic requirement in mice selectively bred for differences in ethanol sensitivity. *British Journal of Anaesthesia*, *53*, 5–10.
Krishnan, K. S., & Nash, H. A. (1990). A genetic study of the anesthetic response: Mutants of Drosophila melanogaster altered in sensitivity to halothane. *Proceedings of the National Academy of Sciences of the United States of America*, *87*, 8632–8636.
Lee, J., Nam, S., Hwang, S. B., Hong, M., Kwon, J. Y., et al. (2004). Functional genomic approaches using the nematode Caenorhabditis elegans as a model system. *Journal of Biochemistry and Molecular Biology*, *37*, 107–113.
Leibovitch, B. A., Campbell, D. B., Krishnan, K. S., & Nash, H. A. (1995). Mutations that affect ion channels change the sensitivity of Drosophila melanogaster to volatile anesthetics. *Journal of Neurogenetics*, *10*, 1–13.
Mali, P., Esvelt, K. M., & Church, G. M. (2013). Cas9 as a versatile tool for engineering biology. *Nature Methods*, *10*, 957–963.
Mali, P., Yang, L., Esvelt, K. M., Aach, J., Guell, M., et al. (2013). RNA-guided human genome engineering via Cas9. *Science*, *339*, 823–826.
McCrae, A. F., Gallaher, E. J., Winter, P. M., & Firestone, L. L. (1993). Volatile anesthetic requirements differ in mice selectively bred for sensitivity or resistance to diazepam: Implications for the site of anesthesia. *Anesthesia and Analgesia*, *76*, 1313–1317.
Metz, L. B., Dasgupta, N., Liu, C., Hunt, S. J., & Crowder, C. M. (2007). An evolutionarily conserved presynaptic protein is required for isoflurane sensitivity in Caenorhabditis elegans. *Anesthesiology*, *107*, 971–982.
Montgomery, M. K., Xu, S., & Fire, A. (1998). RNA as a target of double-stranded RNA-mediated genetic interference in Caenorhabditis elegans. *Proceedings of the National Academy of Sciences of the United States of America*, *95*, 15502–15507.
Morgan, P. G., & Cascorbi, H. F. (1985). Effect of anesthetics and a convulsant on normal and mutant Caenorhabditis elegans. *Anesthesiology*, *62*, 738–744.
Morgan, P. G., Sedensky, M., & Meneely, P. M. (1990). Multiple sites of action of volatile anesthetics in Caenorhabditis elegans. *Proceedings of the National Academy of Sciences of the United States of America*, *87*, 2965–2969.
Nadeau, J. H., Singer, J. B., Matin, A., & Lander, E. S. (2000). Analysing complex genetic traits with chromosome substitution strains. *Nature Genetics*, *24*, 221–225.
Palmer, L. K., Shoemaker, J. L., Baptiste, B. A., Wolfe, D., & Keil, R. L. (2005). Inhibition of translation initiation by volatile anesthetics involves nutrient-sensitive GCN-independent and -dependent processes in yeast. *Molecular Biology of the Cell*, *16*, 3727–3739.
Palmiter, R. D. (2011). Dopamine signaling as a neural correlate of consciousness. *Neuroscience*, *198*, 213–220.
Quintana, A., Morgan, P. G., Kruse, S. E., Palmiter, R. D., & Sedensky, M. M. (2012). Altered anesthetic sensitivity of mice lacking Ndufs4, a subunit of mitochondrial complex I. *PLoS One*, *7*, e42904.
Rapp, J. P. (2000). Genetic analysis of inherited hypertension in the rat. *Physiological Reviews*, *80*, 135–172.
Rossant, J., & Nagy, A. (1995). Genome engineering: The new mouse genetics. *Nature Medicine*, *1*, 592–594.

Sedensky, M. M., & Meneely, P. M. (1987). Genetic analysis of halothane sensitivity in Caenorhabditis elegans. *Science*, *236*, 952–954.

Silva, J. V., Freitas, M. J., Felgueiras, J., & Fardilha, M. (2015). The power of the yeast two-hybrid system in the identification of novel drug targets: Building and modulating PPP1 interactomes. *Expert Review of Proteomics*, *12*, 147–158.

Singaram, V. K., Somerlot, B. H., Falk, S. A., Falk, M. J., Sedensky, M. M., et al. (2011). Optical reversal of halothane-induced immobility in C. elegans. *Current Biology*, *21*, 2070–2076.

Sonner, J. M., Gong, D., & Eger, E. I., 2nd. (2000). Naturally occurring variability in anesthetic potency among inbred mouse strains. *Anesthesia and Analgesia*, *91*, 720–726.

Sonner, J. M., Werner, D. F., Elsen, F. P., Xing, Y., Liao, M., et al. (2007). Effect of isoflurane and other potent inhaled anesthetics on minimum alveolar concentration, learning, and the righting reflex in mice engineered to express alpha1 gamma-aminobutyric acid type A receptors unresponsive to isoflurane. *Anesthesiology*, *106*, 107–113.

Speca, D. J., Chihara, D., Ashique, A. M., Bowers, M. S., Pierce-Shimomura, J. T., et al. (2010). Conserved role of unc-79 in ethanol responses in lightweight mutant mice. *PLoS Genetics*, *6*, e1001057.

St Johnston, D. (2002). The art and design of genetic screens: Drosophila melanogaster. *Nature Reviews Genetics*, *3*, 176–188.

Taylor, N. E., Van Dort, C. J., Kenny, J. D., Pei, J., Guidera, J. A., et al. (2016). Optogenetic activation of dopamine neurons in the ventral tegmental area induces reanimation from general anesthesia. *Proceedings of the National Academy of Sciences of the United States of America*, *113*, 12826–12831.

van Swinderen, B., Galifianakis, A., & Crowder, M. (1998). Common genetic determinants of halothane and isoflurane potencies in Caenorhabditis elegans. *Anesthesiology*, *89*, 1509–1517.

van Swinderen, B., Saifee, O., Shebester, L., Roberson, R., Nonet, M. L., et al. (1999). A neomorphic syntaxin mutation blocks volatile-anesthetic action in Caenorhabditis elegans. *Proceedings of the National Academy of Sciences of the United States of America*, *96*, 2479–2484.

van Swinderen, B., Shook, D. R., Ebert, R. H., Cherkasova, V. A., Johnson, T. E., et al. (1997). Quantitative trait loci controlling halothane sensitivity in Caenorhabditis elegans. *Proceedings of the National Academy of Sciences of the United States of America*, *94*, 8232–8237.

White, J. G., Southgate, E., Thomson, J. N., & Brenner, S. (1986). The structure of the nervous system of the nematode Caenorhabditis elegans. *Philosophical Transactions of the Royal Society of London. Series B, Biological Sciences*, *314*, 1–340.

Wolfe, D., Hester, P., & Keil, R. L. (1998). Volatile anesthetic additivity and specificity in Saccharomyces cerevisiae: Implications for yeast as a model system to study mechanisms of anesthetic action. *Anesthesiology*, *89*, 174–181.

Yoganantharjah, P., & Gibert, Y. (2017). The use of the zebrafish model to aid in drug discovery and target validation. *Current Topics in Medicinal Chemistry*, *17*, 2041–2055.

Zalucki, O. H., Menon, H., Kottler, B., Faville, R., Day, R., et al. (2015). Syntaxin1A-mediated resistance and hypersensitivity to isoflurane in Drosophila melanogaster. *Anesthesiology*, *122*, 1060–1074.

Zarnowska, E. D., Rodgers, F. C., Oh, I., Rau, V., Lor, C., et al. (2015). Etomidate blocks LTP and impairs learning but does not enhance tonic inhibition in mice carrying the N265M point mutation in the beta3 subunit of the GABA(A) receptor. *Neuropharmacology*, *93*, 171–178.

Zimin, P. I., Woods, C. B., Quintana, A., Ramirez, J. M., Morgan, P. G., et al. (2016). Glutamatergic neurotransmission links sensitivity to volatile anesthetics with mitochondrial function. *Current Biology*, *26*, 2194–2201.

CHAPTER SEVEN

Approaches to Anesthetic Mechanisms: The *C. elegans* Model

Louise M. Steele*, Margaret M. Sedensky†,‡,1
*Kent State University at Salem, Salem, OH, United States
†Center for Integrative Brain Research, Seattle Children's Research Institute, Seattle, WA, United States
‡University of Washington, Seattle, WA, United States
[1]Corresponding author: e-mail address: margaret.sedensky@seattlechildrens.org

Contents

Abstract

Understanding the mechanisms of volatile anesthetics has been a complex problem that has intrigued investigators for decades. Through the use of relatively simple model organisms—including the nematode *Caenorhabditis elegans*—progress has been made. Like any model system, *C. elegans* has both advantages and disadvantages, which are discussed in this chapter. Methods are provided for exposing worms to volatile anesthetics in airtight glass chambers, and for measuring the concentrations of anesthetic in the chambers by gas chromatography. In addition, various behavioral assays are described for characterizing the worms' responses to anesthetics. *C. elegans* identified proteins that play a role in anesthetic sensitivity that are highly conserved in other organisms, including humans. With precisely characterized neural development,

Methods in Enzymology, Volume 602
ISSN 0076-6879
https://doi.org/10.1016/bs.mie.2018.01.006

C. elegans has also afforded an excellent opportunity to study anesthetic-induced neurotoxicity. Continued progress in understanding anesthetic action is anticipated from the ongoing study of *C. elegans* and other animal models.

1. INTRODUCTION

The invertebrate model *Caenorhabditis elegans* has many advantages to further the understanding of anesthetic mechanisms. Exploiting a nematode to probe some of science's deepest questions relating to consciousness may certainly appear to be problematic. However, the advantages of this genetic model organism are many and may reveal a great deal about processes highly conserved across the animal kingdom.

1.1 Advantages of the Model Organism, *C. elegans*

C. elegans, a nonparasitic metazoan approximately a millimeter long, consists primarily of a cuticle, gut, gonad, muscle, and nerves. Nerves comprise nearly a third of its somatic cells. Each of its 302 neurons has been mapped in space and time, with all of their relatively simple connections clearly diagrammed (Varshney, Chen, Paniagua, Hall, & Chklovskii, 2011; White, Southgate, Thomson, & Brenner, 1986). Neurotransmission is well characterized via genetics (reviewed by Richmond, 2007). The worm is clear, with distinct developmental stages over a 3-day generation time. Culture is simple and inexpensive.

C. elegans usually exists as a self-fertilizing hermaphrodite; males exist, which makes genetic crosses possible. The wild-type hermaphrodite will produce 300 isogenic offspring over a period of about 2 days. Interesting strains can be frozen in liquid nitrogen and recovered for later study. Approximately 40% of its ~20,000 genes have homologs in the human genome (Shaye & Greenwald, 2011). An array of powerful molecular tools is available. For a full review of worms as a model, as well as techniques for culture of worms, mutagenesis and forward genetic screens, application of RNAi, etc. see Brenner (1974), Stiernagle (2006), Kutscher and Shaham (2014), and Ahringer (2006).

A very important resource is the *Caenorhabditis* Genetics Center (CGC; University of Minnesota, St. Paul)—a repository of mutants, special transgenic strains, tools, and a wealth of shared information. In addition, an extensive database is publicly accessible at http://www.wormbase.org.

1.2 Disadvantages of the Model Organism, *C. elegans*

Although offering an array of powerful molecular techniques, *C. elegans* is not the perfect model in which to study anesthetic action. The worm possesses an outer cuticle that makes delivery of agents other than volatile anesthetics difficult to quantify. There is no ability to isolate individual organs other than the gonad. There is no worm brain or heart. Neurons in the worm are extremely small; the entire length of a worm is less than that of cortical neurons in a human. Electrophysiological measurements, therefore, are technically challenging and have been undertaken by very few investigators. The same is true of cell culture.

Perhaps most obviously, choosing an endpoint in worms that is analogous to the state of general anesthesia in humans must be addressed. Different endpoints have identified changes in different putative targets. Similarities among the putative targets in a variety of model organisms and humans (Humphrey, Sedensky, & Morgan, 2002; Liem et al., 2004; Morgan, Hoppel, & Sedensky, 2002; Steele, Morgan, & Sedensky, 2007) suggest that many of the findings are generalizable across the animal kingdom.

2. EXPOSING WORMS TO VOLATILE ANESTHETICS

2.1 Using Anesthetic Chambers

To assay worm behavior, it is necessary to examine the animals in a clear chamber that is airtight, to have the ability to sample the chamber to determine the concentration of anesthetic within, and to establish a calibration standard. Ordinarily, one simultaneously compares the behavior of mutants to that of the wild-type animal, N2. Mutants are obtained either in forward screens for changed behaviors in an anesthetic, by feeding RNAi-containing bacteria to interrogate relevant genes, or by ordering interesting mutants from the CGC.

Since volatile anesthetics readily pass through plastics, glass chambers with metal ports for injection and sampling are optimal. The chambers must fit easily under a dissecting microscope with illumination from below. Fig. 1 shows our standard anesthetic chamber, fashioned from glass lids for dishes designed to microwave food. They have a ground-glass lip and custom-made covers for an airtight seal, as well as a hole into which we have affixed a metal-lined port made from a spinal needle.

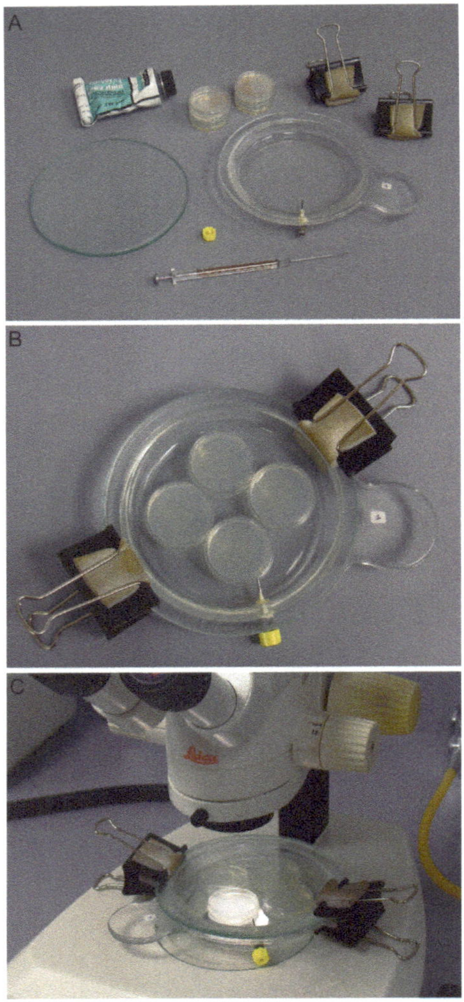

Fig. 1 Photographs of an anesthetic chamber. (A) An open chamber with accessories (from *top left*) vacuum grease, worm plates, padded binder clips, luer-lock cap, and Hamilton syringe. (B) Worm plates, without their lids, are placed in the chamber. The chamber is sealed with vacuum grease, and binder clips are placed on opposite sides of the chamber to hold the cover in place. Immediately after anesthetic is dispensed through the injection port with a Hamilton syringe, a cap is placed on the luer lock. The chamber is shaken from side to side to distribute the anesthetic and then incubated until steady state is achieved. (C) The chamber fits on the stage of a dissecting microscope. The transparency and uniform thickness of the glass chamber allow the worms to be viewed.

2.1.1 Equipment and Reagents

See Table 1.

2.1.2 Procedure

1. Synchronized young adults are optimal for most assays. We commonly place 10 young adult wild-type hermaphrodites on a plate for 4 h or less, then pick them off, leaving eggs behind. This method gives about 100 synchronized young adults ready for exposure to anesthetic 3.5 days later (see Notes 1 and 2). Numbers and times may vary for mutants.
2. Remove the lids from the worm plates and place them in the anesthetic chamber.
3. Seal the chamber, applying vacuum grease as needed, and clamp the cover with two padded 2″ binder clips.
4. Open the injection port and use a Hamilton syringe to inject a dose of liquid anesthetic.
5. Immediately seal the inlet with a luer-lock cap. Cover the inlet with your thumb while exchanging the syringe for the cap.
6. Invert and shake the chamber from side to side immediately and then every half hour to disperse the anesthetic.
7. Incubate the worms in the sealed chambers until steady state is achieved (see Note 3).
8. Perform a behavioral assay as described in Section 3. Most assays will be performed without opening the chamber.
9. Assess the concentration of anesthetic delivered to the animals as described in Sections 2.2 and 2.3.

Table 1 Using Anesthetic Chambers

Equipment	Healthy worm cultures, usually on 35-mm nematode growth medium (NGM) plates
	Airtight chambers and accessories (Fig. 1; Section 2; and Steele, Sedensky, & Morgan, 2010)
	100-μL glass Hamilton syringe for dispensing anesthetic
	Dissecting microscope with illumination from below
Reagents	Standard reagents for worm culture, including 35-mm NGM plates, as described in Stiernagle (2006)
	Vacuum grease
	Anesthetics, sold as liquids, such as isoflurane

2.1.3 Notes

1. To obtain consistent measurements, do not use starved or crowded worms.
2. Unless developmental effects are being studied, the animals being scored should be young adults, i.e., within the first day of egg laying. Sensitivity to volatile anesthetics varies with age.
3. Volatile anesthetics differ in the time necessary to reach steady state. Time courses should be performed to determine when steady-state behavioral effects are reached.
4. Unless repeated anesthetic exposures are necessary to pick and propagate a desired mutant, avoid using worms that have been anesthetized for further culture.

2.2 Determining Anesthetic Concentration: Collection of Gas Samples

The volumes of anesthetic chambers differ. Therefore, for experiments in which the concentration of anesthetic in a chamber must be known precisely, gas chromatography is used to analyze samples of gas taken from each chamber. The concentration of anesthetic is determined by comparison to a known standard. A rule-of-thumb for starting conditions is that 1 μL of any volatile anesthetic yields about 200 μL of vapor. For example, a 500-mL chamber will require ~25 μL of liquid anesthetic to achieve a ~1% v/v anesthetic concentration. For experiments involving desflurane, the Hamilton syringes must be prechilled to <4°C, or the liquid will immediately vaporize making volume determination impossible. Agent-specific vaporizers can also be used, but it takes an exceedingly long time to equilibrate the agar plates when delivering only through the gas phase.

2.2.1 Equipment

See Table 2.

Table 2 Determining Anesthetic Concentrations—Collection of Gas Samples

Equipment	Airtight chambers and accessories (Fig. 1; Section 2; and Steele et al., 2010)
	30-mL gas chromatograph syringes with stopcocks for sampling chambers
	Gas chromatograph and a hydrophobic column of sufficient length to separate the very lipophilic anesthetics

2.2.2 *Procedure*

1. Remove the luer-lock cap from the chamber inlet and immediately cover the inlet with your thumb.
2. Remove your thumb from the inlet and immediately connect a 30-mL gastight syringe into the inlet.
3. Move the syringe plunger in and out a few times to mix the gas in the chamber again.
4. Draw a 10-mL sample of gas into the syringe and close the valve on the syringe.
5. Remove the syringe from the chamber and immediately cover the inlet with your thumb.
6. Remove your thumb from the inlet and immediately replace the luer-lock cap.
7. Analyze the sample according to the operating instructions for the particular gas chromatograph that is used.
8. Expect a large peak of nitrogen and oxygen (air) followed by a smaller peak of anesthetic.
9. Note the area under the anesthetic peak. This area is proportional to the concentration of anesthetic and represents the value of A_{unknown} in the equation in Section 2.3.2, step 20.

2.2.3 *Note*

Less hydrophobic anesthetics (diethyl ether) will elute from the column more quickly than gases like halothane or isoflurane and may be difficult to separate from the large air peak.

2.3 Determining Anesthetic Concentration: Preparation of a Known Standard

2.3.1 *Equipment and Reagents*

See Table 3.

2.3.2 *Procedure*

1. Choose a rubber stopper that will fit a sidearm vacuum flask (we use a 4-L flask).
2. Insert two 3.5-in. spinal needles through the rubber stopper and attach stainless steel three-way stopcocks to the needles.
3. Use parafilm to seal the sidearm of the flask.

Table 3 Determining Anesthetic Concentrations—Preparation of a Known Standard

Equipment	4-L sidearm flask with rubber stopper
	Two 3.5-in. spinal needles
	Two three-way stopcocks, preferably metal with female luer to male luer to sidearm female luer
	Parafilm
	Graduated cylinder
	Vacuum tubing
	Vacuum source
	Hemostats
	100-μL glass Hamilton syringe for dispensing anesthetic
	30-mL gas chromatograph syringes with stopcocks for sampling chambers
	Gas chromatograph and a hydrophobic column of sufficient length to separate the very lipophilic anesthetics
Reagents	Water
	Anesthetics, sold as liquids, such as isoflurane

4. Determine the volume of the flask by completely filling it with water and capping the flask with the rubber stopper. Measure the amount of water to the nearest milliliter.
5. Allow the flask to dry completely and remove the parafilm from the sidearm.
6. Use a piece of vacuum tubing to connect the sidearm of the flask to a vacuum source.
7. Insert the stopper, close the stopcocks, and apply a vacuum to the flask.
8. Use hemostats to clamp off the vacuum tubing as close to the sidearm as possible.
9. Turn off the vacuum and disconnect the tubing from the vacuum source.
10. Use a Hamilton syringe to measure the volume of anesthetic that will give a concentration of gaseous anesthetic in the flask that is in the range of that being used in the chambers during that day's experiment.

The gas concentration in the flask corresponds to $C_{standard}$ in the equation in step 20.

11. Hold the syringe needle in the opening of one of the stopcocks in the rubber stopper.
12. Open that stopcock, immediately insert the needle, and immediately inject the anesthetic. Keep the syringe in place for a second or two. Because of the lower air pressure in the flask, air will enter the flask and draw in the anesthetic.
13. Remove the needle, immediately close the stopcock, and allow the anesthetic to volatilize.
14. After the anesthetic has volatilized, unclamp the tubing to allow the flask to reach atmospheric pressure. When air is no longer entering the flask, immediately reclamp the tubing as close to the sidearm as possible.
15. Invert the flask several times to mix the gases inside to achieve a uniform concentration of anesthetic.
16. Open the other stopcock on the flask and take a 10-mL sample of gas from the flask using a gas chromatograph syringe fitted with a stopcock. Close the stopcock on the syringe.
17. Remove the syringe from the flask and immediately close the stopcock on the flask.
18. Analyze the sample by following the operating instructions for the particular gas chromatograph that is used.
19. Expect a large peak of nitrogen and oxygen (air), followed by a smaller peak of anesthetic. Note the area under the anesthetic peak. This is $A_{standard}$ in step 20.
20. Calculate the concentration of anesthetic in experimental chambers by comparison with the standard. The area under the anesthetic peak is proportional to the concentration of anesthetic in the sample.

$$C_{standard}/A_{standard} = C_{unknown}/A_{unknown}$$

where C equals the concentration of anesthetic, and A equals the area under the anesthetic peak observed on the gas chromatogram. Determine the concentration of anesthetic in each chamber by measuring the area for the experimental samples and solving the equation for $C_{unknown}$.

2.3.3 Note

We make a fresh standard for each day of experimental assays.

3. BEHAVIORAL ASSAYS

A variety of behavioral assays are currently used to measure the anesthetic response in *C. elegans*, including assays for immobility, uncoordination, radial dispersion, mating efficiency, body bends, velocity, and chemotaxis (Crowder, Shebester, & Schedl, 1996; Hawasli, Saifee, Liu, Nonet, & Crowder, 2004; Morgan & Cascorbi, 1985; van Swinderen, Metz, Shebester, & Crowder, 2002; van Swinderen et al., 1999). Additional assays for measuring the anesthetic response have been described in detail elsewhere, including defecation, pharyngeal pumping, egg laying, and mechanosensation (Crowder et al., 1996).

3.1 Immobility

A standard way to refer to the EC_{50} of a volatile anesthetic is by the minimum alveolar concentration necessary to keep 50% of a human population immobile in response to a surgical stimulus. Therefore, reversible immobility of *C. elegans*—an animal which normally moves continuously in a sinuous motion across an agar plate—is an intuitively attractive analogous behavior and an easy endpoint to score.

3.1.1 Equipment and Reagents

As in Table 1.

3.1.2 Procedure

1. Prepare cultures of about 100 synchronized young adults as described in Section 2.1.
2. Expose the worms to appropriate concentration(s) of anesthetic in air-tight chambers until steady state is achieved as described in Section 2.1.
3. Without opening the chambers, score the worms as mobile or immobile (see Note).
4. Assess the concentration of anesthetic delivered to the animals as described in Sections 2.2 and 2.3.

3.1.3 Note

A worm is considered immobile if it does not propagate a wave of body motion during a count of 10 s. We do not count head tossing (characteristic in diethyl ether) as mobility. If a worm writhes in place, it is mobile.

3.2 Uncoordination

Another commonly used endpoint is uncoordination upon anesthetic exposure (Crowder et al., 1996). For a given anesthetic, uncoordination is typically observed at lower concentrations than immobility (see Morgan, Kayser, & Sedensky, 2007).

3.2.1 Equipment and Reagents

As in Table 1.

3.2.2 Procedure

1. Prepare cultures of synchronized young adults as described in Section 2.1.
2. Expose the worms to appropriate concentration(s) of anesthetic in airtight glass chambers until steady state is achieved as described in Section 2.1.
3. Without opening the chambers, an observer who is blinded to the anesthetic concentration should score at least 10 worms as coordinated or uncoordinated by examining them at about 160 × with a dissecting microscope (see Note).
4. Assess the concentration of anesthetic delivered to the animals as described in Sections 2.2 and 2.3.

3.2.3 Note

Wild-type worms exhibit a smooth, sinuous pattern of movement on an agar surface. A worm that displays any defect in this normal pattern of movement is considered "uncoordinated" (Brenner, 1974). Uncoordination may include a variety of movements, such as twitching, halting, jerking, curling, kinking, shrinking, and expanding lengthwise without progress of the body, or partial or complete paralysis (Herndon & Hall, 2010).

3.3 Radial Dispersion

Radial dispersion assays measure the worms' ability to move from a starting point at the center of an agar plate out to a peripheral ring of food (Crowder et al., 1996; van Swinderen et al., 2002, 1999).

3.3.1 Equipment and Reagents

See Table 4.

Table 4 Radial Dispersion

Equipment	1.5-mL microcentrifuge tubes
	Pasteur pipettes
	Micropipettes and tips
	Inoculating loop
	Airtight chambers and accessories (Fig. 1; Section 2; and Steele et al., 2010)
	100-μL glass Hamilton syringe for dispensing anesthetic
	Dissecting microscope with illumination from below
Reagents	S. Basal buffer (5.85 g NaCl, 1 g K_2HPO_4, 6 g KH_2PO_4, 1 mL cholesterol (5 mg/mL in ethanol), H_2O to 1 L. Sterilized by autoclaving)
	Distilled water
	100-mm NGM plates
	Overnight culture of *E. coli*
	Anesthetics, sold as liquids, such as isoflurane

3.3.2 Procedure

1. Use 1 mL of S. Basal to wash well-fed, young adult worms off plates into 1.5-mL microcentrifuge tubes.
2. Rinse the worms twice with 1 mL of S. Basal.
3. Rinse the worms once with 1 mL of distilled water.
4. Resuspend the worms in 100 μL of distilled water.
5. Dispense 10-μL aliquots (containing 50–100 worms) onto the center of a 100-mm plate seeded with a narrow ring (about 5 mm) of *Escherichia coli* along the outer edge of the plate (see Note 1).
6. Immediately place the plates in an airtight chamber and add an appropriate amount of anesthetic (see Note 2).
7. After the droplet containing the worms has dried (usually within 5 min), shake the chamber from side to side.
8. Incubate the worms in the chamber for 45 min at room temperature.
9. Determine the percent of worms reaching the ring of bacteria.
10. Compare the results in step 9 to a control plate treated identically but without anesthetic added (see Note 3).

3.3.3 Notes

1. Use an inoculating loop to spread the ring the night before performing the test so that a visible lawn is present.
2. Only one 100-mm plate will fit in our chamber. Do not shake the chamber until the water droplet containing the worms has dried.
3. Comparison with an unexposed control of the same strain is especially important when testing mutant strains that are uncoordinated.

3.4 Mating Efficiency

Mating efficiencies can be estimated by performing the following assay (Crowder et al., 1996).

3.4.1 Equipment and Reagents

As in Table 1.

3.4.2 Procedure

1. Spread 35-mm plates with a relatively small bacterial lawn.
2. On five plates, plate 10 young adult males and two hermaphrodites with a visible genetic marker such as *dpy-11(e224)*.
3. Place the plates in an airtight chamber and add an appropriate concentration of anesthetic.
4. Take a sample of gas from the chamber to measure the anesthetic concentration at the beginning of the experiment by gas chromatography as described in Sections 2.2 and 2.3.
5. Ensure that the chamber is well sealed.
6. Incubate the chamber for 24 h at 20°C.
7. To verify that the chamber did not leak during the incubation, take another sample of gas from the chamber to measure the anesthetic concentration at the end of the experiment.
8. Open the chambers and remove the male worms from the plates.
9. Allow the offspring to develop for 3 or 4 days at 20°C in room air.
10. Calculate the "mating index" for each chamber as the fraction of plates with cross progeny.

3.5 Body Bends

It has been suggested that the body bend assay (Hawasli et al., 2004) is minimally influenced by factors such as the number of eggs carried or the number of worm reversals (Hart, 2006).

3.5.1 Equipment and Reagents

See Table 5.

3.5.2 Procedure

1. To avoid transferring bacteria to the assay plate, move young adults to an unseeded NGM plate with a platinum wire.
2. Then use the platinum wire (without bacteria) to move the worms to an unseeded NGM assay plate.
3. Incubate the worms in a clear glass chamber containing volatile anesthetic for 30 min (see Section 2.1).
4. Vigorously shake the chamber for ~5 s.
5. After one more minute, and without opening the chambers, count the number of body bends for 2 min (see Note).
6. Count at least 10 animals per strain and condition.
7. Without opening the chambers, assess the concentration of anesthetic delivered to the animals as described in Sections 2.2 and 2.3.
8. Calculate the average number of body bends per minute for the 10 individual worms measured.

3.5.3 Note

One body bend consists of a complete sinusoidal waveform of the nematode's forward motion. Backward body bends are not counted.

3.6 Speed of Locomotion

Some researchers have assessed worms' anesthetic sensitivity by measuring the speed of locomotion in the presence of volatile anesthetics (Wang, Saifee, Nonet, & Salkoff, 2001).

Table 5 Body Bends

Equipment	Platinum wire pick
	Airtight chambers and accessories (Fig. 1; Section 2; and Steele et al., 2010)
	100-μL glass Hamilton syringe for dispensing anesthetic
	Dissecting microscope with illumination from below
Reagents	Unseeded 35-mm NGM plates
	Anesthetics, sold as liquids, such as isoflurane

3.6.1 Equipment and Reagents

See Table 6.

3.6.2 Procedure

1. Transfer L4 animals to fresh NGM plates containing a thin *E. coli* lawn and incubate them until they reach adulthood.
2. Place the worm plates in a clear glass chamber containing volatile anesthetic for 30 min (see Section 2.1).
3. Move the chamber to a microscope and leave the plates undisturbed for 5–10 min.
4. Without opening the chambers, collect photographic images at 5-s intervals.
5. Without opening the chambers, assess the concentration of anesthetic delivered to the animals as described in Sections 2.2 and 2.3.
6. Calculate locomotion velocities by measuring the distance that the tail of each animal moves between successive images.
7. Calculate and average the velocities from three pairs of images.

3.7 Anesthetic-Induced Neurotoxicity (Chemotaxis)

Chemotaxis assays measure the ability of worms to move from a starting point to a spot of chemoattractant (Bargmann & Horvitz, 1991; Crowder et al., 1996). If exploring the notion of anesthetics as developmental neurotoxins, exposure to anesthetic occurs early in development and subsequent behavior changes are measured in adults.

3.7.1 Equipment and Reagents

See Table 7.

Table 6 Speed of Locomotion

Equipment	Airtight chambers and accessories (Fig. 1; Section 2; and Steele et al., 2010)
	100-μL glass Hamilton syringe for dispensing anesthetic
	Dissecting microscope with illumination from below
	Camera
Reagents	Fresh 35-mm NGM plates with a thin lawn of *E. coli*
	Anesthetics, sold as liquids, such as isoflurane

Table 7 Anesthetic-Induced Neurotoxicity (Chemotaxis)

Equipment	Airtight chambers and accessories (Fig. 1; Section 2; and Steele et al., 2010)
	Felt-tipped pen
	100-μL glass Hamilton syringe for dispensing anesthetic
	Micropipette and tips
	Pasteur pipettes
	Dissecting microscope with illumination from below
Reagents	35-mm NGM plates
	Anesthetics, sold as liquids, such as isoflurane
	100-mm unspread NGM plates
	Overnight culture of *E. coli*
	1 *M* sodium azide
	Chemotaxis buffer (5 m*M* potassium phosphate, 1 m*M* calcium chloride, 1 m*M* magnesium sulfate)

3.7.2 Procedure

1. Synchronize cultures by placing 10 young adult wild-type hermaphrodites per 35-mm NGM plate for 4 h or less, and then pick them off, leaving eggs behind. This method gives about 100 synchronized offspring per plate (see Note 1).
2. Incubate the plates for 20 h at 15°C or 20°C as necessary to obtain newly hatched L1 larvae.
3. Expose experimental plates of newly hatched L1 animals to anesthetic (see Note 2).
4. Hold control plates of newly hatched L1 animals in room air at 15°C to slow development to match the anesthetic-exposed cohorts.
5. Check the anesthetic concentration by gas chromatography as described in Sections 2.2 and 2.3.
6. After exposure, grow the experimental and control animals in room air at 20°C until they reach day 1 of adulthood (3–5 days depending on the strain).
7. Use a felt-tipped pen to mark the underside of a 100-mm unspread NGM plate to divide it into three regions as shown in Fig. 2 in a modification of the technique described by Bargmann and Horvitz (1991) and Gentry, Steele, Sedensky, and Morgan (2013).

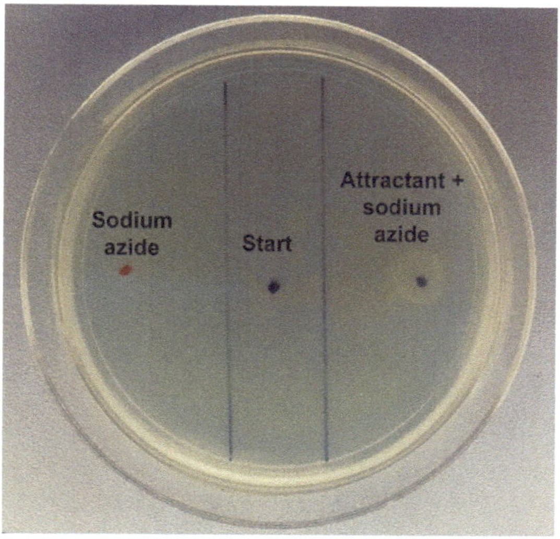

Fig. 2 Photograph of a plate to be used for a chemotaxis assay. Worms will be spotted in the center of the plate. A spot of *E. coli* and sodium azide is seen at the *right* (*black dot*); equidistant from the center is a spot of sodium azide alone on the *left* (*red dot*).

8. In one region, spot an attractant (e.g., 20 μL *E. coli* OP50) with 1 μL of 1 *M* sodium azide.
9. In the opposite region, spot 1 μL of 1 *M* sodium azide (see Fig. 2 and Note 3).
10. Wash young adult worms three times in chemotaxis buffer.
11. Transfer at least 100 washed worms in a minimal volume of chemotaxis buffer to the middle region of the assay plate.
12. Count the worms in each region 1 h after transfer.
13. Calculate a chemotaxis index (CI) using the formula: CI = 100 × (worms at food side − worms at control side)/total number of worms on the plate.

3.7.3 Notes

1. Two to three plates of both control (unexposed) and exposed animals should be assayed in parallel on a given day.
2. *C. elegans* L1 larvae are generally exposed to anesthetic at hours 4–8 of L1 development. For mutants, the time of anesthetic exposure was adjusted to approximate hours 4–8 of normal L1 development.
3. The sodium azide immobilizes worms so they will remain in place to be counted. The middle region serves as the starting point for the animals.

4. SUMMARY AND CONCLUSION

Multiple assays have been used to exploit *C. elegans* as a model for studying mechanisms of action of volatile anesthetics. The advantages of this simple animal are many, and continued progress in understanding anesthetic mechanisms is anticipated from their ongoing study.

ACKNOWLEDGMENT

M.M.S. was supported by NIH GM10569.

REFERENCES

Ahringer, J. (Ed.), (2006). *Reverse genetics*. WormBook, ed. The *C. elegans* Research Community, WormBook. https://doi.org/10.1895/wormbook.1.47.1, http://www.wormbook.org.

Bargmann, C. I., & Horvitz, H. R. (1991). Chemosensory neurons with overlapping functions direct chemotaxis to multiple chemicals in *C. elegans*. *Neuron*, *7*, 729–742.

Brenner, S. (1974). The genetics of *Caenorhabditis elegans*. *Genetics*, *77*, 71–94.

Crowder, C. M., Shebester, L. D., & Schedl, T. (1996). Behavioral effects of volatile anesthetics in *Caenorhabditis elegans*. *Anesthesiology*, *85*, 901–912.

Gentry, K. R., Steele, L. M., Sedensky, M. M., & Morgan, P. G. (2013). Early developmental exposure to volatile anesthetics causes behavioral defects in *Caenorhabditis elegans*. *Anesthesia & Analgesia*, *116*, 185–189.

Hart, A. C. (Ed.), (2006). *Behavior*. WormBook, ed. The *C. elegans* Research Community, WormBook. https://doi.org/10.1895/wormbook.1.87.1, http://www.wormbook.org.

Hawasli, A. H., Saifee, O., Liu, C., Nonet, M. L., & Crowder, C. M. (2004). Resistance to volatile anesthetics by mutations enhancing excitatory neurotransmitter release in *Caenorhabditis elegans*. *Genetics*, *168*, 831–843.

Herndon, L. A., & Hall, D. H. (2010). Glossary U. In *WormAtlas*. https://doi.org/10.3908/wormatlas.6.21.

Humphrey, J. A., Sedensky, M. M., & Morgan, P. G. (2002). Understanding anesthesia: Making genetic sense of the absence of senses. *Human Molecular Genetics*, *11*, 1241–1249.

Kutscher, L. M., & Shaham, S. (2014). *Forward and reverse mutagenesis in C. elegans*. WormBook, ed. The *C. elegans* Research Community, WormBook. https://doi.org/10.1895/wormbook.1.167.1, http://www.wormbook.org.

Liem, E. B., Lin, C. M., Suleman, M. I., Doufas, A. G., Gregg, R. G., Veauthier, J. M., et al. (2004). Anesthetic requirement is increased in redheads. *Anesthesiology*, *101*, 279–283.

Morgan, P. G., & Cascorbi, H. F. (1985). Effect of anesthetics and a convulsant on normal and mutant *Caenorhabditis elegans*. *Anesthesiology*, *62*, 738–744.

Morgan, P. G., Hoppel, C. L., & Sedensky, M. M. (2002). Mitochondrial defects and anesthetic sensitivity. *Anesthesiology*, *96*, 1268–1270.

Morgan, P. G., Kayser, E.-B., & Sedensky, M. M. (2007). C. elegans *and volatile anesthetics*. WormBook, ed. The *C. elegans* Research Community, WormBook. https://doi.org/10.1895/wormbook.1.140.1, http://www.wormbook.org.

Richmond, J. (2007). *Synaptic function*. WormBook, ed. The *C. elegans* Research Community, WormBook. https://doi.org/10.1895/wormbook.1.69.1, http://www.wormbook.org.

Shaye, D. D., & Greenwald, I. (2011). OrthoList: A compendium of *C. elegans* genes with human orthologs. *PLoS One*, *6*, e20085.

Steele, L. M., Morgan, P. G., & Sedensky, M. M. (2007). Genetics and the mechanisms of action of inhaled anesthetics. *Current Pharmacogenomics*, *5*, 125–141.
Steele, L. M., Sedensky, M. M., & Morgan, P. G. (2010). Alternatives to mammalian pain models 1: Use of *C. elegans* for the study of volatile anesthetics. *Methods in Molecular Biology*, *617*, 1–17.
Stiernagle, T. (2006). *Maintenance of C. elegans*. WormBook, ed. The *C. elegans* Research Community, WormBook. https://doi.org/10.1895/wormbook.1.101.1, http://www.wormbook.org.
van Swinderen, B., Metz, L. B., Shebester, L. D., & Crowder, C. M. (2002). A *Caenorhabditis elegans* pheromone antagonizes volatile anesthetic action through a go-coupled pathway. *Genetics*, *161*, 109–119.
van Swinderen, B., Saifee, O., Shebester, L., Roberson, R., Nonet, M. L., & Crowder, C. M. (1999). A neomorphic syntaxin mutation blocks volatile-anesthetic action in Caenorhabditis elegans. *Proceedings of the National Academy of Sciences of the United States of America*, *96*, 2479–2484.
Varshney, L. R., Chen, B. L., Paniagua, E., Hall, D. H., & Chklovskii, D. B. (2011). Structural properties of the *Caenorhabditis elegans* neuronal network. *PLoS Computational Biology*, *7*, e1001066.
Wang, Z. W., Saifee, O., Nonet, M. L., & Salkoff, L. (2001). SLO-1 potassium channels control quantal content of neurotransmitter release at the *C. elegans* neuromuscular junction. *Neuron*, *32*, 867–881.
White, J. G., Southgate, E., Thomson, J. N., & Brenner, S. (1986). The structure of the nervous system of the nematode *Caenorhabditis elegans*. *Philosophical Transactions of the Royal Society, B: Biological Sciences*, *314*, 1–340.

CHAPTER EIGHT

Using *Drosophila* to Understand General Anesthesia: From Synapses to Behavior

Shanker Karunanithi*,†, Michael Troup*, Bruno van Swinderen*,1
*Queensland Brain Institute, The University of Queensland, St Lucia, QLD, Australia
†School of Medical Science and Menzies Health Institute Queensland, Griffith University Gold Coast Campus, Gold Coast, QLD, Australia
[1]Corresponding author: e-mail address: b.vanswinderen@uq.edu.au

Contents

Abstract

Investigating mechanisms of general anesthesia requires access to multiple levels of neuronal function, from effects at individual synapses to responses in behaving animals. *Drosophila melanogaster* provides an excellent model to test different theories for general anesthesia because it offers robust methods for testing local as well as global target processes, in an animal that is similarly impacted by these diverse drugs as humans. Here, we outline methods to quantify two such endpoints, neurotransmission and behavioral responsiveness, focusing on the intravenous drug propofol.

Methods in Enzymology, Volume 602
ISSN 0076-6879
https://doi.org/10.1016/bs.mie.2018.02.003

1. MULTIPLE LEVELS OF GENERAL ANESTHETIC ACTION: FROM PROTEINS TO CIRCUITS TO CONNECTIVITY

Our understanding of general anesthesia has largely paralleled our growing understanding of how the brain works. This may not be surprising, because general anesthesia is fundamentally a whole-brain phenomenon measured by behavioral readouts (Zalucki & van Swinderen, 2016). Thus, the first theories of general anesthesia, developed over a century ago, were centered on the observation that these drugs were lipophilic (Meyer, 1899; Overton, 1891), and a link was then made upon learning that the basic unit of brain function, neurons, was largely made of lipids. The subsequent understanding of neuronal action potential physiology and its dependence on lipid insulation provided an obvious (but wrong) explanation for how general anesthetics might work: by nonspecific effects on neuronal membrane physiology. Subsequent progress in understanding neuronal function and communication largely involved the study of proteins embedded in the lipid bilayer of neurons, such as ion channels. Anesthesia research then followed quite logically in this direction, leading to our current understanding that many of these drugs specifically potentiate inhibitory chloride channels, such as GABA receptors (Franks & Lieb, 1994). Following these mostly biochemical studies, general anesthesia was further explained by placing the receptor effects in the context of circuits in the brain, such as hypothesized arousal pathways (Saper, Chou, & Scammell, 2001), to provide a broader explanation for how these diverse drugs might potentiate endogenous sleep mechanisms in the brain (Franks, 2008; Moore et al., 2012; Ueda, Hirakawa, Arakawa, & Kamaya, 1986). Most recently, there has been much interest in studying the brain by moving beyond hardwired circuits and into the realm of functional connectivity (Boly et al., 2012; Chennu, O'Connor, Adapa, Menon, & Bekinschtein, 2016), and general anesthesia research has accordingly followed in this direction as well, with current theories mirroring the latest approaches to understanding consciousness (Mashour & Hudetz, 2017). Rather than being a criticism, the parallel history of anesthesia and brain research illustrates how general anesthesia, like the brain, is complex problem unlikely to be explained by a single mechanism.

All animals with brains are rendered unresponsive by general anesthetics, often at surprisingly similar drug concentrations (Zalucki & van Swinderen, 2016). This suggests that the brain processes targeted are likely to be similar in different animal brains, irrespective of the specific neuroanatomy or size.

This view also implies that all brains work in fundamentally the same way, and that studying general anesthesia could actually provide general insights into brain functions. A difficulty arises, however, when making comparisons among different species, namely, that it is not always evident which behaviors to measure to quantify loss of responsiveness under anesthesia, and whether these are relevant to human endpoints such as loss of pain perception (termed analgesia) or loss of consciousness (termed sedation or hypnosis). Closer examination of general anesthesia in animals reveals a variety of endpoints across different drug concentrations (Zalucki & van Swinderen, 2016), much like sedation and analgesia occur successively within a patient exposed to increasing isoflurane concentrations, for example (Brown, Purdon, & Van Dort, 2011). Calculating the EC_{50} for different behavioral endpoints in animal models such as nematodes, flies, and mice has revealed intriguing trends which might suggest common mechanisms: except for the nematode model (*Caenorhabditis elegans*), animals are sedated (i.e., rendered quiescent) by volatile anesthetics at similar drug concentrations; however, effects on behavioral responsiveness (potentially providing insight into analgesia) seem to depend on the complexity of the endpoint being assayed (Zalucki & van Swinderen, 2016). This suggests at least two distinct processes targeted by these drugs: one linked to conserved sleep/wake mechanism possibly common to most animals, and another correlated to circuit complexity or functional connectivity. The first proposed target (endogenous sleep/wake pathways) has good support in animal models, at least in rodents and flies (Kottler et al., 2013; Moore et al., 2012). The second target process, however, might appear to involve a grab bag of every other likely protein target (Antkowiak, 2001) and their combined effects on brain function, scaling with the complexity of the processing involved. Thus, in humans, memory formation is more sensitive to general anesthetics than responses to pain (Chortkoff, Bennett, & Eger, 1993), and in nematodes chemotaxis is more sensitive than pharyngeal pumping (Crowder, Shebester, & Schedl, 1996). Beyond the realm of sleep/wake circuits (which remain unidentified in *C. elegans*), is there another target process which might explain this complexity effect seen in animals with very different nervous systems?

Neurotransmission, and specifically the vesicle release machinery, provides a likely alternate target for general anesthetics. There is already strong evidence that both volatile as well as intravenous drugs impair neurotransmission, now documented in a variety of preparations ranging from neurosecretory cells (Herring et al., 2011) to hippocampal and cortical cultures

(Baumgart et al., 2015) to the neuromuscular junctions (NMJs) of amphibians (Leite, Gomez, Fonseca Mde, Gomez, & Guatimosim, 2011) and flies (Zalucki et al., 2015). Together, this new body of work shows that general anesthetics target both pre- and postsynaptic mechanisms. Early work in the *C. elegans* model uncovered the presynaptic SNARE machinery, and particularly the proteins syntaxin1A and *unc-13*, as potential targets for volatile anesthetics (Metz, Dasgupta, Liu, Hunt, & Crowder, 2007; van Swinderen et al., 1999), since mutations in these proteins conferred an astonishing level of resistance to the drugs. Since then, the effect of some of these mutations have been confirmed in flies (Zalucki et al., 2015) and mammalian cell culture preparations (Herring et al., 2011), arguing for a conserved target mechanism centered on the exocytosis machinery in general and syntaxin1A specifically (van Swinderen & Kottler, 2014). A recent study examined the mobility of syntaxin1A on the plasma membrane of cells exposed to clinically relevant concentrations (3 μ*M*) of propofol (Bademosi et al., 2018). Single-molecule imaging of syntaxin1A revealed an effect that had remained invisible until the advent of superresolution microscopy: lateral diffusion of the protein on the membrane is significantly impaired under propofol anesthesia, leading to an increase in the cluster size and density (Bademosi et al., 2018), and suggesting halted cycling into release-ready SNAREs. This effect of propofol requires interaction between syntaxin1A and SNAP25, but not VAMP2, suggesting a propofol effect before the formation of the SNARE complex. Other research in hippocampal cultures exposed to volatile anesthetics points to an even earlier presynaptic process linked to calcium influx (Baumgart et al., 2015), but together these findings are consistent with the view that at least some of these lipophilic drugs, even canonical GABA agonists such as propofol, also impair presynaptic release mechanisms. The likelihood that neurotransmission is targeted by many general anesthetics calls for an array of readouts relevant to this effect, ranging from synaptic release assays to whole-brain imaging and behavior. Thus, the ideal model organism to study general anesthesia would be one where sleep/sedation-related effects are readily addressed alongside presynaptic effects.

2. *DROSOPHILA MELANOGASTER*: A MODEL FOR ANESTHESIA RESEARCH

Drosophila is proving to be an excellent model for anesthesia research because genetic manipulations can be easily interrogated across a variety of anesthesia endpoints. Thus, effects on single molecules and quantal release of neurotransmitter can be addressed at larval synapses (Bademosi et al., 2018); electrophysiology or imaging under anesthesia can be performed at a circuit level (Walcourt & Nash, 2000) or a whole-brain level (Cohen, Zalucki, van Swinderen, & Tsuchiya, 2016); and genetic lesions and circuit manipulations can be tested behaviorally (Allada & Nash, 1993; Kottler et al., 2013). In addition, *Drosophila* provides one of the smallest animal brains where sleep/wake circuits have been identified (Griffith, 2013), thereby providing a likely substrate for investigating at least one of the hypothesized target processes involved in general anesthesia. Together with its highly conserved ion channels and synaptic release mechanisms, this makes the fly a powerful model for testing theories of general anesthetic action, especially more complex theories involving multiple target processes.

The first *Drosophila* approaches to general anesthesia naturally exploited the power of this model for conducting genetic screens, especially using whole-animal behaviors as readouts. Such screens were hugely successful in revealing other key biological processes, such as the molecular mechanisms underlying circadian rhythms, so the hope was that a similar approach might prove successful for general anesthesia. Most of the mutations that were uncovered early on (only volatile anesthetics were assayed) were hypersensitive rather than resistant to these drugs (Krishnan & Nash, 1990). While not necessarily uninteresting, there are multiple nonspecific paths to hypersensitivity (a sick disposition, for starters), while resistance is much more likely to reflect a change that is closer to the drug's target mechanism. Most mutations uncovered in these first behavioral screens were in ion channels that controlled neuronal excitability, such as *shaker* (Weber et al., 2009) and *narrow abdomen* (Humphrey et al., 2007). Behavioral endpoints (reviewed extensively elsewhere Zalucki & van Swinderen, 2016) typically involved fly locomotion assays, sometimes in response to a stimulus such as a tap (Guan, Scott, & Nash, 2000) or a beam of light (Campbell & Nash, 1994). Results in wild-type flies generally supported the view that more complex behaviors (i.e., requiring some level of perceptual integration) were more sensitive than less complex behaviors, such as walking (Friedman et al., 2010) or the visually induced jump reflex (Campbell & Nash, 1998).

The discovery that flies sleep (Shaw, Cirelli, Greenspan, & Tononi, 2000) provided a novel direction to *Drosophila* anesthesia research. This is because research in mammalian systems was increasingly pointing to endogenous sleep/wake circuits as likely processes targeted by most commonly used general anesthetics (Franks, 2008). The capacity to exquisitely manipulate neuronal circuits is a cornerstone of current *Drosophila* neuroscience research, and a number of circuits have been uncovered relevant to fly sleep (Griffith, 2013; Pimentel et al., 2016). These provided excellent candidates for further exploring general anesthetic mechanisms, with some sleep-promoting circuits such as the dorsal fan-shaped body of the central complex identified as potentially involved (Kottler et al., 2013). This research direction will be crucial for better understanding how these drugs work in the fly brain, and indeed it could be argued that general anesthesia in *Drosophila* could not be understood without first having discovered and explained how sleep is "switched on" in flies. There is much research being currently done in this area (Liu, Liu, Tabuchi, & Wu, 2016; Pimentel et al., 2016), and identifying GABAergic sleep-promoting circuits in the fly brain will also identify the circuits likely to be targeted by many general anesthetics.

Yet, as we have discussed earlier, mapping anesthetic mechanisms onto endogenous sleep pathways may only explain half of the mystery that is general anesthesia—with the other half being potentially an even more complex problem linked to diverse presynaptic effects (van Swinderen & Kottler, 2014). This complementary research direction requires a variety of different techniques, from recording synaptic release events from individual boutons, to whole-brain electrophysiology or imaging. Linking the former with the latter is especially pertinent, because one idea here is that small effects on synaptic release efficacy might translate to major effects on whole-brain function. Readouts such as field-potential coherence (Cohen et al., 2016) may thus reflect global consequences of smaller effects that might not seem important at the level of a single cell, or even a single circuit. There is increasing evidence that this level of analysis might be important for understanding general anesthesia, with functional connectivity studies pointing to whole-brain coherence defects coming into play, especially at surgical drug concentrations slightly higher than those used to achieve sedation and sleep (Chennu et al., 2016). As far as animal models go, *Drosophila* is particularly well poised to address this question as well as for understanding sleep pathways. This is because the same genetic manipulations can be carried through multiple levels of analysis, from synaptic approaches, to whole-brain electrophysiology or imaging, to whole-animal behavior. This genetics-based approach allows us to determine whether a resistance phenotype (for example) carries through different levels, in this way linking the global with the local. In this review, we will be discussing a "global" readout—behavioral responsiveness to mechanical stimuli—and a "local" readout, recordings from individual synaptic boutons. Whereas our focus here is on novel approaches using an intravenous anesthetic, propofol, the same approaches can be adapted to volatile anesthetics, as referenced herein.

3. BEHAVIORAL ENDPOINTS: STARTLE-INDUCED LOCOMOTION

3.1 Introduction

Recent *Drosophila* research on general anesthesia has focussed on volatile agents, using locomotion as a readout (Friedman et al., 2010; Joiner et al., 2013; Kottler et al., 2013; Weber et al., 2009). In particular, behavioral paradigms that were originally adapted to measure sleep in flies were repurposed to measure general anesthesia. This strategy is logical, since sleep in flies is well studied and it has been proposed that sleep and anesthesia may be similar processes. Sleep in *Drosophila* has been defined as a period of 5 min or more of immobility (Shaw et al., 2000), and this is usually measured by recording a fly's activity as it walks back and forth along a small glass tube. Most experiments utilize the *Drosophila* Activity Monitor System (DAMS; Trikinetics, Waltham, MA), where an infrared beam registers activity as a fly crosses through it. Friedman et al. (2010) used this experimental approach to study the effect of the volatile anesthetics halothane and isoflurane, examining transitional states between anesthetic sedation and wakefulness. By monitoring fly locomotion over increasing concentrations of drug during anesthesia induction, they could compare how wakefulness (measured by fly locomotion) returns when the drug concentration is decreased back to zero. An asymmetry between anesthesia induction and recovery suggested distinct mechanisms were involved, with recovery characterized by hysteresis, or inertia. This approach demonstrated the practicality of using established sleep-monitoring devices for assessing anesthesia phenotypes in flies, with only minor modifications such as placing the devices in a gas-tight chamber (Weber et al., 2009) or pumping the gas through the Trikinetics tubes (Friedman et al., 2010; Joiner et al., 2013).

Yet, anesthesia is not the same as sleep, so it is not quite sufficient to use *Drosophila* sleep assays to measure general anesthesia in flies. For instance, if anesthetics indeed act on sleep circuitry, how could one determine if an immobile fly is anesthetized, as opposed to merely asleep? A key endpoint of general anesthesia is loss of behavioral responsiveness; a sleeping fly is less arousable but will respond to a sufficiently salient stimulus, while a deeply anesthetized fly should fail to respond at all. While the DAMS system only monitors activity, attempts have been made to more actively probe for arousal levels in sleeping flies, by introducing ad hoc stimuli to the apparatus, such light flashes or mechanical stimuli (Kume, Kume, Park, Hirsh, & Jackson, 2005). Improved sleep-monitoring systems have integrated stimulus delivery while also tracking fly activity using video recording (Calcagno, Eyles, van Alphen, & van Swinderen, 2013; Faville, Kottler, Goodhill, Shaw, & van Swinderen, 2015). Kottler et al. (2013) used this approach to measure behavioral responsiveness to automated mechanical vibrations delivered to flies housed in glass tubes. Flies respond to vibration stimuli by increasing their locomotion if they are awake, and if they are asleep this provides a way of probing sleep intensity (Calcagno et al., 2013). Adapting this stimulation paradigm to anesthesia studies provided a startle-induced locomotion endpoint for flies (Fig. 1A). Startle-induced locomotion is more sensitive to isoflurane than baseline locomotion (Fig. 1B) which supports the view discussed earlier that more "complex" behaviors are more sensitive to the effects of these drugs. Interestingly, sleep duration in *Drosophila* mutants is correlated with sensitivity to isoflurane measured by the startle endpoint (Kottler et al., 2013), supporting the view that common processes are involved.

General anesthetics vary widely in their size and structure. While all are lipophilic, at room temperature an anesthetic can either be a liquid delivered intravenously or an inhaled gas. In small animals such as flies, most experiments use volatile anesthetics since the delivery method is straightforward: a behavioral apparatus can be placed inside a chamber which can be filled with the anesthetic. Often a behavior is assayed by gradually increasing the dose and monitoring loss of the behavior. However, this is difficult with an intravenous anesthetic, as the delivery system would need to be miniaturized. Such systems, such as tethering a fly and perfusing its brain with an anesthetic, forfeit the advantage of high-throughput screening of large populations in these small animal models. Instead, liquid-soluble drugs have been delivered to flies by mixing them into fly food, and this has also been applied to sleep-promoting agents (Dissel et al., 2015). Gardner et al. (2016) used this approach to deliver the intravenous anesthetic propofol to flies while monitoring their activity in glass tubes. Using activity levels as a measure, the authors confirmed that flies are sensitive to propofol, although it was not clear whether propofol was promoting natural sleep.

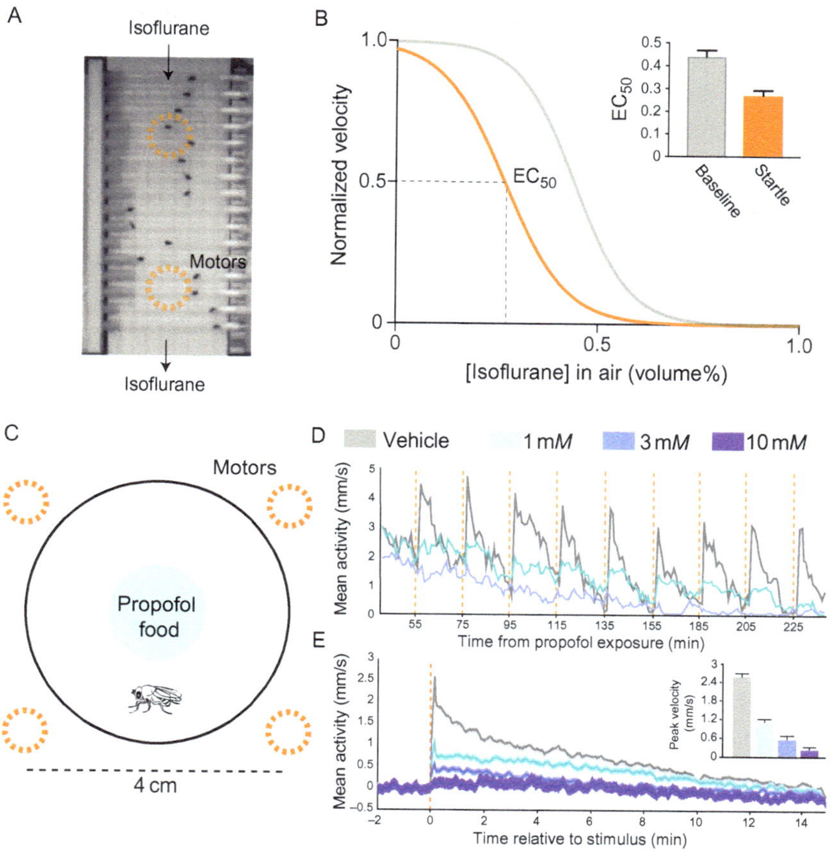

Fig. 1 Measuring anesthesia sensitivity in flies. (A) Flies can be housed in glass tubes, and their activity is tracked from above. Motors underneath provide a mechanical stimulus. A sealed chamber allows a controlled concentration of isoflurane to be delivered to the flies. (B) An example dose–response curve for an experiment. As isoflurane concentration increases, fly activity decreases. Two endpoints can be measured in this assay: prior to a stimulus "baseline" fly activity (*gray*) and following a stimulus, flies increase their activity and this is captured as the "startle" endpoint (*orange*). Often, more complex behaviors such as responding to a stimulus (*orange*) are abolished at lower concentrations, compared to simpler behaviors such as baseline locomotion (*gray*). This is reflected by a lower EC_{50} value, the concentration at which the behavior becomes half-maximal. Error bars are standard errors of the estimate. (C) Flies can be housed in open-field arenas to allow walking in two dimensions. Agar–sucrose medium in the center of the arena contains propofol-laced food which decreases fly activity following ingestion. (D) Fly activity after 45 min in the propofol arena. *Orange dashed line* shows when the stimulus occurs, and a subsequent increase in activity can be seen in wild-type flies that have consumed only the vehicle control (*gray*), compared with propofol doses of 1 and 3 m*M* (*aqua* and *blue*, respectively). (E) Average fly activity following the stimulus (*orange line*), normalized to baseline activity. The peak of this response compared to baseline is decreased as propofol concentration increases (*inset*, vehicle $n = 26$, 1 m*M* $n = 26$, 3 m*M* $n = 26$, 10 m*M* $n = 10$).

High-performance liquid chromatography ascertained that the concentration of propofol in the fly brain was within a clinically relevant range.

We describe here a variation on the propofol anesthesia assay for flies, involving monitoring of startle-induced locomotion in an open field. As discussed earlier, measuring behavioral responsiveness provides an acute endpoint that may be more relevant to general anesthesia than baseline activity or locomotion. Additionally, placing this in an open-field context (Fig. 1C) provides a potentially cleaner measure that is less likely to be affected by other confounding endpoints, such as the ability of a fly to perform coordinated turns in a narrow tube.

3.2 Equipment

In order to track fly activity and deliver vibration stimuli to test behavioral responsiveness, the following pieces of equipment are required:

- Open-field arenas ~50 mm in diameter to house flies, with a clear covering so they may be viewed and filmed from above (Fig. 1C).
- Small shaftless motors (for example, 312-101, Precision Microdrives London, UK) capable of delivering approximately 2.5 G. These are attached equal distance apart underneath the fly housing to deliver stimuli (Fig. 1C).
- A camera for filming flies form above, for example, an inexpensive webcam.
- Tracking software optimized for *Drosophila* such as described in Colomb, Reiter, Blaszkiewicz, Wessnitzer, and Brembs (2012) and Faville et al. (2015).

3.3 Solutions

Propofol–agar solution adapted from Gardner et al. (2016):

- Propofol (≥97% Sigma-Aldrich D126608 or equivalent)
- Polyethylene glycol (Sigma-Aldrich 202398 or equivalent)
- Low melting point agar
- Sucrose

Stock propofol in polyethylene glycol is diluted down to concentrations of 1 and 3 m*M*. This is then added to a 2% agar and 5% sucrose melted medium. A vehicle control of PEG equivalent to the highest concentration is also used. A small amount of food coloring can be added to confirm that flies are ingesting this mixture. A small drop (~0.5 cm diameter) of this propofol-laced media is added to the center of circular chambers (4 cm diameter) and allowed to solidify.

3.4 Experiment

For experiments, use 3–5-day-old female flies that have been starved for 12 h overnight. This ensures approximately equal feeding drive when exposed to propofol media. Load flies into the chambers via mouth pipetting or very briefly anesthetizing flies on CO_2 (<1 min) before loading them into the behavioral chambers. When a mechanical vibration is delivered, flies quickly increase their speed which returns to baseline levels after approximately 10 min (Fig. 1D). Therefore, a dependable stimulus protocol is to deliver a vibration every 20 min for 6 h using the vibrating motors attached underneath the chambers. A force of 1.2 g is sufficient to elicit a reliable response. In flies fed the vehicle control, an increase in speed in response to this stimulus should be maintained throughout the experiment (Fig. 1D, gray). However, note that after being placed in a new environment such as this, flies may initially be briefly hyperactive. Flies take approximately 45 min to consume and be affected by the propofol. As flies become affected by the drug, their peak increase in activity following the stimulations is diminished. To determine the magnitude of the response, baseline activity for 1 min prior to the stimulus is averaged and subtracted from the peak activity following the stimulus (Fig. 1E). 1 m*M* propofol is sufficient to impair responsiveness, and flies are affected in a dose-dependent manner (Fig. 1E).

4. SYNAPTIC ENDPOINTS: RECORDINGS FROM INDIVIDUAL BOUTONS

4.1 Introduction

The *Drosophila* larval NMJ is a highly studied model glutamatergic synapse (Frank, 2014; Harris & Littleton, 2015; Menon, Carrillo, & Zinn, 2013), making it an excellent model to study presynaptic effects of general anesthetics. The ability to manipulate gene expression at this synapse in sophisticated ways, though with relative simplicity, is the primary reason for its popularity. The larval preparation (Fig. 2A) is unique in that each muscle receives multiple inputs from different motor neurons (Hoang & Chiba, 2001). Each neuron forms a discrete NMJ on the same muscle (Hoang & Chiba, 2001). The motor nerve terminals which form the NMJs are made up of several synaptic boutons, reminiscent of beads on strings (Fig. 2A; Hoang & Chiba, 2001; Menon et al., 2013).

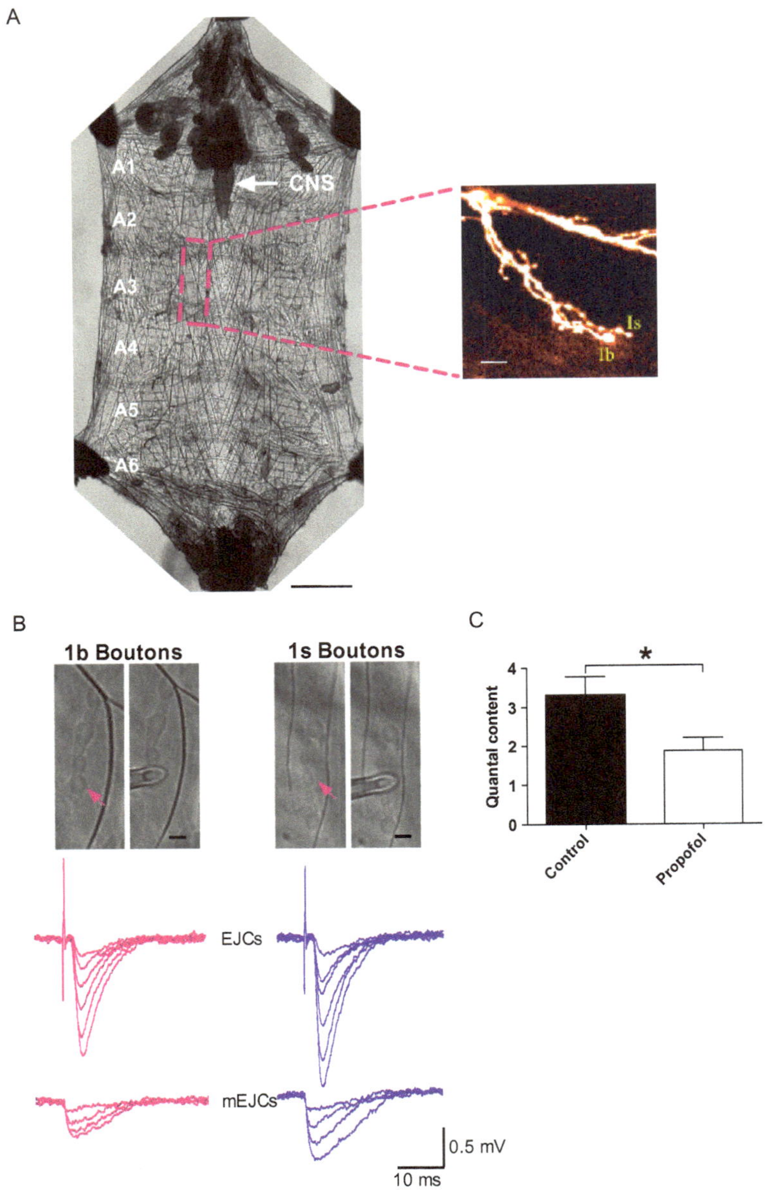

Fig. 2 (A) The *left panel* shows a *Drosophila* third instar larva prepared for electrophysiological recordings. The central nervous system (CNS) and the abdominal segments (A1–6) are indicated. The *magenta rectangle* marks the most highly studied muscles in this preparation, muscles 6 and 7 (shown in this case, in one of the two A3 hemisegments). Scale bar represents 500 μm. The *right panel* shows the motor nerve terminals which form NMJs on each muscle. The motor nerve terminals are composed

"Fast" synaptic transmission on each muscle is mediated by NMJs releasing the excitatory transmitter, glutamate (Harris & Littleton, 2015). Emerging evidence indicates that individual glutamatergic boutons share a number of similarities with their mammalian counterparts (Atwood & Karunanithi, 2002; Menon et al., 2013). However, access to individual boutons for electrical recordings is easier at the larval NMJ than in the mammalian brain (Karunanithi, Barclay, Robertson, Brown, & Atwood, 1999; Karunanithi, Marin, Wong, & Atwood, 2002). Use of the larval NMJ therefore allows us to obtain information about glutamatergic transmission at single-bouton resolution that cannot be readily obtained from mammalian glutamatergic synapses. In this section, we describe the methods for performing the focal macropatch recording technique, used for recording postsynaptic currents from single synaptic boutons (Fig. 2B). We describe the application of this technique to study the actions of propofol at single-bouton resolution, although the same methods should work for volatile anesthetics such as isoflurane (Zalucki et al., 2015).

4.2 Equipment

- *Dissection microscope (Olympus SZX7 or an equivalent)*: Used for larval dissection and the pinning of the preparation.
- *Upright microscope fitted with Nomarski DIC optics for electrophysiology (Nikon Eclipse FN1 fixed-stage microscope or an equivalent)*: DIC optics is required

---

of synaptic boutons. Virtually every muscle receives synaptic inputs from the 1b and the 1s glutamatergic neurons. The motor nerve terminals of 1b neurons possess larger synaptic boutons than those of 1s neurons, and this difference can be clearly seen in this image. Scale bar represents 10 μm. (B) Focal macropatch recordings from individual synaptic boutons. The *top panel* shows DIC images of 1b and 1s motor nerve terminal branches where the individual boutons can be clearly seen. The 1b and 1s boutons chosen for recordings are indicated with *magenta arrows*. The focal macropatch electrode is gently placed on the muscle, enclosing a well-isolated 1b or 1s bouton from which postsynaptic currents are recorded. Scale bar represents 5 μm. The *bottom panel* provides examples of EJC and mEJC traces recorded from a 1b bouton (*magenta traces*) and a 1s bouton (*blue traces*). In an experiment, the average EJC amplitude (resulting from low stimulus frequencies) represents synaptic strength and the average mEJC amplitude represents quantal size. (C) In wild-type Canton-S lines, propofol (3 μ*M*) decreases quantal content at 1s boutons. Quantal content was derived by dividing the mean EJC amplitude by the mean mEJC amplitude in each experiment. The average EJC and average mEJC amplitudes were 0.86 ± 0.41 and 0.27 ± 0.02 mV, respectively, in controls ($n = 12$) and 0.76 ± 0.12 and 0.46 ± 0.05 mV, respectively, in propofol-treated preparations ($n = 11$). *$P < 0.05$, *t*-test. Error bars are s.e.m.

for viewing single boutons and for positioning the focal macropatch electrode (Fig. 2B; Karunanithi et al., 1999, 2002; Kurdyak, Atwood, Stewart, & Wu, 1994). A 60 × water-immersion objective is needed to visualize the boutons. The microscope should be fitted with standard epifluorescence components enabling fluorescence microscopy of living cells, such as to view GFP-labeled NMJs. The microscope should be placed on a vibration isolation table (TMC Micro-g lab table or an equivalent). A Faraday cage enclosing the microscope is needed to shield the system from electrical noise.

- *Low-light intensity video camera (Panasonic WV-CP310 security camera or an equivalent):* To aid positioning of the focal macropatch electrode, electrode movement has to be tracked live using a video camera mounted onto the upright microscope (with additional magnification possibly required) and projected onto a computer screen (Karunanithi et al., 2002). A cheap (~$300) security camera with high sensitivity (for example, 0.008 lx, B/W at F1.4) is sufficient for the purpose. These cameras are also capable of detecting low-level fluorescence (Karunanithi et al., 1999). The live images can be projected onto the computer screen using commercially available video capture software for either Mac or Windows platforms. For some cameras, a bnc to usb video converter may be needed for capturing the video images onto the computer.
- *Amplifier (Axoclamp 900a or an equivalent bridge amplifier)*: Used under bridge mode (or I-clamp on the 900a, with high- and low-pass filtering set at 1 Hz and 3 kHz, respectively) to record the postsynaptic currents. The focal macropatch recording technique is a form of extracellular recording that was originally pioneered by Bernard Katz and colleagues (Del Castillo & Katz, 1956). In this recording configuration, the postsynaptic currents are represented as voltage changes in the extracellular solution. As the currents flow through the resistance of the extracellular solution, they produce voltage drops across that resistance. The amplitudes of the voltage drops are directly proportional to the amplitudes of the postsynaptic currents since the resistance of the extracellular solution is constant.

  The currents are recorded, digitized, and stored onto the computer using a data acquisition system such as PowerLab 4/35 hardware and LabChart software (ADInstruments), or an equivalent.
- *Stimulator (Master-8 and ISO-Flex stimulus isolator, A.M.P·I or an equivalent)*: Synaptic responses are elicited when the cut end of the hemisegmental nerve (containing the motor axons to each abdominal

hemisegment) is sucked into a glass stimulating electrode (~12–16 μm lumen diameter) and stimulated at selected frequencies. The multiple output channels on the Master-8 stimulator can be operated independently or synchronized to produce complex signals, or used to trigger other instruments in synchrony with the stimulus.

- *Micromanipulators (Narishige WR-6 or an equivalent. The WR-6 is fitted with a UST-2 solid universal joint for easy setting of the electrode at chosen angles)*: Two micromanipulators are used to maneuver the stimulating and recording electrodes within the preparation chamber. The manipulators are mounted onto a custom-built microscope stage enabling the electrodes and preparation dish to move together in the same direction as the stage. This ensures that the electrode position relative to the preparation position remains unperturbed.
- *Syringe pump (KDS 100 Legacy Syringe Pump, KD Scientific, or an equivalent)*: The pump is used to perfuse the preparation dish with physiological solution.

  The effects of propofol on synaptic function can be assessed by perfusing physiological solution containing propofol. NMJs are exposed to propofol by dissolving stock solutions of propofol (prepared in DMSO at 0.4% v/v concentration) into the HL3 physiological solution (see below). Perfusion rates at 1–2 mL per min over 30–40 min produce propofol-induced physiological changes.
- *Microelectrode beveler (Sutter Instruments BV-10 microelectrode beveler (fitted with a stereo microscope) or an equivalent)*: The beveler is used to fashion the tip and the lumen of the focal macropatch electrode so that it sits flat on the muscle surface, enclosing the chosen bouton without damaging the muscle (Fig. 2B). The electrode lumen is created by grinding the tip of the focal macropatch electrode on a rotating diamond abrasive plate (104D, Sutter Instruments). The opening of the lumen is approximately ~5 μm in diameter (Karunanithi et al., 1999). Once the electrode tip is beveled, it is washed to remove any attached debris. Bubbles passed through the electrode during the washing helps to dislodge the debris.

  Both the stimulating and the recording electrodes are made from thin-walled borosilicate glass capillaries (1.5 mm OD, 1.17 mm ID, and 100 mm length; 30-0066, Harvard Apparatus or an equivalent), and HL3 is used as the internal solution (see below).
- *Upright microscope used for microelectrode fabrication (Olympus CX23, or an equivalent possessing both 10× and 40× objectives)*: This microscope is used to monitor the fire polishing of the stimulating and the macropatch

electrode tips to smoothen their surfaces. For fire polishing, a simple heating element can be made using nichrome wire and powered with a DC power supply. The heating element is mounted onto a coarse manipulator.

4.3 Solutions

- Schneider's insect medium (S0146-100ML, Sigma-Aldrich or from an equivalent vendor): A few drops of the medium are needed to bathe the larva during dissection and pinning. The medium reduces muscle vaculation and preserves muscle integrity. A newly opened bottle of the medium needs to be refrigerated to prevent its degradation. To prevent contamination of the medium, sterile needles and syringes are needed to draw out the required amounts of medium for dissection. Once the dissection and pinning are complete, the medium can be replaced with physiological solution required for experimentation.
- HL3 hemolymph-like physiological solution for electrophysiological recordings contain (m*M*): 70 NaCl, 5 KCl, 1 $CaCl_2$, 20 $MgCl_2$, 10 $NaHCO_3$, 5 Trehalose, 115 Sucrose, and 5 BES or HEPES (pH 7.2 and ~340 mOsm). $CaCl_2$ and $MgCl_2$ can be varied depending on the type of experiments being performed.

4.4 Dissection

- The dissection and preparation of the larva for electrophysiological recordings has been previously described (Brent, Werner, & McCabe, 2009). The tools needed for this procedure are a dissection microscope, a preparation dish, one spring scissor (15000-00, Fine Science Tools or an equivalent), one fine forceps (11252-20, Fine Science Tools or an equivalent), and a short 30-G needle. The dissection and pinning steps are described as follows:
  - A wandering third instar larva is immobilized with its dorsal end up by pinning its head (near the mouth hooks) and its tail. Slightly stretch the larva. The two primary tracheal branches, which have a silvery appearance, can be seen extending all the way from the head to the tail just under the cuticle.
  - Place a few drops of Schneider's insect medium onto the larva, completely engulfing it. Using the needle, puncture a hole through the cuticle between the primary tracheal branches, close to the tail end.

- Using the spring scissor, cut all the way from the tail to the head along the dorsal midline between the two primary tracheal tubes.
- Use four more pins, pin down both sides of the cut cuticle (two near the head end and two near the tail end) so that the shape of the pinned preparation is reminiscent of a "bear skin rug" (Fig. 2A).
- Cut out the guts, salivary glands, tracheal branches and clear away any loose-hanging imaginal discs. Essentially all that should be left is the larval CNS (reminiscent of a head template of Mickey Mouse!) (Fig. 2A) with the hemisegmental nerves radiating out from it to each of the abdominal hemisegments. At this point (or at any other point during the dissection), if the Schneider's insect medium becomes unclear as a result of dissection debris, replace it with new medium.
- By increasing the dissection microscope magnification, locate and carefully cut the hemisegmental nerve close to the hemisegment possessing the muscle whose synaptic responses you want to record. Using the forceps, gently pull at loosely hanging connective tissue attached to the larval brain in order to best expose the segmental nerve for cutting. To avoid nerve damage, great care should be taken to prevent stretching of the segmental nerve before and after cutting.
- Once the chosen segmental nerve is cut, the brain can be removed by severing the connective tissue attaching it to the ventral surface and cutting the remaining segmental nerves closer to the hemisegments.
- Once the dissection is complete, the preparation can be further stretched out, but not tightly. Finally, the Schneider's insect medium is replaced with the physiological solution by thoroughly rinsing the preparation with HL3.

4.5 Recordings

- Once the dissection is complete, the preparation dish is fixed onto the upright microscope stage. Using low magnification (10 × objective), the stimulating and the recording electrodes are lowered into the preparation dish and appropriately positioned. Switch to high magnification (using the 60 × water-immersion objective) and then start to perfuse the preparation with the perfusate using the syringe pump. The cut end of the hemisegmental nerve is gently sucked into the stimulating electrode. Avoid stretching the nerve. The focal macropatch electrode is then positioned close to the muscle from which recordings are to be made.

- Live images of the synaptic boutons innervating the muscle are captured with the low-light video camera mounted on the microscope. The images are projected onto the computer screen; the same computer which is used for data acquisition. The live images on the computer screen allow for selection of a bouton and for positioning of the focal macropatch electrode. Virtually all the larval muscles are innervated by types 1b and 1s glutamatergic motor neurons (Atwood, Govind, & Wu, 1993; Hoang & Chiba, 2001). The synaptic boutons formed by the 1b neuron are larger than those formed by the 1s neuron (Fig. 2A; Atwood et al., 1993; Kurdyak et al., 1994). We can readily differentiate between the larger 1b and smaller 1s boutons under Nomarski optics. Well-isolated 1b or 1s boutons are selected for recordings (Fig. 2B). Once a bouton is chosen for recording, that bouton's position is marked onto the computer screen with a nonpermanent felt-tip pen, along with prominent landmarks near that bouton in the event that repositioning is required. The focal macropatch electrode is gently placed onto the muscle surface, enclosing the bouton from which nerve-evoked (EJCs) and spontaneously occurring (mEJCs) excitatory junctional currents are to be recorded (Fig. 2B). The position of the electrode lumen should also be marked on the computer screen following electrode placement to ensure that the electrode lumen has not been displaced from the recording site. Once the macropatch electrode has been correctly placed, recordings can commence.

  In our experiments, we recorded from 1b and 1s boutons formed on muscle 6 in the third abdominal segment (Fig. 2A). In each experiment, 200 stimuli were delivered to the segmental nerve at a frequency of 1 Hz and the resulting EJCs recorded. 25–30 mEJCs were collected in the absence of nerve stimulation (Karunanithi et al., 1999).

4.6 Assessing the Quality of Recordings

- The ring of contact formed between the rim of focal macropatch electrode and the muscle surface is tight enough to reduce propofol-containing perfusate moving into the electrode lumen and affecting the synaptic bouton. For this reason, preparations need to be exposed to propofol-containing perfusate before recording electrode placement. Similarly, in separate control experiments, the preparations also need to be bathed with perfusate (lacking propofol) before placement of the recording electrode. Despite the tightness of contact between the electrode rim and the muscle surface, no contact resistance is formed to impede the flow of currents across electrode rim (Karunanithi et al., 2002).

- The segmental nerve is stimulated well above threshold. Otherwise, no EJCs will be produced.
- Throughout the recording session, there should be no movement which causes the bouton to be displaced from the recording site or no movement of the recording electrode, or both. If movement occurs, the recording has to be abandoned and restarted from another bouton.
- Blockages in the recording electrode can significantly increase high frequency noise and produce baseline fluctuations. There will be an increase in the electrode resistance. A new electrode is required.
- Baseline fluctuations can also arise if the flow of the perfusate is impeded. The impedance needs to be cleared and the smooth flow of perfusate restored.
- The presence of line frequency noise in the traces indicates that there is a breach in the shielding, or the silver wires in the electrode holder or ground wire, or both, require rechloriding. Removal of this type of noise will ensure more accurate measures of synaptic parameters.

4.7 Results

Key synaptic parameters derived from EJC and mEJC measurements being: (a) synaptic strength, the average size of the postsynaptic response following nerve stimulation at low stimulus frequencies (mean EJC amplitude); (b) quantal size, the average size of the postsynaptic response produced by individual vesicles releasing transmitter in the absence of nerve stimulation (mean mEJC amplitude); and (c) quantal content, the average number of vesicles releasing transmitter in response to a nerve stimulus (calculated by dividing the mean EJC amplitude by the mean mEJC amplitude) (Fig. 2C; Bykhovskaia & Vasin, 2017; Karunanithi et al., 1999).

Synaptic strength is the product of quantal size and quantal content (Atwood & Karunanithi, 2002). Changes in quantal size or quantal content, or both, therefore affect synaptic strength. Determination of the key parameters would enable us to interpret the actions of the propofol in the following ways: (a) changes in synaptic strength indicate changes in the size of the postsynaptic response, which in turn affects the ability of the postsynaptic cell to fire an action potential (Atwood & Karunanithi, 2002). For example, if propofol reduces synaptic strength, the ability to fire an action potential is diminished; (b) changes in quantal size indicate alterations in the amount of released transmitter from individual vesicles, or in the number of activated postsynaptic receptors by the released transmitter, or both (Atwood & Karunanithi, 2002; Lisman, Raghavachari, & Tsien, 2007). Some of the factors which could change the amount of released transmitter and could also be targets of propofol action include the amount of transmitter packaged into vesicles, the size of synaptic vesicles, and the fusion pore size and kinetics. Changes in the number of activated postsynaptic receptors may arise through the antagonistic actions of propofol; and (c) changes in quantal content indicate alterations in the amount of released transmitter following nerve stimulation. In fact, propofol (3 μ*M*) decreases quantal content at 1s boutons (Fig. 2C). Some of the factors which could affect quantal content and could also be targets of propofol action include the calcium dependence of transmitter release, the pool of synaptic vesicles, and the SNARE proteins which regulate exocytosis (Atwood & Karunanithi, 2002; Harris & Littleton, 2015; Lisman et al., 2007).

5. CONCLUSION

General anesthesia cannot be fully understood without linking local effects at the synapse with global consequences on brain function and behavior. Thus, some purported effects which may seem relatively minor when assayed in vitro may turn out to have major effects on whole-animal behavior. Neurotransmission would appear to be one such target mechanism. Minor impairments in synaptic release efficacy—such as decreased quantal content or mistimed release—may be unlikely to register as important in a cell culture preparation while having profound consequences in a live brain comprised of trillions of synapses. Similarly, dissecting the effect of general anesthetics on a sleep circuit remains incomplete without being able to assay sleep-like readouts in an animal. The fruit fly model *Drosophila melanogaster* provides an excellent halfway point between the efficiency required for performing molecular manipulations and the behavioral readouts needed to actually assess general anesthesia effects. We outlined here two levels of investigation, measuring effects on neurotransmission and behavioral responsiveness, which should ideally allow a thorough investigation of presynaptic hypotheses for general anesthesia. Combined with growing knowledge about postsynaptic effects, especially as they relate to endogenous sleep pathways, *Drosophila* research again proves advantageous: by using combinatorial genetic approaches (such as split LexA and

Gal4 systems; Pfeiffer et al., 2010), it should be possible to simultaneously manipulate different mechanisms in order to better understand the relative contribution of distinct target processes. Thus, it seems plausible to test in flies whether anesthesia induction might be primarily a GABAergic sleep process, whereas anesthesia maintenance and recovery could involve presynaptic mechanisms (van Swinderen & Kottler, 2014). Simultaneous control (for example, via optogenetics; Klapoetke et al., 2014) of each process should in principle generate flies that are either behaviorally resistant or hypersensitive to different general anesthetics, depending on their respective proposed mechanisms of action. The next level of understanding (not covered in this review) will be to relate these behavioral effects to novel approaches to measuring whole-brain activity, such as multichannel electrophysiology (Cohen et al., 2016) or whole-brain calcium imaging (Mann, Gallen, & Clandinin, 2017). This should provide a further level of understanding on how local effects on different proteins at synapses translate to that amalgam of different phenomena we call general anesthesia.

REFERENCES

Allada, R., & Nash, H. A. (1993). Drosophila melanogaster as a model for study of general anesthesia: The quantitative response to clinical anesthetics and alkanes. *Anesthesia and Analgesia*, *77*, 19–26.

Antkowiak, B. (2001). How do general anaesthetics work? *Naturwissenschaften*, *88*, 201–213.

Atwood, H. L., Govind, C. K., & Wu, C. F. (1993). Differential ultrastructure of synaptic terminals on ventral longitudinal abdominal muscles in Drosophila larvae. *Journal of Neurobiology*, *24*, 1008–1024.

Atwood, H. L., & Karunanithi, S. (2002). Diversification of synaptic strength: Presynaptic elements. *Nature Reviews. Neuroscience*, *3*, 497–516.

Bademosi, A. T., Steeves, J., Karunanithi, S., Zalucki, O. H., Gormal, R. S., Liu, S., et al. (2018). Trapping of syntaxin1A in presynaptic nanoclusters by a clinically relevant general anesthetic. *Cell Reports*, *22*(2), 427–440.

Baumgart, J. P., Zhou, Z. Y., Hara, M., Cook, D. C., Hoppa, M. B., Ryan, T. A., et al. (2015). Isoflurane inhibits synaptic vesicle exocytosis through reduced Ca2+ influx, not Ca2+-exocytosis coupling. *Proceedings of the National Academy of Sciences of the United States of America*, *112*, 11959–11964.

Boly, M., Moran, R., Murphy, M., Boveroux, P., Bruno, M. A., Noirhomme, Q., et al. (2012). Connectivity changes underlying spectral EEG changes during propofol-induced loss of consciousness. *The Journal of Neuroscience*, *32*, 7082–7090.

Brent, J. R., Werner, K. M., & McCabe, B. D. (2009). Drosophila larval NMJ dissection. *Journal of Visualized Experiments: JoVE* (24), pii: 1107.

Brown, E. N., Purdon, P. L., & Van Dort, C. J. (2011). General anesthesia and altered states of arousal: A systems neuroscience analysis. *Annual Review of Neuroscience*, *34*, 601–628.

Bykhovskaia, M., & Vasin, A. (2017). Electrophysiological analysis of synaptic transmission in Drosophila. *Wiley Interdisciplinary Reviews: Developmental Biology*, *6*(5).

Calcagno, B., Eyles, D., van Alphen, B., & van Swinderen, B. (2013). Transient activation of dopaminergic neurons during development modulates visual responsiveness, locomotion and brain activity in a dopamine ontogeny model of schizophrenia. *2*, e2026.

Campbell, D. B., & Nash, H. A. (1994). Use of Drosophila mutants to distinguish among volatile general anesthetics. *Proceedings of the National Academy of Sciences of the United States of America*, *91*, 2135–2139.
Campbell, J. L., & Nash, H. A. (1998). The visually-induced jump response of Drosophila melanogaster is sensitive to volatile anesthetics. *Journal of Neurogenetics*, *12*, 241–251.
Chennu, S., O'Connor, S., Adapa, R., Menon, D. K., & Bekinschtein, T. A. (2016). Brain connectivity dissociates responsiveness from drug exposure during propofol-induced transitions of consciousness. *PLoS Computational Biology*, *12*, e1004669.
Chortkoff, B. S., Bennett, H. L., & Eger, E. I., 2nd (1993). Subanesthetic concentrations of isoflurane suppress learning as defined by the category-example task. *Anesthesiology*, *79*, 16–22.
Cohen, D., Zalucki, O. H., van Swinderen, B., & Tsuchiya, N. (2016). Local versus global effects of isoflurane anesthesia on visual processing in the fly brain. *eNeuro*, *3*(4).
Colomb, J., Reiter, L., Blaszkiewicz, J., Wessnitzer, J., & Brembs, B. (2012). Open source tracking and analysis of adult Drosophila locomotion in buridan's paradigm with and without visual targets. *PLoS One*, *7*, e42247.
Crowder, C. M., Shebester, L. D., & Schedl, T. (1996). Behavioral effects of volatile anesthetics in Caenorhabditis elegans. *Anesthesiology*, *85*, 901–912.
Del Castillo, J., & Katz, B. (1956). Localization of active spots within the neuromuscular junction of the frog. *The Journal of Physiology*, *132*, 630–649.
Dissel, S., Angadi, V., Kirszenblat, L., Suzuki, Y., Donlea, J., Klose, M., et al. (2015). Sleep restores behavioral plasticity to Drosophila mutants. *Current Biology: CB*, *25*, 1270–1281.
Faville, R., Kottler, B., Goodhill, G. J., Shaw, P. J., & van Swinderen, B. (2015). How deeply does your mutant sleep? Probing arousal to better understand sleep defects in Drosophila. *Scientific Reports*, *5*, 8454.
Frank, C. A. (2014). Homeostatic plasticity at the Drosophila neuromuscular junction. *Neuropharmacology*, *78*, 63–74.
Franks, N. P. (2008). General anaesthesia: From molecular targets to neuronal pathways of sleep and arousal. *Nature Reviews Neuroscience*, *9*, 370–386.
Franks, N. P., & Lieb, W. R. (1994). Molecular and cellular mechanisms of general anaesthesia. *Nature*, *367*, 607–614.
Friedman, E. B., Sun, Y., Moore, J. T., Hung, H. T., Meng, Q. C., Perera, P., et al. (2010). A conserved behavioral state barrier impedes transitions between anesthetic-induced unconsciousness and wakefulness: Evidence for neural inertia. *PLoS One*, *5*, e11903.
Gardner, B., Strus, E., Meng, Q. C., Coradetti, T., Naidoo, N. N., Kelz, M. B., et al. (2016). Sleep homeostasis and general anesthesia: Are fruit flies well rested after emergence from propofol? *Anesthesiology*, *124*, 404–416.
Griffith, L. C. (2013). Neuromodulatory control of sleep in Drosophila melanogaster: Integration of competing and complementary behaviors. *Current Opinion in Neurobiology*, *23*, 819–823.
Guan, Z., Scott, R. L., & Nash, H. A. (2000). A new assay for the genetic study of general anesthesia in Drosophila melanogaster: Use in analysis of mutations in the X-chromosomal 12E region. *Journal of Neurogenetics*, *14*, 25–42.
Harris, K. P., & Littleton, J. T. (2015). Transmission, development, and plasticity of synapses. *Genetics*, *201*, 345–375.
Herring, B. E., McMillan, K., Pike, C. M., Marks, J., Fox, A. P., & Xie, Z. (2011). Etomidate and propofol inhibit the neurotransmitter release machinery at different sites. *The Journal of Physiology*, *589*, 1103–1115.
Hoang, B., & Chiba, A. (2001). Single-cell analysis of Drosophila larval neuromuscular synapses. *Developmental Biology*, *229*, 55–70.
Humphrey, J. A., Hamming, K. S., Thacker, C. M., Scott, R. L., Sedensky, M. M., Snutch, T. P., et al. (2007). A putative cation channel and its novel regulator: Cross-species conservation of effects on general anesthesia. *Current Biology*, *17*, 624–629.

Joiner, W. J., Friedman, E. B., Hung, H. T., Koh, K., Sowcik, M., Sehgal, A., et al. (2013). Genetic and anatomical basis of the barrier separating wakefulness and anesthetic-induced unresponsiveness. *PLoS Genetics, 9*, e1003605.

Karunanithi, S., Barclay, J. W., Robertson, R. M., Brown, I. R., & Atwood, H. L. (1999). Neuroprotection at Drosophila synapses conferred by prior heat shock. *The Journal of Neuroscience, 19*, 4360–4369.

Karunanithi, S., Marin, L., Wong, K., & Atwood, H. L. (2002). Quantal size and variation determined by vesicle size in normal and mutant Drosophila glutamatergic synapses. *The Journal of Neuroscience, 22*, 10267–10276.

Klapoetke, N. C., Murata, Y., Kim, S. S., Pulver, S. R., Birdsey-Benson, A., Cho, Y. K., et al. (2014). Independent optical excitation of distinct neural populations. *Nature Methods, 11*, 338–346.

Kottler, B., Bao, H., Zalucki, O., Troupe, M., Imlach, W., Paulk, A., et al. (2013). A sleep/wake circuit controls isoflurane sensitivity in Drosophila. *Current Biology: CB, 23*, 594–598.

Krishnan, K. S., & Nash, H. A. (1990). A genetic study of the anesthetic response: Mutants of Drosophila elanogaster altered in sensitivity to halothane. *Proceedings of the National Academy of Sciences of the United States of America, 87*, 8632–8636.

Kume, K., Kume, S., Park, S. K., Hirsh, J., & Jackson, F. R. (2005). Dopamine is a regulator of arousal in the fruit fly. *The Journal of Neuroscience: The Official Journal of the Society for Neuroscience, 25*, 7377–7384.

Kurdyak, P., Atwood, H. L., Stewart, B. A., & Wu, C. F. (1994). Differential physiology and morphology of motor axons to ventral longitudinal muscles in larval Drosophila. *The Journal of Comparative Neurology, 350*, 463–472.

Leite, L. F., Gomez, R. S., Fonseca Mde, C., Gomez, M. V., & Guatimosim, C. (2011). Effect of intravenous anesthetic propofol on synaptic vesicle exocytosis at the frog neuromuscular junction. *Acta Pharmacologica Sinica, 32*, 31–37.

Lisman, J. E., Raghavachari, S., & Tsien, R. W. (2007). The sequence of events that underlie quantal transmission at central glutamatergic synapses. *Nature Reviews. Neuroscience, 8*, 597–609.

Liu, S., Liu, Q., Tabuchi, M., & Wu, M. N. (2016). Sleep drive is encoded by neural plastic changes in a dedicated circuit. *Cell, 165*, 1347–1360.

Mann, K., Gallen, C. L., & Clandinin, T. R. (2017). Whole-brain calcium imaging reveals an intrinsic functional network in Drosophila. *Current Biology: CB, 27*, 2389–2396. e2384.

Mashour, G. A., & Hudetz, A. G. (2017). Bottom-up and top-down mechanisms of general anesthetics modulate different dimensions of consciousness. *Frontiers in Neural Circuits, 11*, 44.

Menon, K. P., Carrillo, R. A., & Zinn, K. (2013). Development and plasticity of the Drosophila larval neuromuscular junction. *Wiley Interdisciplinary Reviews: Developmental Biology, 2*, 647–670.

Metz, L. B., Dasgupta, N., Liu, C., Hunt, S. J., & Crowder, C. M. (2007). An evolutionarily conserved presynaptic protein is required for isoflurane sensitivity in Caenorhabditis elegans. *Anesthesiology, 107*, 971–982.

Meyer, H. (1899). Theorie der alkoholnarkose. *Archiv für Experimentalle Pathologie and Pharmakologie, 42*, 109–118.

Moore, J. T., Chen, J., Han, B., Meng, Q. C., Veasey, S. C., Beck, S. G., et al. (2012). Direct activation of sleep-promoting VLPO neurons by volatile anesthetics contributes to anesthetic hypnosis. *Current Biology: CB, 22*, 2008–2016.

Overton, C. E. (1891). *Studies of narcosis* (1st ed.). London: Chapman & Hall.

Pfeiffer, B. D., Ngo, T. T., Hibbard, K. L., Murphy, C., Jenett, A., Truman, J. W., et al. (2010). Refinement of tools for targeted gene expression in Drosophila. *Genetics, 186*, 735–755.

Pimentel, D., Donlea, J. M., Talbot, C. B., Song, S. M., Thurston, A. J., & Miesenbock, G. (2016). Operation of a homeostatic sleep switch. *Nature, 536*, 333–337.
Saper, C. B., Chou, T. C., & Scammell, T. E. (2001). The sleep switch: Hypothalamic control of sleep and wakefulness. *Trends in Neurosciences, 24*, 726–731.
Shaw, P. J., Cirelli, C., Greenspan, R. J., & Tononi, G. (2000). Correlates of sleep and waking in Drosophila melanogaster. *Science, 287*, 1834–1837.
Ueda, I., Hirakawa, M., Arakawa, K., & Kamaya, H. (1986). Do anesthetics fluidize membranes? *Anesthesiology, 64*, 67–71.
van Swinderen, B., & Kottler, B. (2014). Explaining general anesthesia: A two-step hypothesis linking sleep circuits and the synaptic release machinery. *BioEssays, 36*, 372–381.
van Swinderen, B., Saifee, O., Shebester, L., Roberson, R., Nonet, M. L., & Crowder, C. M. (1999). A neomorphic syntaxin mutation blocks volatile-anesthetic action in Caenorhabditis elegans. *Proceedings of the National Academy of Sciences of the United States of America, 96*, 2479–2484.
Walcourt, A., & Nash, H. A. (2000). Genetic effects on an anesthetic-sensitive pathway in the brain of Drosophila. *Journal of Neurobiology, 42*, 69–78.
Weber, B., Schaper, C., Bushey, D., Rohlfs, M., Steinfath, M., Tononi, G., et al. (2009). Increased volatile anesthetic requirement in short-sleeping Drosophila mutants. *Anesthesiology, 110*, 313–316.
Zalucki, O. H., Menon, H., Kottler, B., Faville, R., Day, R., Bademosi, A. T., et al. (2015). Syntaxin1A-mediated resistance and hypersensitivity to isoflurane in Drosophila melanogaster. *Anesthesiology, 122*, 1060–1074.
Zalucki, O., & van Swinderen, B. (2016). What is unconsciousness in a fly or a worm? A review of general anesthesia in different animal models. *Consciousness and Cognition, 44*, 72–88.

CHAPTER NINE

High-Throughput Screening to Identify Anesthetic Ligands Using *Xenopus laevis* Tadpoles

Kellie A. Woll, Roderic G. Eckenhoff[1]
University of Pennsylvania Perelman School of Medicine, Philadelphia, PA, United States
[1]Corresponding author: e-mail address: roderic.eckenhoff@uphs.upenn.edu

Contents

Abstract

General anesthetics are considered among the most significant advances in modern medicine; however, they are also some of the most dangerous commonly administered drugs. Despite this, the discovery of novel anesthetics has been slow, with few clinically used agents regardless of their nearly 200-year history. *Xenopus laevis* frogs have a long history as a model organism and provide a vital bridge between in vitro and preclinical mammalian assays. The provided protocols are efficient and cost-effective and therefore readily amendable for high-throughput evaluation of novel anesthetic ligands. By using the *X. laevis* bioassay, a researcher is capable of determining relative general anesthetic tolerance and/or cross-tolerance for candidate nonvolatile and/or volatile ligands.

1. INTRODUCTION

The tadpole bioassay to determine anesthetic potency has a long history within anesthesiology research, including application by Meyer and Overton (Downes & Courogen, 1996; Meyer, 1899; Overton, 1901).

Methods in Enzymology, Volume 602
ISSN 0076-6879
https://doi.org/10.1016/bs.mie.2018.01.007

Amphibians are, as a group, phylogenetically closer in relation to humans compared to other simple model organisms including zebrafish (*Danio rerio*), roundworm (*Caenorhabditis elegans*), and fruit fly (*Drosophila melanogaster*) (Wheeler & Brandli, 2009). The genomes of some species have been fully sequenced and showed striking similarity to human (Hellsten et al., 2010; Session et al., 2016). The premise for validity of the tadpole assay is that neurobiologic pathways underlying aversive behavior are conserved, most molecular pathways and neurotransmitter pathways are similar to mammals, and most importantly, the relative EC_{50} for most anesthetics is strikingly similar to mammals, including humans (Fig. 1A). The practical advantage of the tadpole bioassay is that, with adequate time for equilibration, the anesthetic ligand concentration within the surrounding aqueous environment equals (or is very near to) the concentration within the blood of the tadpole. As a result, the bioassay allows for a researcher to apply a constant and known concentration of anesthetic ligand and thereby associate overt behavioral changes to that value. In addition, the tadpole bioassay is a cost-effective means to determine relative drug tolerance and cross-tolerance before embarking upon more expensive mammalian assays. This chapter will provide protocols and points for discussion regarding the application of *Xenopus laevis* tadpole bioassay for the screening of anesthetic ligands.

2. *X. laevis* MODEL ORGANISM

The *Xenopus* genus, commonly known as the African clawed frogs, is fully aquatic and contains two species, *X. laevis* and *Xenopus tropicalis*. Both species have an extensive history as model organisms in biomedical research.

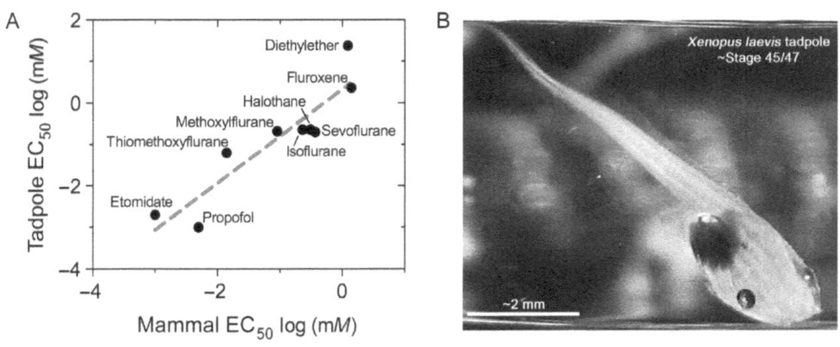

Fig. 1 Comparison of reported potencies of various clinical anesthetics within tadpoles and mammals (Hall et al., 2010; Husain et al., 2003; Roth & Miller, 1986; Woll et al., 2017) (A). An image of an albino *Xenopus laevis* tadpole, approximate stage of 45–47 (B).

The *Xenopus* embryo develops externally directly after fertilization and the tadpole has a complete set of organs that are fully functional (Nieuwkoop & Faber, 1994). Furthermore, both species have fully sequenced genomes that show startling structural similarity with the human genome allowing for *Xenopus* to act as a potential model for human conditions as well as in drug-discovery assays (Session et al., 2016; Wheeler & Brandli, 2009). Advantages of these species within the anesthetic tadpole bioassay include the easy relative maintenance as well as the size of the tadpole (~1 cm). Ultimately the bioassay allows for a robust dose–response curve for a ligand to be completed within a matter of days. Finally, *Xenopus* tadpoles are commercially available year round, with multiple vendors that provide a variety of stages. Generally for 50 tadpoles, the costs range between 20 and 40 USD at the current 2017 exchange. It should be noted, however, that some regions do require a permit before shipment.

Comparing the two species, *X. laevis* has a much longer life cycle, requiring 12 months to reach adulthood, compared to *X. tropicalis* that requires just 4 months (Wheeler & Brandli, 2009). The slower development displayed by *X. laevis* is an advantage given that it provides a larger window to execute the tadpole bioassay. A consistent stage is important to consider given that there are shifts in potency based on the development of the animal. As a whole, the bioassay has been conducted on tadpoles within the Weisz's first-form tadpole, specifically between the Nieuwkoop and Faber stages ~45–47 (Nieuwkoop & Faber, 1994) (Fig. 1B). This stage range for *X. laevis* lasts ~2 weeks at an optimal temperature of 23°C and represents a fully formed tadpole, before hind limb development (Nieuwkoop & Faber, 1994). *The husbandry for the X. laevis tadpoles is ultimately at the discretion and approval of the Institution Animal Care and Use Committee (IACUC).* As a whole, commercially available tadpoles can be maintained on dry feed and clean artificial pond solution (APS), or similar solution, at a temperature between ~20°C and 25°C. Previous investigations have used buffered solutions (Downes & Courogen, 1996) or dechlorinated tap water (Krasowski et al., 2001) as suitable aqueous environments for short-term housing. We apply commercially available APS as it gives consistent results and is favored by our institution. For long-term housing (>24 h) approximately 50 tadpoles should be kept in no less than 1 L of aqueous solution. We suggested at least 24 h acclimatization after arrival of the tadpole shipment and a 12 h light–dark cycle.

The personnel training and protocols are at the discretion and must be approved by the IACUC. The following care and protocols are based on our previous experience and our own IACUC. Aquariums are cleaned daily using serological pipettes to remove debris. To transfer tadpoles from the aquarium, a modified 25-mL serological pipette that is carefully cut 1 cm from the tip of the pipette, with sharp edges removed, can be used. Due to their small size, *X. laevis* tadpoles have the propensity to get tangled in most aquarium nets. We find that using the modified serological pipette for transfer results in less injury and/or loss of tadpoles. To establish a basic dose–response curve (i.e., sigmoidal with a variable Hill slope) a range of at least five solubilized ligand concentrations would be required, with one dosage reaching the maximum efficacy. For each tested dose, at least three individual population sets of 10 *X. laevis* tadpoles are considered necessary. The average percent response of the population set plus/minus the standard error of the mean provides the total result for the examined concentration. It is important to note that the tadpoles are not reused over the course of multiple experiments given that there is potential for retaining ligand within the animal. Euthanization and disposal should be performed as dictated by the IACUC. For euthanization, we apply a prolonged exposure to a buffered overdose (>3 g/L) of tricaine methanesulfonate (MS-222), a commonly used fish and amphibian anesthetic.

3. LIGAND PHYSICOCHEMICAL PROPERTIES AND DOSE PREPARATION

Anesthetic ligands are unique in that they are a remarkably chemically diverse class of drugs, ranging from inert gases to large steroidal ligands. Significant factors to consider prior to starting the high-throughput assay are the physicochemical limitations of the ligands that are to be examined. The physicochemical properties (stability, solubility, vehicles, concentrations, etc.) will dictate the experimental setup for the *X. laevis* biological activity assay as well as the ability of the ligand to be an anesthetic agent.

A key physicochemical property to take into account is the volatility of the agent at the ambient temperature (~20–25°C) because the assay is conducted on the bench within a designated facility. This can be established using a variety of methods from the quantitative to the qualitative by determining the vapor pressure or through general observation. Volatile ligands require a gas-tight experimental setup explained in Section 5. It is recommended to lean toward the side of caution when establishing the volatile nature for a set of compounds given that during the nonvolatile assay the ligand will be exposed to air for an extended amount of time (+30 min) (see Section 4). Additionally, it is essential that when examining multiple ligands that all are compared using the same assay setup as the two protocols contain significant differences.

The bioassay is conducted in aqueous solution; therefore, a degree of solubility is required for the experimental ligand. Hydrophobic agents often require solubilization within a biologically compatible vehicle (a cosolvent), generally dimethyl sulfoxide (DMSO), prior to further dilution into the aqueous assay solutions. The concentration of this vehicle should remain as low as possible in order to prevent interference with the end points, for example, <0.01% (v/v%) DMSO. Vortexing and/or sonication should be used to insure solubilization within the aqueous solution. Finally, the exposure container needs to be compatible with the hydrophobicity of the ligands of interest. Often glass or Teflon are optimal for exposure chambers, because these materials adsorb less hydrophobic compound than plastic.

Finally, when establishing the potency of a novel ligand, the results in the high-throughput *X. laevis* bioassay should be compared to those from a clinically used anesthetic. The appropriate clinical anesthetic(s) for comparison can be determined based on similarity of structure and/or the range of potency. In general, the EC_{50}s for intravenous anesthetics (e.g., propofol and etomidate) and volatile anesthetics (e.g., isoflurane and sevoflurane) are 0.5–3 μ*M* (Eckenhoff et al., 2010; Woll et al., 2017) and 100–300 μ*M* (Hall et al., 2010; Zhong, Rusch, & Forman, 2008), respectively, for *X. laevis* tadpoles stages 40–47.

4. BEHAVIORAL ASSESSMENTS

In general there are two forms of behavioral assessments to determine anesthetic potency within *X. laevis* tadpoles: the loss of *spontaneous* movement and the loss of *elicited* movement. Normal *X. laevis* tadpole behavior is characterized as continuous slow swimming with brief pauses (<5 s) while keeping an upright orientation (Hall et al., 2010; Hewapathirane, Dunfield, Yen, Chen, & Haas, 2008). The loss of this spontaneous movement provides a quantitative behavioral end point for potency. Under normal conditions (i.e., appropriate aqueous environment and with ambient light and temperature (20–25°C)), *X. laevis* tadpoles will exhibit a form of spontaneous movement that includes full body displacement from one area to another, either by "swimming" or "darting" as well as minor "flickering" of the tail. We typically monitor spontaneous movement for 30 s, and thus to meet the "loss of spontaneous movement" end point, the tadpole must not show any type of movement over the full 30-s time period. In addition, tadpoles often adopt a straight prone position and a slight opaque coloration of the eyes. As a result of this overt change in behavior, "loss of spontaneous movement" is both a consistent and simple end point.

There are a few variations to determine loss of *elicited* movement to evaluate *X. laevis* tadpole immobility. *X. laevis* tadpoles demonstrate a C-start reflex, also called a fast escape or startle reflex, that qualitatively appears as a "darting" movement upon external stimuli (Zarei, Elliott, Zarei, Fritzsch, & Buchholz, 2017). This external stimulus can be gentle physical prodding (with a smooth glass rod) or sharp taps upon the exposure chamber (McKinstry-Wu et al., 2015; Milutinovic, Zhao, & Sonner, 2009). Within a population, the average loss of startle reflex over the three repetitive taps would be quantified as the percent of immobilized tadpoles. It should be recognized that the potential for inconsistency between experimental groups and individuals might exist given that the strength of external stimuli may vary. Finally, the loss of righting reflex has also been used to determine loss of elicited movement (Downes & Courogen, 1996). The loss of righting reflex in tadpoles has a more specific definition being the lack of purposeful and sustained swimming response after a gentle and complete inversion using a smooth glass rod or similar tool (Downes & Courogen, 1996; Zhong et al., 2008).

Finally, other behaviors might be observed when utilizing *X. laevis* as a model organism. These include continuous rapid and irregular swimming, a tight circular body shape, and obvious convulsive behavior, all of which may reflect pain or seizures (Hewapathirane et al., 2008). The final behavioral assessment to address is the toxicity of a given agent. Obvious fatalities should be recorded over the course of exposure to the ligand and in the ensuing 18–24-h period after placing the tadpoles in fresh pond water. A dead tadpole typically will be lying on the bottom of the exposure chamber or tank, without movement, and will have an opaque coloration.

5. SCREENING OF NONVOLATILE ANESTHETICS

For nonvolatile anesthetics chambers of sufficient size that are exposed to open air can be applied. Our lab generally uses Petri dishes at least 30 mm diameter and 10 mm in height (Fig. 2A); the Petri dish (glass preferred) must be able to hold enough aqueous solution to reach greater than 5 mm from the bottom of the dish. The following contains our basic step-by-step method for exposure of a *X. laevis* population at a given concentration of ligand:

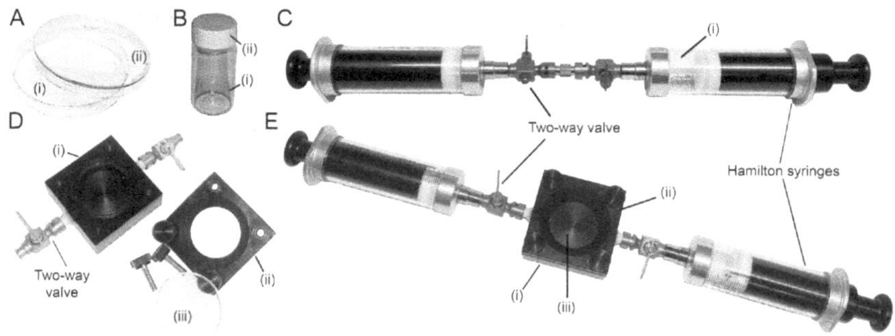

Fig. 2 Images of sufficient *Xenopus laevis* exposure chambers for nonvolatile and volatile anesthetic ligands. Nonvolatile ligands can use chambers that are open to air like a Petri dish (A). Volatile anesthetics require gas-tight chambers such as glass vials (B), joint Hamilton syringes (C), or homemade exposure chambers (D, E). Tadpoles are placed in chambers (i) with or without lid (ii), and, if present, glass or quartz viewing window (iii).

(a) Transfer 10 tadpoles into a Petri dish using modified (see Section 2) 25-mL serological pipette (end cutoff to enlarge opening, with edges smoothed) and allow tadpoles to adapt to the chamber for at least 10 min within APS.

(b) Using a nonmodified serological pipette carefully remove the APS without injuring the tadpoles. Immediately and gently pour in prepared ligand-containing APS (see Section 3) with a sufficient working volume to reach a height at least 5 mm from the bottom of the Petri dish.

(c) Set the timer for 30 min for equilibration/exposure time (see Section 7). Keep record of abnormal behavior and intervene as outlined by protocol approved by the Institution's Animal Care Committee.

(d) After 30 min immediately initiate behavioral assessment(s) chosen for evaluation and record results (see Section 4).

(e) Using a serological pipette, carefully remove ligand-containing APS without injuring tadpoles. Immediately and gently pour in fresh APS to a volume that reaches at least 5 mm from the bottom of the Petri dish.

(f) Monitor for reversible recovery then, as outlined by the protocol approved by the Institution's Animal Care Committee, carefully transfer the population to their own chamber to monitor for 18–24 h toxicity.

6. SCREENING OF VOLATILE ANESTHETICS

Volatile ligands provide a unique challenge given that they require gas-tight conditions to maintain a constant drug concentration for exposure to *X. laevis*. Various exposure chambers have been applied, including chambers fitted for constant gas flow and equipped with vaporizers (Milutinovic et al., 2009); however, this method may result in slow and unknown kinetics of equilibration with the APS. We have elected to use sealed glass vials, Hamilton gas-tight syringes, and homemade exposure chamber to overcome these challenges. These chambers are shown in Fig. 2B–E (Eckenhoff et al., 2010; Woll et al., 2017). These chambers and vials have little-to-no gas space, and therefore the advantage of maintaining fixed, known drug concentrations. Moreover, with suitable volumes, the metabolic activity of the tadpoles does not cause a detectable decline in the solution oxygen partial pressure over the 5-min exposure period. The shorter exposures are possible as the smaller volatile anesthetics equilibrate much more quickly than the nonvolatile ones. Each chamber has its own limitation(s); therefore, it is important for application of the experimental ligand to be performed in tandem with control clinical anesthetic(s) of a similar chemotype. As a result, anesthetic potency can be compared in *X. laevis* bioassays for volatile ligands vs nonvolatile ligands assayed under open Petri dish conditions (Section 5). The following describes the step-by-step method using our lab-made exposure chamber (Fig. 2D and E) that holds 5.6 mL aqueous solution.

a. Prepare ligand-containing APS within a 10-mL working volume glass Hamilton syringe.

b. Close both two-way valves on the exposure chamber and gently, using modified serological pipette, add tadpoles into chamber with a final aqueous volume to reach at least 5 mm from bottom of the chamber.

c. Place glass or quartz cover window over the chamber. Place on the lid and close chamber tightly, connect one Hamilton syringe (>10 mL working volume) containing 5 mL of APS with a two-way valve and open connected valve.

d. Gently invert chamber, with connected Hamilton syringe on bottom and the closed two-way valve on top. Open the upper second two-way valve and gently, using the bottom Hamilton syringe, push out the remaining air trapped in the chamber without harming the contained tadpoles.

e. Close the top and bottom two-way valves, return chamber back to its noninverted orientation, and remove connected Hamilton syringe. The chamber is now gas tight. The following steps should be performed quickly but carefully.

f. To one two-way valve connect the prepared ligand-containing APS and to the second two-way valve connect an empty Hamilton syringe.

g. Open both valves and, using a gentle push–pull method, push in ALL of the ligand-containing APS while drawing out excess APS using both the Hamilton syringes. Close both two-way valves and set timer for 5-min equilibration/exposure time. The ligand dose equals the initial ligand APS concentration diluted by the working volume of the chamber (e.g., 5.6 mL).

h. After 5 min immediately initiate appropriate behavioral assessment.

i. Immediately open the chamber, remove glass or quartz window, and transfer tadpoles using a serological pipette into a Petri dish containing fresh APS.

j. Using a serological pipette, carefully remove the diluted ligand-containing APS from the dish and gently pour in fresh APS to a volume that reaches at least 0.5 cm from the bottom of the Petri dish.

k. Monitor for reversible recovery then, as outlined by the protocol approved by the IACUC, carefully transfer the population to their own chamber to monitor for 18–24 h toxicity.

7. ADDITIONAL CONSIDERATIONS FOR NOVEL AND/OR CHEMICALLY ACTIVE LIGANDS

A major advantage of using *X. laevis* tadpole bioassay is the ability to integrate these organisms efficiently in high-throughput workflows to evaluate novel and/or chemically active anesthetic ligands. Some additional considerations need to be taken into account when applying this bioassay within these investigations. While similar to mammals, it should still be recognized that there are subtle genomic differences; therefore, anesthetic activity, or lack thereof, may not be fully reflected in higher-order vertebrates. Furthermore, the conditions of the bioassay need to be considered to either negate or enable chemically active ligands. For example, the *X. laevis* tadpole bioassay has been often applied to evaluate anesthetic photolabels and/or photoactivated anesthetic ligands (Stein et al., 2012; Weiser, Kelz, & Eckenhoff, 2012). Therefore to determine baseline tolerance, the exposure chamber should have minimal exposure to activating wavelength(s). Given that viable albino *X. laevis* tadpoles are also commercially available, researchers have the opportunity to evaluate the photoactivity in vivo, resulting in light-induced immobilization (Stein et al., 2012) or prolonged emergence (also known as "optoanesthesia") (Weiser et al., 2012) upon photoactivation.

8. CONCLUSIONS

The behavioral assessments and experimental procedures for the *X. laevis* tadpole bioassay to determine the potency of anesthetic ligands have been described. The tadpole bioassay has been a useful tool to evaluate the relative potency of clinical, novel, and/or chemically active anesthetic ligands, and the application of *X. laevis* as the model organism allows for the easy integration into high-throughput workflows. The two procedures provided permit the assessment of either nonvolatile or volatile ligands as well as key points to consider for design of the method in practice. The genome of *X. laevis* displays striking similarity to humans and can also be genetically manipulated (Session et al., 2016; Wheeler & Brandli, 2009). Therefore, the *X. laevis* tadpole bioassay would allow for a ready model to determine the impact of human conditions and diseases on anesthetic potency. The *X. laevis* tadpole bioassay provides a convenient and efficient first-line animal model for the discovery and evaluation of anesthetic ligands.

ACKNOWLEDGMENT

This work was supported by the National Institutes of Health (GM055876, GM107117).

REFERENCES

Downes, H., & Courogen, P. M. (1996). Contrasting effects of anesthetics in tadpole bioassays. *The Journal of Pharmacology and Experimental Therapeutics*, *278*(1), 284–296.

Eckenhoff, R. G., Xi, J., Shimaoka, M., Bhattacharji, A., Covarrubias, M., & Dailey, W. P. (2010). Azi-isoflurane, a photolabel analog of the commonly used inhaled general anesthetic isoflurane. *ACS Chemical Neuroscience*, *1*(2), 139–145. https://doi.org/10.1021/cn900014m.

Hall, M. A., Xi, J., Lor, C., Dai, S. P., Pearce, R., Dailey, W. P., et al. (2010). m-Azipropofol (AziPm) a photoactive analogue of the intravenous general anesthetic propofol. *Journal of Medicinal Chemistry*, *53*(15), 5667–5675. https://doi.org/10.1021/jm1004072.

Hellsten, U., Harland, R. M., Gilchrist, M. J., Hendrix, D., Jurka, J., Kapitonov, V., et al. (2010). The genome of the western clawed frog Xenopus tropicalis. *Science (New York, N.Y.)*, *328*(5978), 633–636. https://doi.org/10.1126/science.1183670.

Hewapathirane, D. S., Dunfield, D., Yen, W., Chen, S., & Haas, K. (2008). In vivo imaging of seizure activity in a novel developmental seizure model. *Experimental Neurology*, *211*(2), 480–488.

Husain, S. S., Ziebell, M. R., Ruesch, D., Hong, F., Arevalo, E., Kosterlitz, J. A., et al. (2003). 2-(3-Methyl-3H-diaziren-3-yl)ethyl 1-(1-phenylethyl)-1H-imidazole-5-carboxylate: A derivative of the stereoselective general anesthetic etomidate for photolabeling ligand-gated ion channels. *Journal of Medicinal Chemistry*, *46*(7), 1257–1265. https://doi.org/10.1021/jm020465v.

Krasowski, M. D., Jenkins, A., Flood, P., Kung, A. Y., Hopfinger, A. J., & Harrison, N. L. (2001). General anesthetic potencies of a series of propofol analogs correlate with potency for potentiation of gamma-aminobutyric acid (GABA) current at the

GABA(A) receptor but not with lipid solubility. *The Journal of Pharmacology and Experimental Therapeutics*, *297*(1), 338–351.

McKinstry-Wu, A. R., Bu, W., Rai, G., Lea, W. A., Weiser, B. P., Liang, D. F., et al. (2015). Discovery of a novel general anesthetic chemotype using high-throughput screening. *Anesthesiology*, *122*(2), 325–333.

Meyer, H. (1899). Zur theorie der alkoholnarkose. *Archiv Für Experimentelle Pathologie und Pharmakologie*, *42*(2–4), 109–118. https://doi.org/10.1007/bf01834479.

Milutinovic, P. S., Zhao, J., & Sonner, J. M. (2009). Tolerance to isoflurane does not occur in developing Xenopus laevis tadpoles. *Anesthesia and Analgesia*, *108*(1), 176–180. https://doi.org/10.1213/ane.0b013e31818ca33e.

Nieuwkoop, P. D., & Faber, J. (1994). *Normal table of Xenopus laevis (Daudin): A systematical and chronological survey of the development from the fertilized egg till the end of metamorphosis.* New York: Garland Publishing.

Overton, C. E. (1901). *Studien über die narkose zugleich ein beitrag zur allgemeinen pharmakologie.* Jena, Switzerland: Gustav Fischer.

Roth, S. H., & Miller, K. W. (1986). In S. H. Roth, K. W. Miller, & Plenum Publishing Corporation (Eds.), *Molecular and cellular mechanisms of anesthetics* (1st ed.). New York: Springer US.

Session, A. M., Uno, Y., Kwon, T., Chapman, J. A., Toyoda, A., Takahashi, S., et al. (2016). Genome evolution in the allotetraploid frog Xenopus laevis. *Nature*, *538*(7625), 336–343. https://doi.org/10.1038/nature19840.

Stein, M., Middendorp, S. J., Carta, V., Pejo, E., Raines, D. E., Forman, S. A., et al. (2012). Azo-propofols: Photochromic potentiators of $GABA_A$ receptors. *Angewandte Chemie International Edition*, *51*(42), 10500–10504. https://doi.org/10.1002/anie.201205475.

Weiser, B. P., Kelz, M. B., & Eckenhoff, R. G. (2012). In vivo activation of azi-propofol prolongs anesthesia and reveals synaptic targets. *The Journal of Biological Chemistry*, *26*, 26.

Wheeler, G. N., & Brandli, A. W. (2009). Simple vertebrate models for chemical genetics and drug discovery screens: Lessons from zebrafish and Xenopus. *Developmental Dynamics: An Official Publication of the American Association of Anatomists*, *238*(6), 1287–1308. https://doi.org/10.1002/dvdy.21967.

Woll, K. A., Peng, W., Liang, Q., Zhi, L., Jacobs, J. A., Maciunas, L., et al. (2017). Photoaffinity ligand for the inhalational anesthetic sevoflurane allows mechanistic insight into potassium channel modulation. *ACS Chemical Biology*, *12*, 1353–1362. https://doi.org/10.1021/acschembio.7b00222.

Zarei, K., Elliott, K. L., Zarei, S., Fritzsch, B., & Buchholz, J. H. J. (2017). A method for detailed movement pattern analysis of tadpole startle response. *Journal of the Experimental Analysis of Behavior*, *108*(1), 113–124. https://doi.org/10.1002/jeab.263.

Zhong, H. J., Rusch, D., & Forman, S. A. (2008). Photo-activated azi-etomidate, a general anesthetic photolabel, irreversibly enhances gating and desensitization of gamma-aminobutyric acid type A receptors. *Anesthesiology*, *108*(1), 103–112.

CHAPTER TEN

Zebrafish: A Pharmacogenetic Model for Anesthesia

Victoria Bedell*[,1], Elena Buglo[§,¶,||], Daniel Marcato[†], Christian Pylatiuk[‡], Ralf Mikut[‡], Johannes Stegmaier[‡], Will Scudder[||], Maxwell Wray[||], Stephan Züchner[§,¶], Uwe Strähle[†], Ravindra Peravali[†], Julia E. Dallman[||]

*Perelman School of Medicine at the University of Pennsylvania, Philadelphia, PA, United States
[†]Institute of Toxicology and Genetics, Karlsruhe Institute of Technology, Karlsruhe, Germany
[‡]Institute of Applied Computer Science, Karlsruhe Institute of Technology, Karlsruhe, Germany
[§]John P. Hussman Institute for Human Genomics, University of Miami, Miami, FL, United States
[¶]Dr. John T. MacDonald Foundation, University of Miami, Miami, FL, United States
[||]University of Miami, Coral Gables, FL, United States
[1]Corresponding author: e-mail address: Victoria.bedell@uphs.upenn.edu

Contents

Abstract

General anesthetics are small molecules that interact with and effect the function of many different proteins to promote loss of consciousness, amnesia, and sometimes, analgesia. Owing to the complexity of this state transition and the transient nature of these drug/protein interactions, anesthetics can be difficult to study. The zebrafish is an emerging model for the discovery of both new genes required for the response

Methods in Enzymology, Volume 602
ISSN 0076-6879
https://doi.org/10.1016/bs.mie.2018.02.004

to and side effects of anesthesia. Here we discuss the tools available to manipulate the zebrafish genome, including both genetic screens and genome engineering approaches. Additionally, there are various robust behavior assays available to study anesthetic and other drug responses. These assays are available for single-gene study or high throughput for genetic or drug discovery. Finally, we present a case study of using propofol as an anesthetic in the zebrafish. These techniques and protocols make the zebrafish a powerful model to study anesthetic mechanisms and drug discovery.

Key Points

1. Genetic tools available to knockout or overexpress genes include both transient and stably inheritable approaches. Specific base pair changes can be engineered within the genome. Additionally, forward genetic random mutation screening is readily available.
2. Multiple complex, validated behavior assays are available to study drug response, including the photomotor and startle responses.
3. The zebrafish is an idea model for high-throughput studies to discover new genes for anesthetic response or drugs that induce anesthesia.
4. Owing to its hydrophobic nature, use of propofol with the zebrafish requires use of glass equipment and careful mixing into solution to maintain a consistent concentration.

1. INTRODUCTION

Despite the widespread use of anesthetics for surgical practice, the contribution of genetics to variability in individual responses to anesthesia is still poorly understood (Garcia, Rothman, & Fitzpatrick, 2017; Mashour, Woodrum, & Avidan, 2015; Saczynski et al., 2012). For example studies of red-haired women, who anecdotally had required increased anesthetic, actually verified an association with increased anesthetic requirement over their dark haired matched counterparts (Liem et al., 2004). This illustrates a genetic component to anesthetic response that is not well understood. Additionally, classic, large-scale studies show that, irrespective of age, 3%–8% of people experience prolonged grogginess for more than 2h after surgery with unknown cause (compared to ~80% that recover within 15min) (Aldrete & Kroulik, 1970; Philip, Kallar, Bogetz, Scheller, & Wetchler, 1996). Postoperative cognitive issues are both more severe and more prevalent in patients over the age of 60, as up to 40% experience delirium associated with negative prognoses after undergoing inpatient surgeries (Giattino et al., 2017). Other complications include prolonged hospital stays as well as increased morbidity and mortality (Mashour et al., 2015;

Neufeld et al., 2013; Saczynski et al., 2012). It has also been shown that reduced cognitive ability before anesthesia correlates with greater risk for cognitive decline postanesthesia (Giattino et al., 2017). To address the contributions of genetics and environment to postoperative delirium, a large twin study of middle aged and elderly patients strongly suggest that surgery/anesthesia alone are not sufficient to explain such complications (Dokkedal, Hansen, Rasmussen, Mengel-From, & Christensen, 2016; Terrando et al., 2011). Taken together these data support the possibility of an individual's genetic make-up can profoundly impact in their response to anesthetics and that inherited neurological disorders contribute their perioperative experiences.

Since the sequencing of the human genome in 2001, thousands of genetic variants that cause inherited disorders have been identified (Huguet, Benabou, & Bourgeron, 2016; Puschmann, 2013; Rudnik-Schoneborn, Auer-Grumbach, & Senderek, 2017). Case studies link rare inherited nervous system disorders such as narcolepsy and glycine encephalopathy to complications after anesthesia that include cataplexy and trouble breathing (Liu & Fan, 2006; Mesa, Diaz, & Frosth, 2000). Perioperative risks are also well established for more common inherited disorders such as epilepsy and Parkinson's disease (Dhallu, Baiomi, Biyyam, & Chilimuri, 2017). Therefore, it has been hypothesized that at least a portion of an individual's sensitivity to anesthetics and postoperative complications could result from preclinical inherited neurological disorders (Dhallu et al., 2017; Dokkedal et al., 2016).

Zebrafish provide a promising emerging model in which to better understand genetic variants that confer risk with anesthesia. Established as a new model organism in the 1970s and 80s by George Streisinger at the University of Oregon (Streisinger, Walker, Dower, Knauber, & Singer, 1981), zebrafish (*Brachydanio rerio*) combine the advantages of two more established model organisms, the fruit fly and the mouse. Like fruit flies, a single zebrafish female produces hundreds of embryos at a time that develop externally and rapidly acquire a diversity of sensory-evoked locomotor behaviors (Orger & de Polavieja, 2017) that can be used to assess depth of sedation. In addition, embryos/larvae are transparent making it possible to track neuronal activity in vivo (Ahrens, Huang, Narayan, Mensh, & Engert, 2013) to determine neural circuit correlates of anesthesia. Like mice and humans, zebrafish are vertebrates with largely conserved genes (Howe et al., 2013) and similar physiology, including ion channels, neurotransmitters, and receptors (Renier et al., 2007; Rihel & Schier, 2013) which are known to be among the targets of anesthetic agents (Forman & Miller, 2011; Kretschmannova et al., 2013; Orser, Wang, Pennefather, & MacDonald, 1994). Combined with flexible methods for mutagenizing the genome that includes both forward (Lawson & Wolfe, 2011) and reverse genetic approaches (Hsu, Lander, & Zhang, 2014) and high-throughput methods for behavioral screening (Kokel & Peterson, 2011; McCarroll, Gendelev, Keiser, & Kokel, 2016; Rihel et al., 2010), we suggest that zebrafish is an excellent model for elucidating the genetic contribution to complications associated with anesthesia.

2. GENETIC MANIPULATION IN THE ZEBRAFISH

In recent years, there has been an explosion of technologies that make engineering the zebrafish genome both simple and reliable. These technologies are discussed below.

2.1 Reverse Genetics: Recapitulating Human Mutations in Zebrafish

Many inherited disorders are caused by nonsense, loss-of-function mutations in a specific gene. To generate a zebrafish model of this type of disorder, both knockdown and knockout approaches are used. To knockdown gene expression, morpholinos were developed in the late 1990s (Summerton & Weller, 1997). Morpholinos are antisense nucleotides primarily used to study genes expressed within the first 2–3 days postfertilization (dpf). At later stages, they are unreliable due to degradation and dilution. They are also not ideal for studying the p53 pathway and neurological development due to their nonspecific upregulation of p53 causing nonspecific neurologic damage. They are, however, very simple to design and to use as reviewed in Bedell, Westcot, and Ekker (2011) and Stainier et al. (2017).

Better suited to anesthesia studies are genome engineering approaches that make stably inherited changes to the genome. These methods create double-stranded breaks at a designated site in the genome (Fig. 1). The break is then repaired by endogenous machinery using nonhomologous end joining, which causes a small number of base pairs to be trimmed off or added to the ends of the break before those loose ends are fused back together (Bibikova, Golic, Golic, & Carroll, 2002). This generally creates a small insertion or deletion (an "indel"). These indels can cause loss-of-function, nonsense mutations resulting in the introduction of a stop codon and premature protein truncation. While it is possible to use zinc finger endonucleases (Doyon et al., 2008) or transcription activator-like effector nucleases (TALENs) (Huang et al., 2011), the most common approach currently is

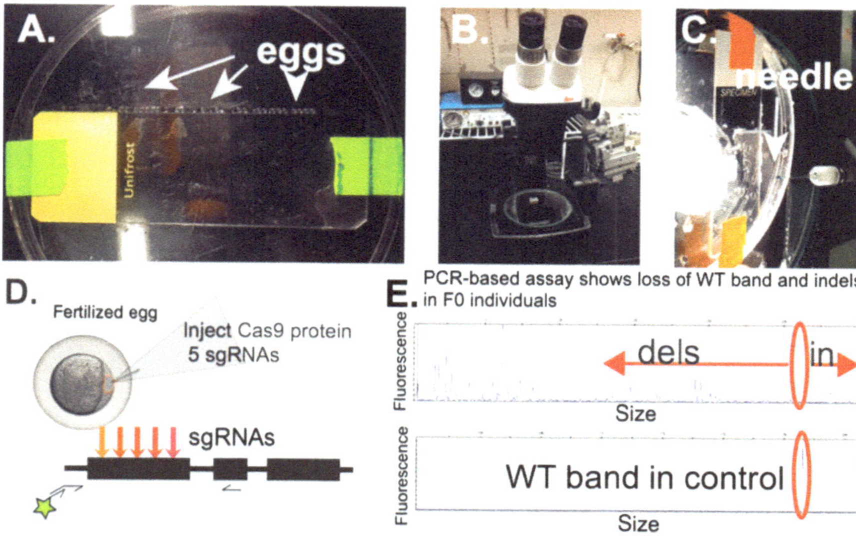

Fig. 1 Genetic manipulation of zebrafish by injecting 1-cell stage embryos. Fertilized eggs surrounded by an acellular chorion are collected and aligned against a glass slide taped to a petri plate (A). These are injected on a dissecting microscope equipped with a foot pedal-driven pressure injector and manipulator (B). For CRISPR/Cas9, small guide RNAs and Cas9 protein are mixed and loaded into an injection needle that is then mounted on the manipulator (C and D). Depending on the topic of study, different stable transgenic lines can be injected. As in Varshney et al. (2016), five guide RNAs are designed against a single exon of the target gene and coinjected with Cas9 protein. Subsequently, embryos are tested for genome editing by using a three-primer polymerase chain reaction to make fluorescent amplicons of the edited site. These are then run on a sequencing capillary to see either a series of insertions (in) and deletions (dels) in the embryo injected with the guide RNAs or a single wild-type band in the case of the embryos that are only injected with cas9 (E).

based on CRISPRs (clustered regularly interspaced short palindromic repeats)/Cas9. For this method, an endonuclease protein, Cas9, binds to a small guide RNA (sgRNA; Fig. 1D) to target a specific genomic sequence and create double-stranded breaks: reviewed in Hsu et al. (2014) and Varshney et al. (2016). The workflow for disrupting a gene is shown in Fig. 1.

It is also possible to specifically engineer missense mutations that change a single, targeted base pair change resulting in a specific change in amino acid sequence. This was first accomplished in zebrafish using TALENs (Bedell et al., 2012) and then CRISPRs (Hoshijima, Jurynec, & Grunwald, 2016; Ran et al., 2013). The TALENs/CRISPRs are designed as described earlier to target a specific genetic locus. Then a single-stranded DNA oligo is designed with the mutation of interest flanked by homologous arms that are around 10 base pairs, with the entire oligo being around 20 base pairs. The ssDNA provides a template for the repair machinery of the cell to undergo homology directed repair rather than nonhomologous end joining (Bedell et al., 2012). Alternatively, a donor plasmid can be engineered with longer homologous arms (Hoshijima et al., 2016). In these ways, specific, disease-causing human mutations can be recapitulated in a zebrafish model.

2.2 Expression of Transgenes

In addition to loss-of-function nonsense and missense alleles, gain-of-function or overexpression experiments can also be used to study missense alleles that produce dominant gain-of-function phenotypes (Koenighofer et al., 2016). Once again, there are both transient and permanent methods. To transiently overexpress, mRNA of a gene is injected into the single cell of the zebrafish embryo. This mRNA is translated, depending on the design of the untranslated regions, for as long as 2 dpf (Hyatt & Ekker, 1999).

If experiments require longer duration of expression or for the expression to be in a specific tissue, transposons can be used to insert DNA into the genome. Transposons are proteins that interact with a pair of specific repeats to excise a piece of DNA from one place and insert that DNA randomly into another. Any sequence including transgenes with transcriptional activator and promoter sequences as well as the coding sequence of the protein to be expressed can be placed between the two transposase recognition sequences (Clark et al., 2011). This approach results in stable expression of the transgene.

2.3 Zebrafish Models of Disorders Linked to Anesthetic Complications

Many zebrafish models have already been made, among them models of narcolepsy (Prober, Rihel, Onah, Sung, & Schier, 2006; Yokogawa et al., 2007), glycine encephalopathy (Cui et al., 2005; Mongeon et al., 2008), epilepsy/autism (Grone et al., 2016; Grone, Qu, & Baraban, 2017; Hoffman et al., 2016; Scheldeman et al., 2017), and Parkinson's (Soman et al., 2017). These and other models can be leveraged to better understand the genetic contributions to complications associated with anesthesia.

2.4 Forward Genetics: Unbiased Discovery of Genetic Variants That Confer Risk for Anesthesia

To identify new genetic variants that alter responses to anesthetics, there are several methods regularly employed by the zebrafish community to generate mutant lines in an unbiased way. These lines are then studied by investigators interested in specific phenotypes (Mullins, Hammerschmidt, Haffter, & Nusslein-Volhard, 1994). Traditionally, mutations are generated with chemicals, such as N-ethyl-N-nitrosourea (ENU), that create single base pair changes: reviewed in Knapik (2000). This method creates base pair changes in up to 1 in every 100 base pairs. Therefore, it can be difficult to assess which mutation is causative for the phenotype. Additionally, it can be difficult to find the mutation of interest because it is a single base pair change in an unknown gene. To address these issues, insertional mutagenesis has also been used including DNA viruses (Amsterdam et al., 1999) and transposons (Clark et al., 2011; Varshney et al., 2013). Similar forward genetic approaches in invertebrate models have defined new anesthetic targets (Morgan & Sedensky, 2013).

Several distinct large collections of mutant zebrafish lines are maintained at large facilities including the National Institutes of Health (Varshney et al., 2013), European Zebrafish Resource Center (EZRC) (Geisler, Borel, Ferg, Maier, & Strahle, 2016), Chinese Zebrafish Resource Center (https://zfin.org/ZDB-LAB-130226-1), and Zebrafish International Resource Center (Varga, 2016).

3. BEHAVIORS TO ASSESS SEDATION AND RECOVERY FROM ANESTHESIA

Zebrafish have emerged as the model organism of choice for high-throughput drug screening of behavioral phenotypes (Bruni, Lakhani, & Kokel, 2014; Kokel & Peterson, 2011; Rihel et al., 2010). A number of behaviors have been analyzed and reported using the zebrafish model including locomotion, feeding, cognition, memory, and stress (Kalueff et al., 2013; Orger & de Polavieja, 2017). However, in order to exploit the advantages of the zebrafish model and to achieve rapid and statistically robust readouts, it is essential to phenotype the behavioral patterns in high throughput. The Screening Center at the Karlsruhe Institute of Technology is a transdisciplinary center that specializes in developing automated and intelligent morphological and behavioral screening platforms (e.g., Figs. 3 and 4). Every platform is equipped with automated sampling handling systems, a high-content and high-throughput imaging component, an image analysis pipeline for real-time data analysis and finally a large-scale data storage system for data archival and retrieval. The center is connected with the

EZRC that serves as a hub to obtain wild type, mutant and transgenic zebrafish and medaka. The Screening Center offers options to conduct morphological and behavioral screening for zebrafish and the medaka community. Here we report two behavioral phenotyping platforms that function in high throughput and can be used efficiently to understand and quantify effects of anesthetics.

3.1 Photomotor Response (PMR)

The PMR is one of the earliest manifestations of behavior in zebrafish occurring between 30 and 42 h hpf and does not involve any canonical visual organs (Kokel et al., 2013; Kokel & Peterson, 2011). Between 30 and 42 hpf, zebrafish embryos exhibit a low basal movement. The application of a strong and short pulse of light, however, evokes rapid movement of the embryos within the chorion after a few seconds of latency that lasts for about 5 s. After this, the embryos enter a refractory phase and do not respond to a second pulse of light (Kokel & Peterson, 2011; Marcato et al., 2015). Fig. 2 shows this behavior schematically.

Interestingly, it has been found that this behavior is significantly altered in the presence of neuroactive compounds and anesthetics. Consequently, we have developed a high-throughput PMR platform that is completely automated from the sampling handling stage to the analysis stage. The entire

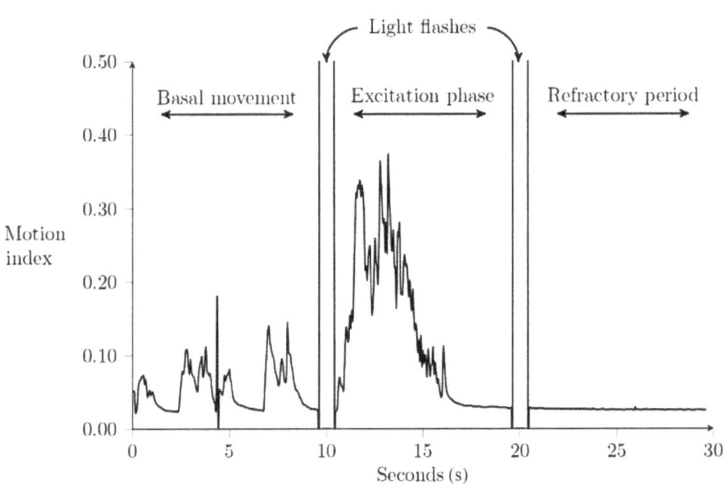

Fig. 2 The photomotor response (PMR) of zebrafish embryos between 30 and 42 hpf.

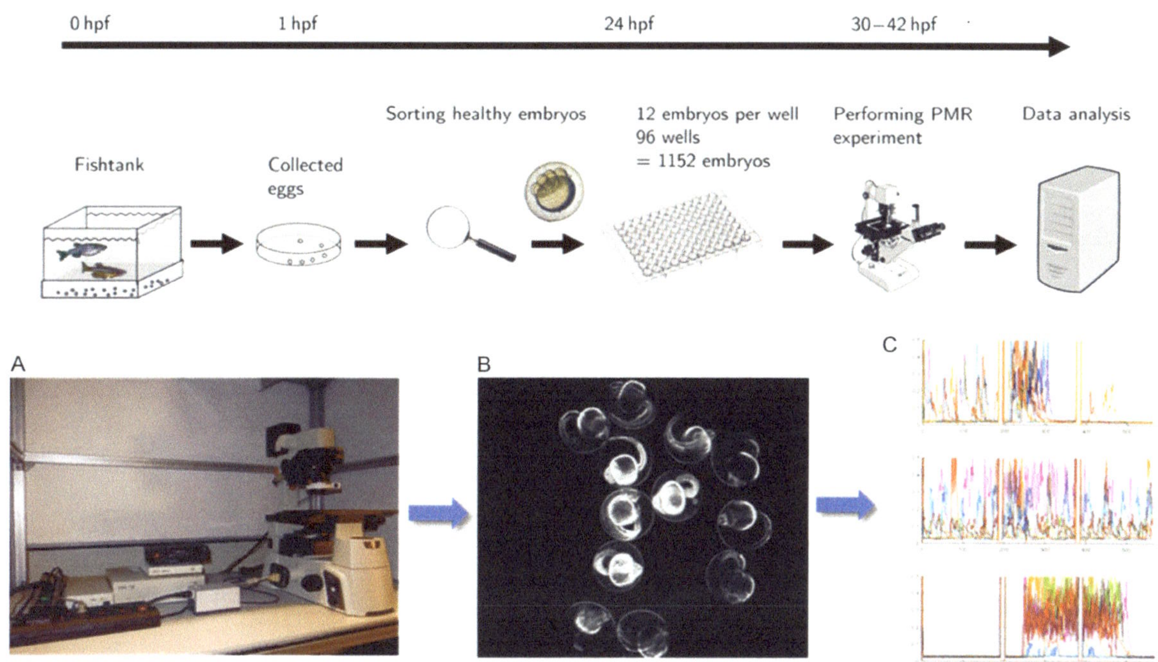

Fig. 3 PMR high-throughput pipeline. The *upper panel* describes the flow of the PMR experiment. (A) The imaging system with A, representing the power source; B, the shutter controller; C, the stage controller; D, the custom IR ring; and E, the microscope itself. (B) A snapshot of a single well with embryos that will be subjected to the light flash. (C) Representative results of the PMR response to 3 different compounds with 10 repeats for each compound showing typical PMR phenotypes: normal response on the *top*, the *middle graph* showing increased activity throughout the different PMR phases, and finally the *bottom graph* showing increased latency and high activity afterward.

platform functions sequentially and is described below and shown in Fig. 3 (Marcato et al., 2015).

Step 1: Sample handling

In the first step, zebrafish embryos are automatically sorted to remove coagulated and unfertilized eggs. After this, 10–12 embryos are automatically pipetted into the wells of a standard 96-well microtiter plate. Furthermore, the entire process works in the absence of ambient white light (thereby avoiding exposure to light before the experiment) as this system operates under infrared illumination. This step helps in avoiding human errors and facilitates the preparation of a large number of plates per day for experimentation.

Step 2: Data acquisition

After sample handling, the automatically prepared plate is transferred to a custom-built imaging system to conduct the actual PMR experiment. A Nikon Eclipse Ti-E inverted microscope equipped with a custom LED ring is programed so that each well containing embryos is successively in the focal plane of the microscope. The light flash is provided by a Nikon 100W Xenon bulb and a controllable shutter (Sutter Instruments, Novato, CA, USA). The video recording is done using a Nikon DS-Qi 1 monochrome camera with a 1280 × 1024 pixel resolution at 19 frames per second and with a CFI Plan Achromat UW 2× objective to get a large field of view to cover the entire well. The entire acquisition and control software is custom developed.

Step 3: Data analysis and visualization

An entire analysis and visualization pipeline were developed to automatically analyze the data acquired from the PMR experiment. The analysis begins with egg detection using a circular Hough Transform and a weighting algorithm that avoids false detections. A motion index which is a numerical value reflecting the changes in consecutive frames of the PMR video is then calculated for each egg. The temporal change in the motion index is then reflective of the test compound on the PMR behavior.

Different anesthetics show different effects on the PMR including decreased activity on the application of a light pulse, increased time to respond (increased latency), and for some doses no response to the application of the light pulse. Taken together, this is a complex, noninvasive behavioral readout that can easily be used to assess anesthetic response and any changes in that response due to genomic changes.

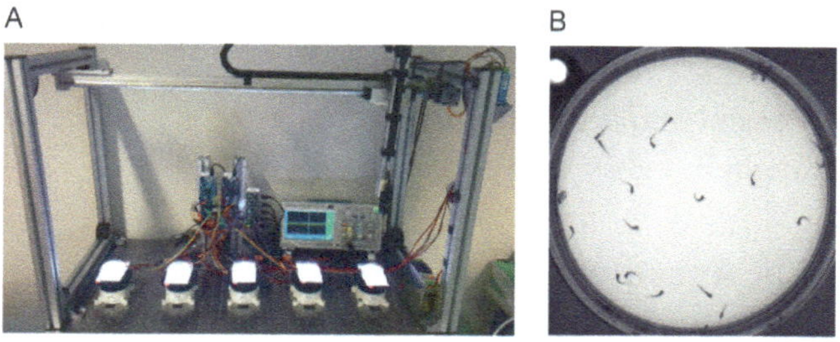

Fig. 4 High-throughput vibration analysis platform. (A) The screening platform showing five different vibration pulse sources mounted below five petri-dishes containing embryos. The robotic arm carries the high-speed camera that moves sequentially over the different source positions. (B) A typical C bend response after the application of the vibration pulse.

3.2 Startle/Vibration Response

A startle response is a fast reaction to an applied vibrational stimulus or sound. At about 4 dpf zebrafish larvae respond to vibration in a characteristic way (Burgess & Granato, 2007). The maturation of the ear/swim bladder to Mauthner neuron escape circuit is believed to be responsible for this behavior: reviewed in Eaton, Lee, and Foreman (2001). With application of a vibration pulse the larvae respond by bending the head all the way to the tail forming a C shape. This is known as the C bend. The larvae then release themselves from this position and propel themselves forward and swim away in an escape response. This behavior is robust and consistent and like the PMR can be modulated by anesthetics and other neuroactive compounds. In the absence of a high-throughput vibration response system, we have developed one that automatically images the response and analyzes the data to provide kinematic parameters. We have used this system to study emergence time of zebrafish larvae subjected to different anesthetics and at different doses.

The system consists of several acoustic speakers that are calibrated to provide a vibration pulse of around 15 g for 1 ms. The response is then recorded using a high-speed camera at 1000 frames per second. The data are then automatically analyzed to extract activity, the bend angles and the distance traveled after the application of the pulse. The system is shown in Fig. 4.

4. ANESTHESIA IN ZEBRAFISH

Methods of delivering anesthetics vary depending on life-history stage. While gavage is used in adult zebrafish and other adult teleosts

(Dang, Henderson, Garraway, & Zon, 2016), anesthetics are more often added directly to the bath (immersion) for larval zebrafish. Here, we focus on larval stages and immersion because these are the stages that are amenable to high-throughput behavioral studies.

4.1 Types of Anesthetics

There is a growing body of literature on the behavioral and physiologic effects of various anesthetics on teleosts (Collymore, Tolwani, Lieggi, & Rasmussen, 2014; Readman, Owen, Knowles, & Murrell, 2017; Sneddon, 2012). The most commonly used anesthetics, mechanisms of action, behavioral and physiologic effects on fish are described in many publications elsewhere and are beyond the scope of this review. Anesthetics evaluated in zebrafish range from nonpharmaceutical grade anesthetics such as MS222/benzocaine (Sneddon, 2012), Eugenol (Javahery, Nekoubin, & Moradlu, 2012), and carbon dioxide administered with ultrafine oxygen bubbles (Kugino, Tamaru, Hisatomi, & Sakaguchi, 2016) and pharmaceutical grade general anesthetics used in humans including isoflurane (Huang et al., 2010) and propofol (Warren, Baker, & Fishman, 2001). Isoflurane is also sometimes used in combination with MS222 in the zebrafish (Huang et al., 2010). The combination is preferable to MS222 alone because it does not depress cardiac function, fish can be sedated for longer periods of time, and recover faster. The same is true for propofol (Warren et al., 2001). The fact that propofol and isoflurane are effective in both zebrafish and humans raises the potential for the discovery of new anesthetics using high-throughput behavioral screening in zebrafish. Such a screening strategy was shown to be effective for identifying novel antipsychotics (Bruni et al., 2016).

4.2 Special Considerations When Making Propofol

Owing to its hydrophobic nature, propofol adheres to plastic and is not particularly water soluble. Therefore, to avoid uncertain concentrations due to loss from absorption, glass is used to pipette and store propofol. To break up micelles that form when propofol is added to aqueous solutions, it is essential to vortex and sonicate the solution (Fig. 5). A previous publication that used propofol in zebrafish larvae (Warren et al., 2001) used a 10-fold higher dose than what is known to anesthetize *Xenopus* tadpoles (Krasowski et al., 2001; Weiser, Kelz, & Eckenhoff, 2013) and humans (Purdon et al., 2013), likely due to these solubility issues. An alternate method is to use DMSO as a cosolvent or to use a

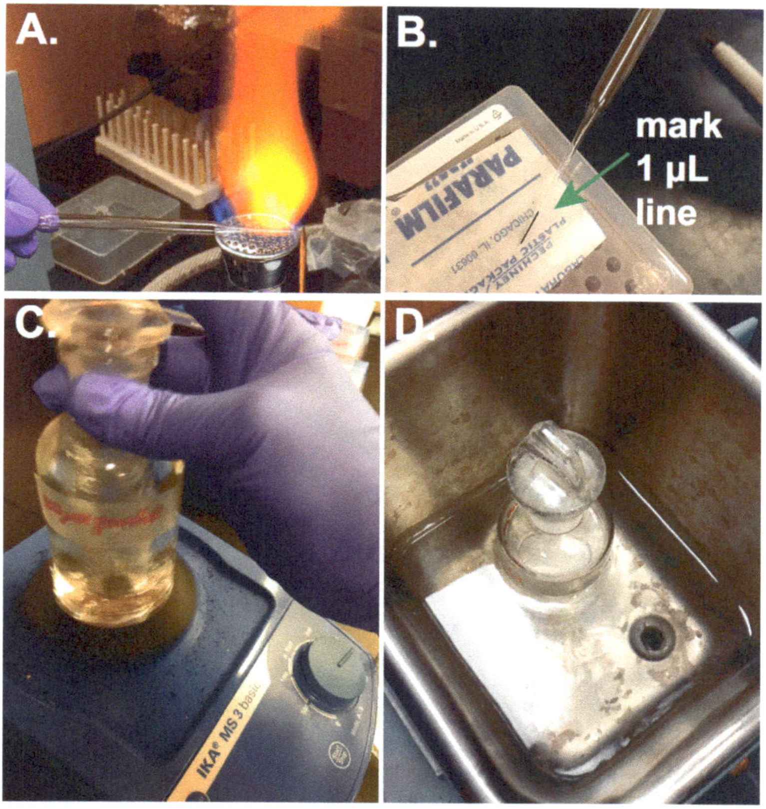

Fig. 5 Diluting propofol for anesthesia. To make propofol, we first use a Bunsen burner (A) to make a drawn out glass pipet that we calibrate to 1 μL (B). We add 1 μL propofol to 100 mL system water in a glass bottle, vortex for 30 s (C), and sonicate for 5 min (D).

formulation of propofol that is more readily soluble in water. When propofol solutions are made using the following protocol (Fig. 5), sedative (~1 μ*M*) and anesthetic (~10 μ*M*) doses of propofol for zebrafish are similar to what have been reported for tadpoles and humans (Fig. 6A and B).

For our purposes, we:

1. Make a drawn out glass pipet using a Bunsen burner (using hemostats; Fig. 5A)
2. Break of the end of the glass pipet and dispose in glass waste
3. Measure 1 μL water with green food coloring onto parafilm using a pipetman
4. Draw up solution into newly fashioned glass pipet and mark the fill line (Fig. 5B)

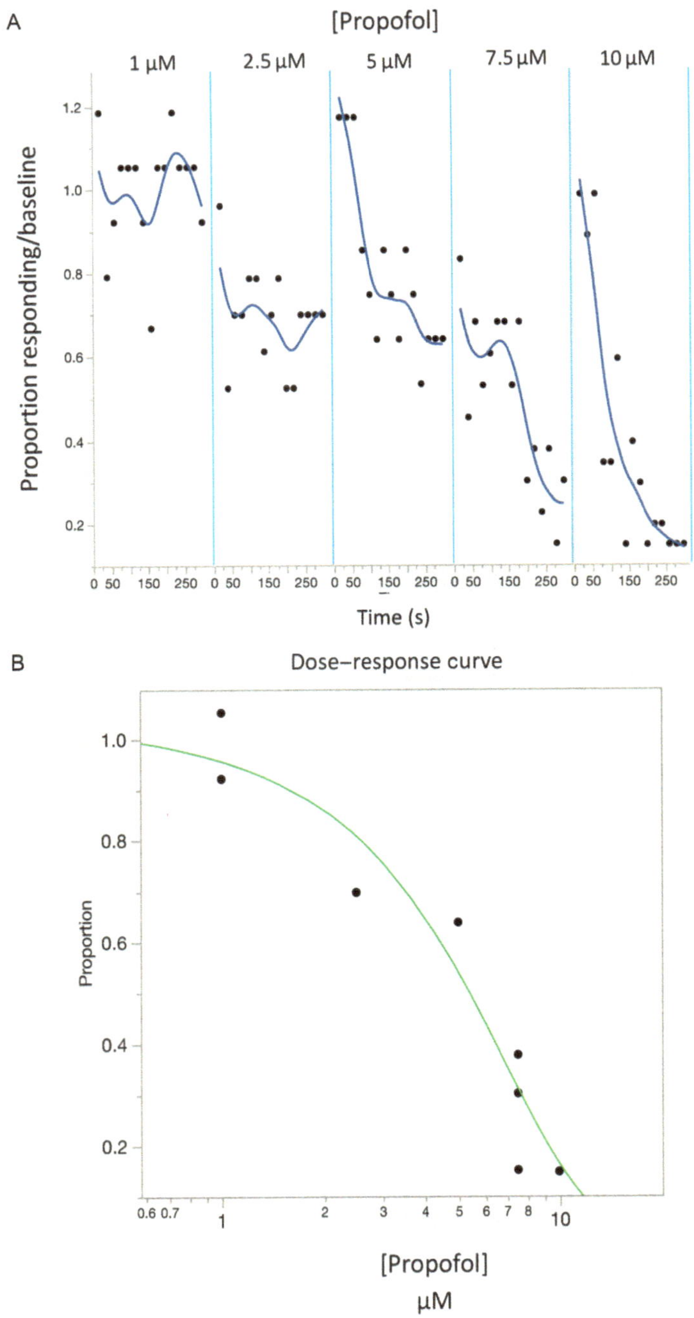

Fig. 6 Vibration response can be used to assess sedation and emergence parameters for anesthesia. (A) 10–20 five-day-old larvae are exposed to vibration for every 20 s for 5 min after addition of propofol and the proportion responding normalized to baseline is plotted vs time for different concentrations of propofol. (B) A dose–response curve is constructed from proportion responding at 5 min. Proportions are fitted with a logistic regression in JMP13 (*green line*).

5. Wash pipet in water and then use to measure propofol
6. Add 100 mL system water to a 100 mL glass bottle
7. Pipet 1 μL propofol into 100 mL
8. Vortex 30 s (Fig. 5C)
9. Sonicate 5 min (Fig. 5D)
10. Measure concentration on a spectrophotometer, measuring absorbance @270 nm. The molar extinction coefficient is 1600/M.

4.3 HPLC to Determine Tissue Concentration of Anesthetic

Because the route of delivery of the propofol anesthetic is immersion, it is important to determine the concentration that has reached the brain. To do this we euthanize fish and then isolate brain tissue for HPLC using the following protocol:

1. Prepare several 35-mm petri plates for surgery arena
   (a) Coat with Sylgard (Corning)
       mix polymers 1:10
       pour ~3 mm depth in each petri
       allow to harden overnight
       soak in distilled water before using for surgeries
2. Expose 5–7-day-old larvae to anesthetic in baskets for ease of exposure to anesthetic and washing
3. After anesthetic exposure, euthanize larvae by placing baskets on dry ice
4. Place surgery arena in 10 cm petri filled with wet ice
5. Transfer basket from dry ice and use forceps to transfer euthanized larva by its tail to the surgery arena so that the head faces the inside of the plate and the notochord is visible
6. Pin the larva to the plate using FST insect pins bent to a right angle with forceps
7. Add a couple drops of system water over the larva
   (a) System water is taken from recirculating aquaria housing adult fish and contains salt for an osmolarity between 400 and 500 mOsm and a neutral pH ~7.4
8. Remove the eyes using the tungsten needles (Electron Microscopy Sciences)
9. Sever the head (usually do this right before the swim bladder)
10. Remove the skin and ears from the head by running over the matter with the tungsten needles

11. Using the tungsten needles drag the brain matter outside of the water using the surface tension to make a bubble on the needle with the brain matter inside
12. Transfer to an Eppendorf and store on dry ice for HPLC as described in Gardner et al. (2016).

4.4 Propofol Sedation Curve

To construct dose–response curves (Fig. 6A and B), 10–20 five-day-old larvae are placed in 1.5 mL system water in a 35 mm petri dish lined with a sylgard ring to keep larvae in the field of view. This dish is placed on an aluminum ring attached to a minishaker by a titanium rod to deliver calibrated vibrational stimuli. Vibration-evoked escape responses (Burgess & Granato, 2007) require sensory/motor integration similar to the righting reflex used in rodents (Campagna, Miller, & Forman, 2003) and are therefore appropriate for assessing depth of sedation. In Fig. 6A, stimuli are delivered every 20 s with responses coming to steady state by 4 min after addition of propofol. The dose–response curve (Fig. 6B) is constructed by taking proportion responding at 5 min normalized to averaged responsiveness of the larvae during a baseline test conducted before addition of propofol. This curve can also be obtained using PMR with zebrafish embryos at very early stages of development. These two assays, PMR (at 30 hpf) and vibration response (at 4 dpf), can help our understanding of how the response to anesthetics changes with age.

5. CONCLUSIONS

The zebrafish has become a powerful model for delineating the genes and neural circuits that contribute to a given behavior. Genome engineering approaches like CRISPR/Cas9 have made it straight forward to generate zebrafish models of human inherited disorders associated with anesthetic risk. In addition, collections of randomly mutated lines of zebrafish enable forward genetic screens for those genes as yet unknown that cause complications. The sensory-evoked and highly stereotyped zebrafish behaviors, such as the PMR and vibration response, can be studied in a high-throughput setting that allows for fast screening of both mutations and drugs. Furthermore, it has been shown that zebrafish larvae respond to propofol and other anesthetics in a similar pattern and concentration as other animals. Taken together, the zebrafish is poised to enhance our understanding of the mechanism of anesthesia and the role genetics plays in the response to anesthesia.

REFERENCES

Ahrens, M. B., Huang, K. H., Narayan, S., Mensh, B. D., & Engert, F. (2013). Two-photon calcium imaging during fictive navigation in virtual environments. *Frontiers in Neural Circuits*, *7*, 104.

Aldrete, J. A., & Kroulik, D. (1970). A postanesthetic recovery score. *Anesthesia and Analgesia*, *49*, 924–934.

Amsterdam, A., Burgess, S., Golling, G., Chen, W., Sun, Z., Townsend, K., et al. (1999). A large-scale insertional mutagenesis screen in zebrafish. *Genes & Development*, *13*, 2713–2724.

Bedell, V. M., Wang, Y., Campbell, J. M., Poshusta, T. L., Starker, C. G., Krug, R. G., 2nd, et al. (2012). In vivo genome editing using a high-efficiency TALEN system. *Nature*, *491*, 114–118.

Bedell, V. M., Westcot, S. E., & Ekker, S. C. (2011). Lessons from morpholino-based screening in zebrafish. *Briefings in Functional Genomics*, *10*, 181–188.

Bibikova, M., Golic, M., Golic, K. G., & Carroll, D. (2002). Targeted chromosomal cleavage and mutagenesis in Drosophila using zinc-finger nucleases. *Genetics*, *161*, 1169–1175.

Bruni, G., Lakhani, P., & Kokel, D. (2014). Discovering novel neuroactive drugs through high-throughput behavior-based chemical screening in the zebrafish. *Frontiers in Pharmacology*, *5*, 153.

Bruni, G., Rennekamp, A. J., Velenich, A., McCarroll, M., Gendelev, L., Fertsch, E., et al. (2016). Zebrafish behavioral profiling identifies multitarget antipsychotic-like compounds. *Nature Chemical Biology*, *12*, 559–566.

Burgess, H. A., & Granato, M. (2007). Sensorimotor gating in larval zebrafish. *The Journal of Neuroscience: The Official Journal of the Society for Neuroscience*, *27*, 4984–4994.

Campagna, J. A., Miller, K. W., & Forman, S. A. (2003). Mechanisms of actions of inhaled anesthetics. *The New England Journal of Medicine*, *348*, 2110–2124.

Clark, K. J., Balciunas, D., Pogoda, H. M., Ding, Y., Westcot, S. E., Bedell, V. M., et al. (2011). In vivo protein trapping produces a functional expression codex of the vertebrate proteome. *Nature Methods*, *8*, 506–515.

Collymore, C., Tolwani, A., Lieggi, C., & Rasmussen, S. (2014). Efficacy and safety of 5 anesthetics in adult zebrafish (Danio rerio). *Journal of the American Association for Laboratory Animal Science*, *53*, 198–203.

Cui, W. W., Low, S. E., Hirata, H., Saint-Amant, L., Geisler, R., Hume, R. I., et al. (2005). The zebrafish shocked gene encodes a glycine transporter and is essential for the function of early neural circuits in the CNS. *The Journal of Neuroscience: The Official Journal of the Society for Neuroscience*, *25*, 6610–6620.

Dang, M., Henderson, R. E., Garraway, L. A., & Zon, L. I. (2016). Long-term drug administration in the adult zebrafish using oral gavage for cancer preclinical studies. *Disease Models & Mechanisms*, *9*, 811–820.

Dhallu, M. S., Baiomi, A., Biyyam, M., & Chilimuri, S. (2017). Perioperative management of neurological conditions. *Health Services Insights*, *10*, (eCollection 2017). https://doi.org/10.1177/1178632917711942.

Dokkedal, U., Hansen, T. G., Rasmussen, L. S., Mengel-From, J., & Christensen, K. (2016). Cognitive functioning after surgery in middle-aged and elderly Danish twins. *Anesthesiology*, *124*, 312–321.

Doyon, Y., McCammon, J. M., Miller, J. C., Faraji, F., Ngo, C., Katibah, et al. (2008). Heritable targeted gene disruption in zebrafish using designed zinc-finger nucleases. *Nature Biotechnology*, *26*, 702–708.

Eaton, R. C., Lee, R. K., & Foreman, M. B. (2001). The Mauthner cell and other identified neurons of the brainstem escape network of fish. *Progress in Neurobiology*, *63*, 467–485.

Forman, S. A., & Miller, K. W. (2011). Anesthetic sites and allosteric mechanisms of action on Cys-loop ligand-gated ion channels. *Canadian Journal of Anaesthesia*, *58*, 191–205.

Garcia, P. S., Rothman, D. L., & Fitzpatrick, S. M. (2017). Waking up to a new model for studying neural systems: What emergence from unconscious states can reveal about brain organization. *Frontiers in Systems Neuroscience*, *11*, 78.

Gardner, B., Strus, E., Meng, Q. C., Coradetti, T., Naidoo, N. N., Kelz, M. B., et al. (2016). Sleep homeostasis and general anesthesia: Are fruit flies well rested after emergence from propofol? *Anesthesiology*, *124*, 404–416.

Geisler, R., Borel, N., Ferg, M., Maier, J. V., & Strahle, U. (2016). Maintenance of zebrafish lines at the European zebrafish resource center. *Zebrafish*, *13*(Suppl. 1), S19–23.

Giattino, C. M., Gardner, J. E., Sbahi, F. M., Roberts, K. C., Cooter, M., Moretti, E., et al. (2017). Intraoperative frontal alpha-band power correlates with preoperative neurocognitive function in older adults. *Frontiers in Systems Neuroscience*, *11*, 24.

Grone, B. P., Marchese, M., Hamling, K. R., Kumar, M. G., Krasniak, C. S., Sicca, F., et al. (2016). Epilepsy, behavioral abnormalities, and physiological comorbidities in Syntaxin-binding protein 1 (STXBP1) mutant zebrafish. *PLoS One*, *11*, e0151148.

Grone, B. P., Qu, T., & Baraban, S. C. (2017). Behavioral comorbidities and drug treatments in a zebrafish scn1lab model of Dravet syndrome. *eNeuro*, *4*.

Hoffman, E. J., Turner, K. J., Fernandez, J. M., Cifuentes, D., Ghosh, M., Ijaz, S., et al. (2016). Estrogens suppress a behavioral phenotype in zebrafish mutants of the autism risk gene, CNTNAP2. *Neuron*, *89*, 725–733.

Hoshijima, K., Jurynec, M. J., & Grunwald, D. J. (2016). Precise editing of the zebrafish genome made simple and efficient. *Developmental Cell*, *36*, 654–667.

Howe, K., Clark, M. D., Torroja, C. F., Torrance, J., Berthelot, C., Muffato, M., et al. (2013). The zebrafish reference genome sequence and its relationship to the human genome. *Nature*, *496*, 498–503.

Hsu, P. D., Lander, E. S., & Zhang, F. (2014). Development and applications of CRISPR-Cas9 for genome engineering. *Cell*, *157*, 1262–1278.

Huang, W. C., Hsieh, Y. S., Chen, I. H., Wang, C. H., Chang, H. W., Yang, C. C., et al. (2010). Combined use of MS-222 (tricaine) and isoflurane extends anesthesia time and minimizes cardiac rhythm side effects in adult zebrafish. *Zebrafish*, *7*, 297–304.

Huang, P., Xiao, A., Zhou, M., Zhu, Z., Lin, S., & Zhang, B. (2011). Heritable gene targeting in zebrafish using customized TALENs. *Nature Biotechnology*, *29*, 699–700.

Huguet, G., Benabou, M., Bourgeron, T., Sassone-Corsi, P., & Christen, Y. (2016). The genetics of autism spectrum disorders. In *A time for metabolism and hormones*. Springer, Cham (CH).

Hyatt, T. M., & Ekker, S. C. (1999). Vectors and techniques for ectopic gene expression in zebrafish. *Methods in Cell Biology*, *59*, 117–126.

Javahery, S., Nekoubin, H., & Moradlu, A. H. (2012). Effect of anaesthesia with clove oil in fish (review). *Fish Physiology and Biochemistry*, *38*, 1545–1552.

Kalueff, A. V., Gebhardt, M., Stewart, A. M., Cachat, J. M., Brimmer, M., Chawla, J. S., et al. (2013). Towards a comprehensive catalog of zebrafish behavior 1.0 and beyond. *Zebrafish*, *10*, 70–86.

Knapik, E. W. (2000). ENU mutagenesis in zebrafish—From genes to complex diseases. *Mammalian Genome*, *11*, 511–519.

Koenighofer, M., Hung, C. Y., McCauley, J. L., Dallman, J., Back, E. J., Mihalek, I., et al. (2016). Mutations in RIT1 cause Noonan syndrome—Additional functional evidence and expanding the clinical phenotype. *Clinical Genetics*, *89*, 359–366.

Kokel, D., Dunn, T. W., Ahrens, M. B., Alshut, R., Cheung, C. Y., Saint-Amant, L., et al. (2013). Identification of nonvisual photomotor response cells in the vertebrate hindbrain. *The Journal of Neuroscience: The Official Journal of the Society for Neuroscience*, *33*, 3834–3843.

Kokel, D., & Peterson, R. T. (2011). Using the zebrafish photomotor response for psychotropic drug screening. *Methods in Cell Biology*, *105*, 517–524.

Krasowski, M. D., Jenkins, A., Flood, P., Kung, A. Y., Hopfinger, A. J., & Harrison, N. L. (2001). General anesthetic potencies of a series of propofol analogs correlate with potency for potentiation of gamma-aminobutyric acid (GABA) current at the GABA(A) receptor but not with lipid solubility. *The Journal of Pharmacology and Experimental Therapeutics*, *297*, 338–351.

Kretschmannova, K., Hines, R. M., Revilla-Sanchez, R., Terunuma, M., Tretter, V., Jurd, R., et al. (2013). Enhanced tonic inhibition influences the hypnotic and amnestic actions of the intravenous anesthetics etomidate and propofol. *The Journal of Neuroscience: The Official Journal of the Society for Neuroscience*, *33*, 7264–7273.

Kugino, K., Tamaru, S., Hisatomi, Y., & Sakaguchi, T. (2016). Long-duration carbon dioxide anesthesia of fish using ultra fine (nano-scale) bubbles. *PLoS One*, *11*, e0153542.

Lawson, N. D., & Wolfe, S. A. (2011). Forward and reverse genetic approaches for the analysis of vertebrate development in the zebrafish. *Developmental Cell*, *21*, 48–64.

Liem, E. B., Lin, C. M., Suleman, M. I., Doufas, A. G., Gregg, R. G., Veauthier, J. M., et al. (2004). Anesthetic requirement is increased in redheads. *Anesthesiology*, *101*, 279–283.

Liu, C. M., & Fan, S. Z. (2006). Glycine encephalopathy and delayed emergence from anesthesia. *Anesthesia and Analgesia*, *103*, 1631.

Marcato, D., Alshut, R., Breitwieser, H., Mikut, R., Strahle, U., Pylatiuk, C., et al. (2015). An automated and high-throughput photomotor response platform for chemical screens. *Conference Proceedings: Annual International Conference of the IEEE Engineering in Medicine and Biology Society*, *2015*, 7728–7731.

Mashour, G. A., Woodrum, D. T., & Avidan, M. S. (2015). Neurological complications of surgery and anaesthesia. *British Journal of Anaesthesia*, *114*, 194–203.

McCarroll, M. N., Gendelev, L., Keiser, M. J., & Kokel, D. (2016). Leveraging large-scale behavioral profiling in zebrafish to explore neuroactive polypharmacology. *ACS Chemical Biology*, *11*, 842–849.

Mesa, A., Diaz, A. P., & Frosth, M. (2000). Narcolepsy and anesthesia. *Anesthesiology*, *92*, 1194–1196.

Mongeon, R., Gleason, M. R., Masino, M. A., Fetcho, J. R., Mandel, G., Brehm, P., et al. (2008). Synaptic homeostasis in a zebrafish glial glycine transporter mutant. *Journal of Neurophysiology*, *100*, 1716–1723.

Morgan, P. G., & Sedensky, M. M. (2013). The power of unbiased genetic screens to discover novel anesthetic targets. *Anesthesiology*, *119*, 995–996.

Mullins, M. C., Hammerschmidt, M., Haffter, P., & Nusslein-Volhard, C. (1994). Large-scale mutagenesis in the zebrafish: In search of genes controlling development in a vertebrate. *Current Biology*, *4*, 189–202.

Neufeld, K. J., Leoutsakos, J. M., Sieber, F. E., Wanamaker, B. L., Gibson Chambers, J. J., Rao, V., et al. (2013). Outcomes of early delirium diagnosis after general anesthesia in the elderly. *Anesthesia and Analgesia*, *117*, 471–478.

Orger, M. B., & de Polavieja, G. G. (2017). Zebrafish behavior: Opportunities and challenges. *Annual Review of Neuroscience*, *40*, 125–147.

Orser, B. A., Wang, L. Y., Pennefather, P. S., & MacDonald, J. F. (1994). Propofol modulates activation and desensitization of GABAA receptors in cultured murine hippocampal neurons. *The Journal of Neuroscience: The Official Journal of the Society for Neuroscience*, *14*, 7747–7760.

Philip, B. K., Kallar, S. K., Bogetz, M. S., Scheller, M. S., & Wetchler, B. V. (1996). A multicenter comparison of maintenance and recovery with sevoflurane or isoflurane for adult ambulatory anesthesia. The Sevoflurane Multicenter Ambulatory Group. *Anesthesia and Analgesia*, *83*, 314–319.

Prober, D. A., Rihel, J., Onah, A. A., Sung, R. J., & Schier, A. F. (2006). Hypocretin/orexin overexpression induces an insomnia-like phenotype in zebrafish. *The Journal of Neuroscience: The Official Journal of the Society for Neuroscience*, *26*, 13400–13410.

Purdon, P. L., Pierce, E. T., Mukamel, E. A., Prerau, M. J., Walsh, J. L., Wong, K. F., et al. (2013). Electroencephalogram signatures of loss and recovery of consciousness from propofol. *Proceedings of the National Academy of Sciences of the United States of America*, *110*, E1142–1151.

Puschmann, A. (2013). Monogenic Parkinson's disease and parkinsonism: Clinical phenotypes and frequencies of known mutations. *Parkinsonism & Related Disorders*, *19*, 407–415.

Ran, F. A., Hsu, P. D., Wright, J., Agarwala, V., Scott, D. A., & Zhang, F. (2013). Genome engineering using the CRISPR-Cas9 system. *Nature Protocols*, *8*, 2281–2308.

Readman, G. D., Owen, S. F., Knowles, T. G., & Murrell, J. C. (2017). Species specific anaesthetics for fish anaesthesia and euthanasia. *Scientific Reports*, *7*, 7102.

Renier, C., Faraco, J. H., Bourgin, P., Motley, T., Bonaventure, P., Rosa, F., et al. (2007). Genomic and functional conservation of sedative-hypnotic targets in the zebrafish. *Pharmacogenetics and Genomics*, *17*, 237–253.

Rihel, J., Prober, D. A., Arvanites, A., Lam, K., Zimmerman, S., Jang, S., et al. (2010). Zebrafish behavioral profiling links drugs to biological targets and rest/wake regulation. *Science*, *327*, 348–351.

Rihel, J., & Schier, A. F. (2013). Sites of action of sleep and wake drugs: Insights from model organisms. *Current Opinion in Neurobiology*, *23*, 831–840.

Rudnik-Schoneborn, S., Auer-Grumbach, M., & Senderek, J. (2017). Hereditary neuropathies: Update 2017. *Neuropediatrics*, *48*, 282–293.

Saczynski, J. S., Marcantonio, E. R., Quach, L., Fong, T. G., Gross, A., Inouye, S. K., et al. (2012). Cognitive trajectories after postoperative delirium. *The New England Journal of Medicine*, *367*, 30–39.

Scheldeman, C., Mills, J. D., Siekierska, A., Serra, I., Copmans, D., Iyer, A. M., et al. (2017). mTOR-related neuropathology in mutant tsc2 zebrafish: Phenotypic, transcriptomic and pharmacological analysis. *Neurobiology of Disease*, *108*, 225–237.

Sneddon, L. U. (2012). Clinical anesthesia and analgesia in fish. *Journal of Exotic Pet Medicine*, *21*, 32–43.

Soman, S., Keatinge, M., Moein, M., Da Costa, M., Mortiboys, H., Skupin, A., et al. (2017). Inhibition of the mitochondrial calcium uniporter rescues dopaminergic neurons in pink1 −/− zebrafish. *The European Journal of Neuroscience*, *45*, 528–535.

Stainier, D. Y. R., Raz, E., Lawson, N. D., Ekker, S. C., Burdine, R. D., Eisen, J. S., et al. (2017). Guidelines for morpholino use in zebrafish. *PLoS Genetics*, *13*, e1007000.

Streisinger, G., Walker, C., Dower, N., Knauber, D., & Singer, F. (1981). Production of clones of homozygous diploid zebra fish (Brachydanio rerio). *Nature*, *291*, 293–296.

Summerton, J., & Weller, D. (1997). Morpholino antisense oligomers: design, preparation, and properties. *Antisense & Nucleic Acid Drug Development*, *7*, 187–195.

Terrando, N., Brzezinski, M., Degos, V., Eriksson, L. I., Kramer, J. H., Leung, J. M., et al. (2011). Perioperative cognitive decline in the aging population. *Mayo Clinic Proceedings*, *86*, 885–893.

Varga, Z. M. (2016). Aquaculture, husbandry, and shipping at the zebrafish international resource center. *Methods in Cell Biology*, *135*, 509–534.

Varshney, G. K., Carrington, B., Pei, W., Bishop, K., Chen, Z., Fan, C., et al. (2016). A high-throughput functional genomics workflow based on CRISPR/Cas9-mediated targeted mutagenesis in zebrafish. *Nature Protocols*, *11*, 2357–2375.

Varshney, G. K., Huang, H., Zhang, S., Lu, J., Gildea, D. E., Yang, Z., et al. (2013). The zebrafish insertion collection (ZInC): A web based, searchable collection of zebrafish mutations generated by DNA insertion. *Nucleic Acids Research*, *41*, D861–864.

Warren, K. S., Baker, K., & Fishman, M. C. (2001). The slow mo mutation reduces pacemaker current and heart rate in adult zebrafish. *American Journal of Physiology. Heart and Circulatory Physiology*, *281*, H1711–1719.

Weiser, B. P., Kelz, M. B., & Eckenhoff, R. G. (2013). In vivo activation of azipropofol prolongs anesthesia and reveals synaptic targets. *The Journal of Biological Chemistry*, *288*, 1279–1285.

Yokogawa, T., Marin, W., Faraco, J., Pezeron, G., Appelbaum, L., Zhang, J., et al. (2007). Characterization of sleep in zebrafish and insomnia in hypocretin receptor mutants. *PLoS Biology*, *5*, e277.

CHAPTER ELEVEN

The Mouse as a Model Organism for Assessing Anesthetic Sensitivity

Andrzej Z. Wasilczuk*, Kaitlyn L. Maier*, Max B. Kelz*,†,1
*University of Pennsylvania, Perelman School of Medicine, Philadelphia, PA, United States
†Center for Sleep and Circadian Neurobiology, University of Pennsylvania, Perelman School of Medicine, Philadelphia, PA, United States
[1]Corresponding author: e-mail address: kelz@pennmedicine.upenn.edu

Contents

Abstract

The mouse has been used in many medical fields as a powerful model to reveal the genetic basis of human physiology and disease. The past two decades have witnessed an enormous wealth of genetic and informatic resources dedicated to this humble organism. With the ongoing revolution in mapping neural circuitry governing behavior, the mouse is an ideal model organism poised to unravel the mysteries of general anesthetic action. This chapter will describe and provide guidelines for anesthetic phenotyping in the mouse including both motor-dependent and motor-independent assessments.

Methods in Enzymology, Volume 602
ISSN 0076-6879
https://doi.org/10.1016/bs.mie.2018.01.008

1. INTRODUCTION

The mouse has served as a valuable organism yielding insights into human health and disease. It remains the premier model for the genetic basis of physiology and pathophysiology because it is small, relatively inexpensive to maintain, and has a short gestation period of just under 3 weeks (Peters et al., 2007). Mice become sexually mature by the age of 6–8 weeks making the generation time roughly 3 months. It is commonly felt that the biochemical, cellular, developmental, and neuronal pathways in mice are highly conserved between mice and humans, further emphasizing the use of the murine model toward translational and reverse translational studies (Baker, 2011; Deussing, 2013). Within the field of anesthesiology, studies in genetically engineered mice have helped to validate the importance of putative molecular targets of anesthetic action and to offer critical glimpses into the translational promiscuity of general anesthetic drugs from the in vitro to the in vivo setting (Behlke et al., 2016; Chen, Shu, & Bayliss, 2009; Heurteaux et al., 2004; Jurd et al., 2003; Rudolph & Antkowiak, 2004). In this chapter, we will focus on the methods used to assess sedation and hypnosis in mice using both customary motor-dependent and motor-independent measures.

2. ANESTHETIC DOSING IN MICE

2.1 Volatile Anesthetics

Delivery of an anesthetic agent to mice merits consideration as every drug must ultimately reach its effect site to elicit its biological actions. Since it is uncommon to intubate or perform a tracheostomy on a mouse, studies using inhaled anesthetics in mice typically occur with the animal placed in a gas-tight chamber. Inhaled anesthetics are delivered using a calibrated vaporizer in an enriched stream of oxygen. In addition to the customary factors (Eger & Saidman, 2005), the kinetics of inhaled anesthetic uptake will also be dependent upon the fresh gas flow rate and the volume of the anesthetizing chamber. Relative to lower flow rates, increasing the fresh gas flow rate to the anesthetized chamber to achieve one volume turnover per minute will decrease the time required for the mouse to equilibrate with a stepwise increasing or decreasing series of inhaled anesthetic doses, but may also induce relative hypothermia in mice that can confound the interpretation of behavioral responses. Without an active warming, mice breathing inhaled anesthetic gases in fresh gas flow rates as little as 100 mL/min can approach room temperature in less than an hour (Sun et al., 2006). Anesthetic-induced hypothermia is often accompanied by other detrimental physiologic effects. Therefore, it is critical to measure and maintain normothermia in mice. Strategies to avoid hypothermia in mice include partially submerging the anesthetizing chamber in a heated water bath or using a servocontrolled radiant heat source such as a lamp or warmed floor that shuts off when continuous measures of core body temperature reach a normothermic range (Caro, Hankenson, & Marx, 2013).

2.2 Injectable Anesthetics

Due to the ease of administration, the most common route to deliver anesthetic drugs in mice is intraperitoneal (IP). Route of anesthetic delivery can dramatically affect experimental results. It is well known that lesions that impair cardiac output will pharmacokinetically slow anesthetic induction of intravenous agents while speeding induction with inhaled anesthetics (Eger & Saidman, 2005; Upton, Ludbrook, Grant, & Martinez, 1999). Similarly, pharmacokinetic factors can affect the apparent anesthetic phenotype of mice dosed with the same anesthetic drug if given by distinct routes. As shown in Fig. 1, intraperitoneal dosing of 400 μg/kg of dexmedetomidine was given to age-matched 4- to 6-month-old mice genetically engineered with a homozygous deletion of the *dopamine β hydroxylase* gene ($D\beta h^{-/-}$) and to their heterozygous siblings that inherit one wild-type copy of the *dbh* gene and consequently retain normal levels of catecholamines (Thomas, Marck, Palmiter, & Matsumoto, 1998). Using the loss of righting reflex as a behavioral measure of hypnosis (described in Section 3.1), neither the latency to loss of righting nor the duration of loss of righting differed between $D\beta h^{-/-}$ and $D\beta h^{+/-}$ with IP dexmedetomidine. This is in sharp contrast with the known hypersensitivity of $D\beta h^{-/-}$ to a variety of anesthetic drugs, including dexmedetomidine when administered intravenously (Hu et al., 2012). Based upon the cardiac compensations in the $D\beta h^{-/-}$ knockouts (Liles et al., 2007; Swoap, Weinshenker, Palmiter, & Garber, 2004), it is believed that the slower uptake of IP dexmedetomidine from the peritoneum masks the hypersensitivity phenotype of reduced latency to loss of righting and prolonged duration of loss of righting (Fig. 1) that is apparent with intravenous dosing. More recent studies of propofol given to wild-type C57BL/6J mice further highlight that even among intravenously dosed anesthetics, delivery via an indwelling tail vein IV catheter does not produce identical results to the same dose of propofol

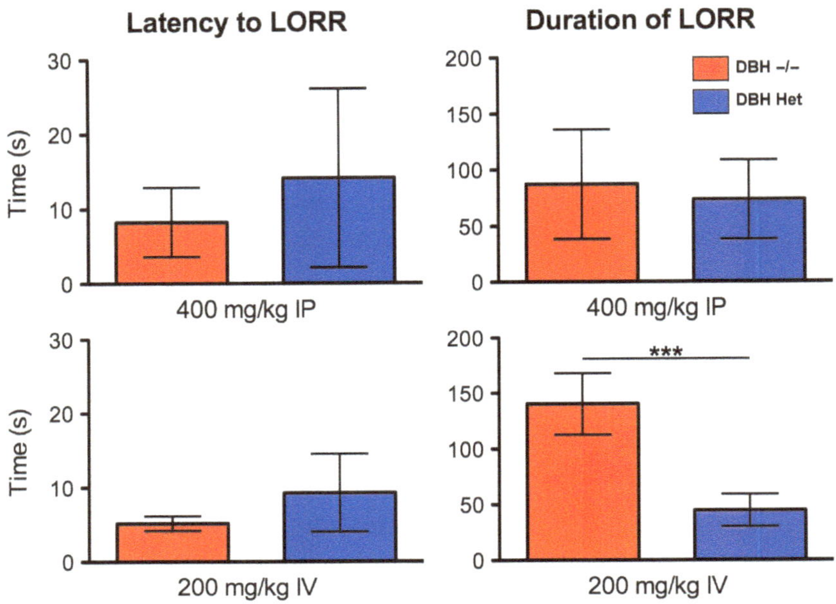

Fig. 1 Route of drug administration plays an integral role in the apparent anesthetic efficacy. When delivered intraperitoneally (IP), dexmedetomidine produces a statistically indistinguishable (mean ± SD) latency to loss of righting (*top left*) and duration of loss of righting (*top right*) in *dopamine β hydroxylase* (*DβH*) knockout mice that are devoid of both epinephrine and norepinephrine ($n = 10$) and *DβH* heterozygous control mice that retain a single wild-type allele and consequently have normal levels of adrenergic ligands ($n = 11$). However, when dexmedetomidine is given intravenously to *DβH* knockout mice and heterozygous control mice, knockouts show a significant dose-dependent increase both the latency and duration of loss of righting. Shown in the *lower panel* are data from an equipotent intravenous dose of 200 mcg/kg ($n = 5$–6/group) in which knockouts show a significantly longer duration of loss of righting. ***$P < 0.001$. *From Hu, F. Y., Hanna, G. M., Han, W., Mardini, F., Thomas, S. A., Wyner, A. J., et al. (2012). Hypnotic hypersensitivity to volatile anesthetics and dexmedetomidine in dopamine beta-hydroxylase knockout mice.* Anesthesiology, 117*(5), 1006–1017.*

given intravenously through a central line (Alex Proekt, University of Pennsylvania, unpublished observations). Hence, when designing and reporting results in mice careful thought must be given to the route of anesthetic drug delivery.

3. DETERMINATION OF HYPNOTIC SENSITIVITY IN MICE

The term anesthesia was first coined by Oliver Wendell Holmes in the 1846 to describe the state that we recognize today as encompassing amnesia, analgesia, motor atonia, and most critically, unconsciousness, which is typically referred to as hypnosis in animals. For a more comprehensive overview of these endpoints, we refer the interested reader to several excellent reviews (Campagna, Miller, & Forman, 2003; Franks, 2008; Hemmings et al., 2005; Mashour, Orser, & Avidan, 2011; Sonner et al., 2003). Each of the anesthetic endpoints, as well as the undesirable autonomic side effects that accompany the anesthetic state, is experimentally approachable in the laboratory mouse.

3.1 Behavioral Testing

The loss and subsequent return of the righting reflex is an established behavioral means to assess behavioral responsiveness in rodents (Franks, 2006). With exposure to hypnotic doses of anesthetic drugs, mice that have been tipped onto their backs in a position of vulnerable dorsal recumbency, lose the protective response to flip back onto all four paws. The loss of the righting reflex is convenient and reliable means to indicate the accompanying loss of consciousness. As rolling a mouse onto its back is sufficient to awaken a sleeping animal but the constant vestibular bombarding inputs of being upside down are not sufficient to antagonize an anesthetic state, investigators regularly use this nonnoxious behavioral paradigm to assess whether an individual animal retains its perceptive awareness of its surroundings or whether the drug delivery has produced a state of hypnosis. By determining the fraction of mice that lose (regain) their righting reflex as a function of increasing (decreasing) anesthetic concentrations, population response curves are easily generated by plotting the fraction of the population that remains responsive vs the log of the anesthetic concentration. A sigmoidal fit yields the population's ED_{50} concentrations for hypnosis as well as its hill slope. The Hill slope term describes the uniformity of the population's state transition from awake to anesthetized. Shallower Hill slopes dictate increased variability, which could be due either to environmental or genetic factors (Sonner, 2002). In the extreme case of an infinite Hill slope, variability would be zero and each member would undergo the state transition at an identical anesthetic dose.

A second means of determining when a mouse has entered a behavioral state of hypnosis can be determined with the tape test. This test has been applied to mice as a sensitive measure of sensorimotor deficits in mice (Bouet et al., 2009). However, as shown in Fig. 2, the test can also be utilized as a sensitive measure of anesthetic hypnosis in mice. Before anesthetic

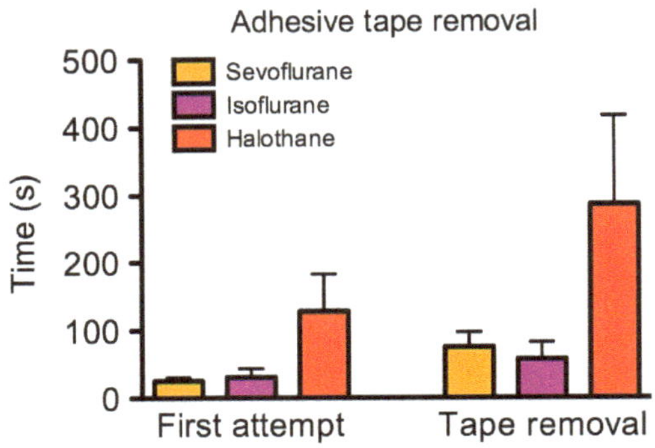

Fig. 2 Adhesive tape removal test. The adhesive tape removal test can be used as a complementary marker for the hypnotic component of anesthesia. When fully trained mice will remove a tape sticker from their nose in under 10 s but show a significant anesthetic-induced impairment in time (mean ± SEM) to their first attempt to removing the tape and in the time to remove the tape following exposure to hypnotic doses of sevoflurane, isoflurane, and halothane ($n = 9$/group).

challenge, mice are trained on the tape test by transiently restraining each animal and placing a round, 0.5-cm diameter sticker on its nose. Each mouse is then transferred into an empty cage. The time to contact and to remove the tape is recorded. These corresponding latencies are measures of intact sensation and motor coordination (Bouet et al., 2009). Mice are considered trained when they remove the tape in fewer than 10 s. Typically, training takes less than four trials. Following a 2-h exposure to 2.3% sevoflurane, 1.2% isoflurane, or 1.0% halothane as previously described (Sun et al., 2006), and after regaining their righting reflexes, mice, respectively, took an additional 25.7 ± 5.1, 30.9 ± 13.1, and 128 ± 54.5 s to first contact the tape and 74.4 ± 21.6, 57.7 ± 25.1, and 286.4 ± 130.6 s to successfully remove the tape. The tape test can be a useful adjunct to testing the righting reflex or a behavioral alternative for assessing the loss of perceptive awareness that defines the hypnotic state in electroencephalography (EEG)-tethered mice in whom testing the righting reflex is not possible.

A third behavioral marker for determining anesthetic-induced loss of consciousness is the forced walking test (Hwang, Kim, Shin, & Choi, 2010). It has the chief advantage of not having to handle the mouse during testing. In this paradigm, mice are placed on a treadmill that moves at 5 cm/s to mimic a spontaneous, comfortable walking pace. Prior to testing, mice are habituated by being suspended in a nonweight bearing position above the treadmill for 15 min and then allowed to walk for another 5 min at 5 cm/s. Through indwelling lines, anesthetic drugs can be remotely infused to induce general anesthesia. Another major advantage of this test is that the precise instant in which gait fails can be continuously measured and determined as the loss of movement. This test consequently yields significant temporal advantages over intermittent assessment of the righting reflex which is impractical to simultaneously assess more frequently than every 30 s in large cohorts of mice. Another advantage of the forced walking test is that it can be simultaneously acquired with continuous recording of EEG in mice (Hwang et al., 2010).

3.2 Motor-Independent Measures

One theoretical problem with assessing hypnotic sensitivity of anesthetics with all behavioral criteria is that these depend upon a motor output. Therefore, a drug that intrinsically affects muscle tone as anesthetics do, could in principle elicit hypnosis, impair movement, or both. For this reason, acquisition of a motor-independent measure of hypnosis is desirable. Experiments in mice typically turn to EEG, electrocorticography, or local field potentials acquired from chronically implanted depth electrodes.

As is true for humans, processed EEG measures can be useful for phenotyping the state of anesthetic hypnosis. Issues in EEG signal acquisition and processing that are true in humans (Proekt, 2018) also apply to murine EEG. One common manipulation often applied to the EEG is a Fourier transformation (Hjorth, 1991). This converts the EEG from a voltage as a function of time into the sum of a series of sine waves of varying amplitudes and frequencies. Such spectral decomposition of the EEG with the Fourier technique has been applied both in humans (Bruhn et al., 2003) and mice (Hwang et al., 2010) to characterize anesthetic-induced waveform changes that reflect varying depth of hypnosis. However, a major advantage of conducting studies of anesthetic hypnosis in mice is the ability to test hypotheses about the contribution of putative molecular targets and neuronal of anesthetic action. For example, positive allosteric modulation of $GABA_A$ receptor signaling by the general anesthetic, etomidate, occurs preferentially at β2- and β3-containing $GABA_A$ receptors and is entirely dependent on the presence of an asparagine at the 265th residue. Reynolds and colleagues used knock-in mice with an N265S point mutation that renders β2 subunits insensitive to etomidate and then dosed β2 N265S knock-in or control animals with intravenous etomidate and assessed both behavioral and EEG measures of sensitivity (Reynolds et al., 2003). They demonstrated that hypnotic actions of etomidate as measured by righting reflex and EEG indices of deep anesthesia, characterized by the appearance of burst suppression, did not differ as a function of genotype. However, etomidate-induced sedation (discussed in Section 4) does indeed depend upon β2-containing $GABA_A$ receptors. In an elegant and complementary study, Jurd and colleagues used β3 N265M $GABA_A$ receptor knock-in mice to demonstrate that these latter $GABA_A$ receptors that are also insensitive to etomidate, do indeed serve as a molecular target that contributes to etomidate-induced hypnosis and immobility (Jurd et al., 2003).

Additional insights into the hypnotic actions of anesthetics have also arisen from studying mice. While it has been commonly assumed that entry into and exit from states of general anesthesia are mirror opposite processes, results obtained from mice with a genetic lesion that ablates all orexin/hypocretin neurons and causes narcolepsy have demonstrated that afflicted animals have a specific deficit exiting states of isoflurane and sevoflurane anesthesia that is not explained by pharmacokinetic factors (Kelz et al., 2008). In behavioral tests of righting, mice that inherited the gene causing narcolepsy exhibited significantly slower return of righting. As shown in Fig. 3, spectral analysis of EEG recorded before, during, and after isoflurane exposure from narcoleptic mice and their wild-type sibling controls reveal similar loss of EEG power with induction of isoflurane, but consistent with

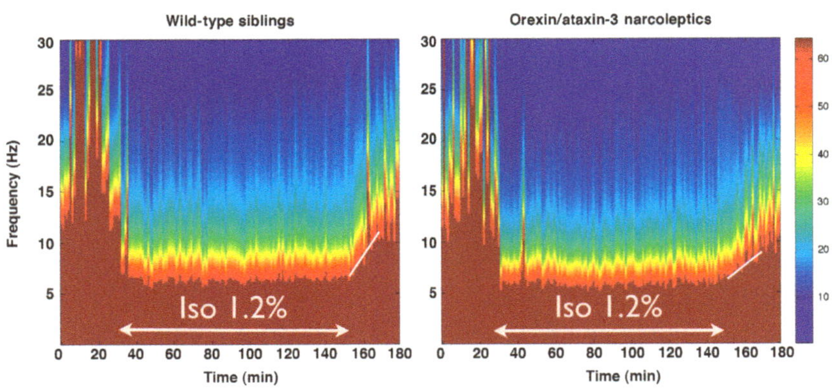

Fig. 3 Spectral analysis of murine EEG shows delayed emergence in narcoleptic orexin/ataxin-3 mice. Compared to their wild-type siblings ($n=4$, *left*), narcoleptic orexin/ataxin-3 ($n=4$, *right*) shows similar baseline spectral power in the 30 min prior to a 2-h 1.2% isoflurane exposure, a similar decrease in higher frequency power during the isoflurane anesthetic, but a delayed return of power in the 30 min after isoflurane is discontinued. Note the shallower slope of the *white line* in the narcoleptic mice to the *right*.

delayed return of righting, narcoleptic orexin–ataxin-3 mice exhibit a slower return of EEG power, confirming a motor-independent delayed emergence.

4. ASSESSING SEDATION IN MICE AS A MEASURE OF ANESTHETIC SENSITIVITY

Sedative effects of anesthetic drugs are also easily modeled in mice. Most commonly, behavioral measures to evaluate the sedative effects of anesthetics depend upon a loss of motor coordination, such as with a reduced ability of mice to remain on a rotating bar in the "rotorod" test (Lakhlani et al., 1997) or with reduced locomotor activity of mice (Drexler, Antkowiak, Engin, & Rudolph, 2011). One could also envision the application of postural sway with a force plate (Zumwalt, Hamrick, & Schmitt, 2006) as another sensitive means of measuring the sedative effects of anesthetic drugs. Given that the molecular targets underlying sedation are dissociable from those underlying hypnosis (Jurd et al., 2003; Reynolds et al., 2003), molecular genetic studies in intact mice have provided valuable insights. More recently, sophisticated hijacking of neuronal activity using viral transformation of discrete populations in mouse brain has pointed to neuronal targets that differentially impact anesthetic sedation from hypnosis (Zhang et al., 2015). In the following sections, we detail affordable strategies to measure anesthetic sedation in mice using video tracking, accelerometry, and electromyographic tone.

4.1 Videographic Analysis of Anesthetic-Induced Reductions in Movement

Video recording of the mouse model is generally considered to be the paragon methodology toward assessment of immobility (Baker, 2011). Video tracking is commonly used as a direct measure of movement and inactivity, where minor image processing allows for automated frame-by-frame detection of a mouse about its home cage or experimental enclosure. Video recording for social and behavioral analysis is extensively used in pharmacological, transgenic, and sleep research studies to track movement and can be applied to classify differences within a variety of behaviors, such as rearing or grooming (Noldus, Spink, & Tegelenbosch, 2001; Ou-Yang, Tsai, Yen, & Lin, 2011). Moreover, video analysis has been used to distinguish wakefulness from NREM and REM sleep (McShane et al., 2012, 2010;

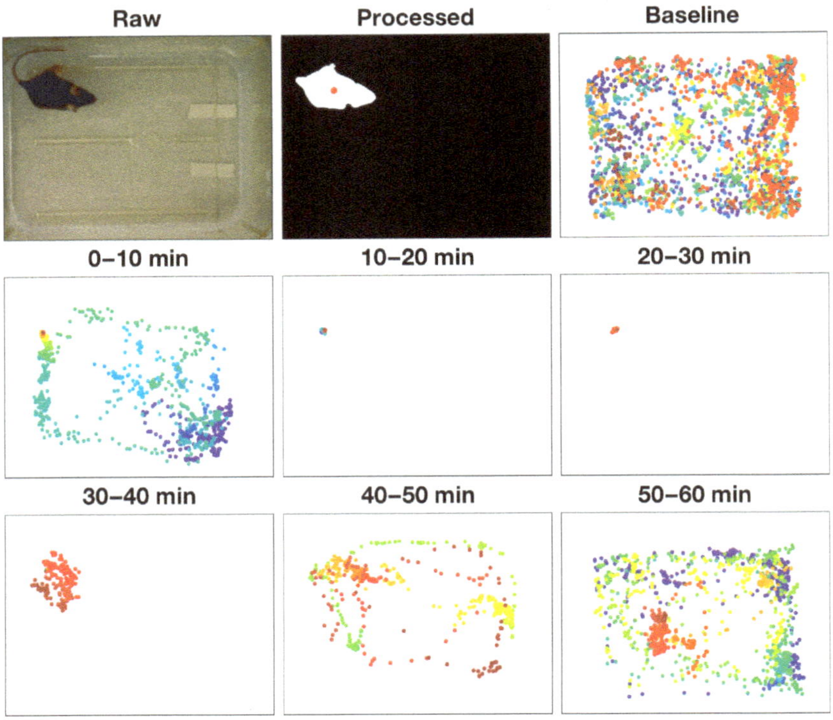

Fig. 4 Video tracking of movement reveals duration of anesthetic sedation. Using the mouse tracking methodology described, a representative experimental paradigm is seen in this figure for a 100 mg/kg IP injection of propofol for a single mouse. Video was captured at 5 fps and was digitally processed to highlight the body of the mouse, with the central point of the mouse being featured in the processed image. Ten minutes epochs are shown for baseline activity, where no drug intervention is present, and for each subsequent 10 min epoch following the 100 mg/kg IP propofol injection up to 60 min after injection. Every point represents a single frame from the video, where time within each epoch is represented by color (0 min = *Blue* → 10 min = *red*).

Pack et al., 2007). Hence, it should not be surprising that video tracking is also useful in the context of studying anesthetics (Fig. 4). Our generalized protocol for creating video tracking of individual mice requires the Image Processing Toolbox within MATLAB® (Mathworks®, Natick, MA) and is described in brief later. Optimal mouse detection occurs with uniform lighting and high color contrast between the mouse and its environment, such as black mouse against a light-colored box. With a recording area that is uniformly lit and a camera positioned in a top-down view, ensure that the entire recording area is in full view. Measure the recording area's length and width, which will establish the physical features of the frames for subsequent distance tracking. Let the mouse habituate to the novel environment and then proceed to record a period of movement activity. Five frames per second afford sufficient temporal resolution while minimizing the movie's file sizes. Load the video file into MATLAB® using a videoreader function and create a structure array in which each individual frame color information will be included in the array. Each frames' physical height and width can be converted into pixel units. The number of frames in the video based upon the movie's duration and frame rate must also be specified to preallocate sufficient memory for the MATLAB® structure array. A "for" loop can be used to load the video data frame-by-frame into the video structure.

Create a temporary variable that contains color information as a matrix for the first frame. Upload this first frame into the Color Thresholder App found within MATLAB® and generate a color threshold mask using. Highlight a portion of the mouse to choose the color of interest. Force a binary output for color thresholding, and export the Color Thresholding parameters using the internal export function feature. Next, produce a region of interest mask using the roipoly function, marking areas of the image that are to be included in the analysis. The entire cage can be selected to be the region of interest. Note that the creation of the color and region of interest masks only need to be created once for a given experiment, as lighting conditions and cage position should remain unchanged between frames.

Filter the first frame with a two-dimensional Gaussian smoothing kernel. This creates a blurred image that helps in the removal of noise in the image acquisition. Proceed by applying both the Color Thresholding mask and the region of interest mask. The resulting image should be binary.

Continue image processing by dilating the image using a morphological structuring element to overlap any coarse features between the foreground and background in the processed image, and to remove fine structures such as the tail and whiskers, since the aim is to determine the central point of the mouse body. Use of imdilate and strel functions is recommended. Subsequently, fill image holes using the imfill function. At this point, image processing is complete. The resulting binary image should convey an isolated mouse.

Determine the centroid point of the mouse using the regionprops function. It is recommended to use the "area" property name when determining the centroid point, due to the irregular shape of the mouse body. The resulting point is the pixel at which the center of the mouse is located as determined using the described image analysis. Repeat the procedure for each frame of the loaded video using a "for" loop. A preallocated matrix can be created where the x–y pixel coordinates for the mouse centroid point can be stored for each frame. Once centroid points are established for each frame, periods of immobility can be defined by total distance traveled or by occupation of a single area over an extended period of time (McShane et al., 2012; Nelson et al., 2002; Noldus et al., 2001; Ou-Yang et al., 2011; Pack et al., 2007). This centroid tracking approach can be applied to both inhalational and intravenous anesthetic challenges.

4.2 Accelerometric Analysis of Anesthetic-Induced Reductions in Movement

The use of accelerometer to directly measure mouse movement can be applied to assess both the sedating and immobilizing properties of anesthetic drugs. Darting is a phenotyping assay routinely used in murine open-field testing, which highlights periods of increased acceleration as a defining feature of mouse behavior (Kafkafi et al., 2003). Due to the physical constraints of experimental space and to the inherent unpredictability in mouse movement, it is improbable that a mouse will be able to sustain a constant velocity for a prolonged period of time (Proekt, Banavar, Maritan, & Pfaff, 2012). Changes in acceleration are a result of fluctuations in velocity, therefore, acceleration can be used as a direct measure of movement. One way to use accelerometer data to determine immobility in anesthetic challenges is by determining prolonged periods where net acceleration is zero. Since it is not possible to establish tilt or orientation on a 3-axis accelerometer without use of a gyroscope, periods of net acceleration equivalent to 1 g (unit of gravity) are interpreted as periods of immobility. During instances of quiescence, the square root of the sum of squares with respect to the 3-axis accelerometer data resembles 1 g. A simple subtraction of 1 g from this cumulative sensor data resulting from periods of motionlessness would yield values near zero. Tri-axis accelerometer voltage data are converted to acceleration in units of gravity by first obtaining the zero-g bias voltage and voltage sensitivity per unit-g. Static accelerometer calibration methods are described in brief later for a RHD-2000 amplifier chip with accelerometer (Intan Technologies, Los Angeles, CA) (Godfrey, Conway, Meagher, & O'Laighin, 2008; Steele et al., 2000; Yang & Hsu, 2010).

Position the accelerometer such that one axis is parallel to vertical and acquire approximately 10 s of acceleration data from the accelerometer in this orientation. Use modeling clay to secure the accelerometer in a fixed position. Reverse the orientation of the accelerometer and record a small amount of acceleration data again. The voltages seen from each orientation correlate to ±1 g. Open-source Open-Ephys hardware (http://www.open-ephys.org) can be used for data acquisition. Take the average of voltage of the two orientations to obtain the zero-g voltage bias, and calculate half of the voltage difference between the two orientations to attain the voltage sensitivity of the vertically oriented axis. Repeat this procedure for the remaining two axes. This concludes calibration of the accelerometer chip. With actual accelerometery data, subtract zero-g voltage bias with respect to each axis for all data points, and then divide by the respective voltage sensitivity per axis to convert raw voltages into units of gravity. Perform a square root sum of squares with the converted axes data per time point to obtain cumulative sensor acceleration. Subtract 1 g from this time series data vector to readjust the data such that the value of zero coincides with no movement. Smooth the time series data with a nonoverlapping 100 ms root mean square (RMS) moving window. Fig. 5 shows prototypic results of experiments in mice in which an RHD-2000 amplifier chip is used in conjunction with

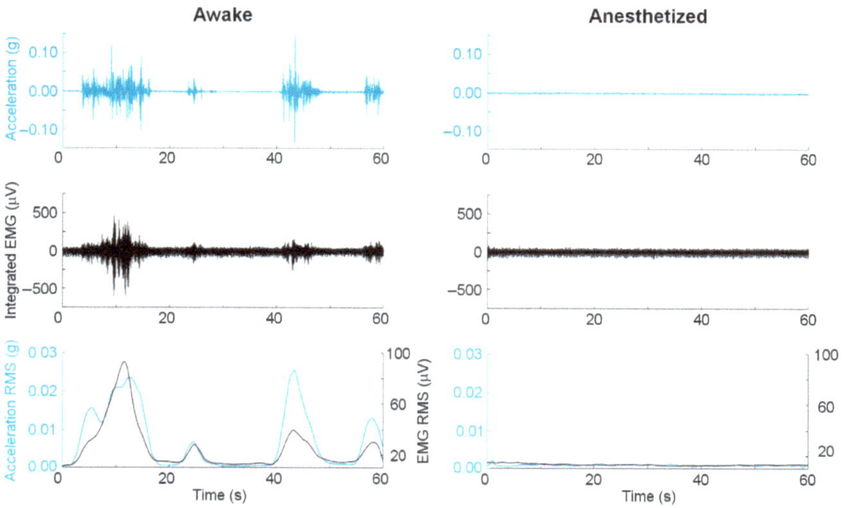

Fig. 5 Analysis of acceleration and EMG are useful modalities to determine duration of anesthetic sedation. One-minute raw acceleration (*Row 1*) and electromyographic traces (*Row 2*) are shown before (*Left*) and after (*Right*) a 1.2% isoflurane anesthetic exposure for an individual mouse. RMS moving average filtering of the raw data shows high correlation (>0.7) between accelometry and electromyography for movement detection (*Row 3*).

mice that have been surgically implanted with EEG and electromyography (EMG) electrodes (Wasilczuk, Proekt, Kelz, & McKinstry-Wu, 2016).

4.3 Electromyographic Analysis of Anesthetic-Induced Reductions in Motor Tone

Another methodology that can be used to determine anesthetic reductions in movement is through EMG analysis (Fig. 5). This technique delivers information about the electrical activity of skeletal muscles. Typically for EMG recordings, intramuscular electrodes are used to create muscle contraction; however, passive recording of EMG reveals spontaneous muscle tone and contractility characteristics that are correlated to muscle activity (De Luca, Gilmore, Kuznetsov, & Roy, 2010). Biopotentials are generated when electrical activity is recorded between two electrodes. Strategic placement of EMG electrodes in cervical muscles provides insight into murine polysomnography, discerning wakefulness, nonREM, and REM sleep (Fisher et al., 2012; Pack et al., 2007). Similarly, EMG can provide a gradient for anesthetic phenotyping, ranging from the high motor tone seen in active waking state to the profound drop in tone that commonly characterizes the unconscious state. A succinct protocol for the analysis of EMG activity can be found below, highlighting two common methods of signal processing, rectified and integrated EMG (Oishi et al., 2016). The aim of these methodologies is to combat the ephemeral oscillations of the raw EMG signal, where smoothing would provide uninterpretable data.

Implant intramuscular electrodes into nuchal muscle as previously described, and allow 2 weeks of recovery time before the animal is used in experimental studies (Oishi et al., 2016; Wasilczuk et al., 2016). For the integrated EMG analysis method, apply a high-pass filter on the EMG biopotential with cutoff frequency of 20 Hz. This removes any DC offset in the signal and movement artifact that may be present (van Boxtel, 2001). If the raw data sampling frequency is greater than 1000 Hz, additional low-pass filtering should be performed, as the optimal low-pass cutoff frequency for muscle activity is approximately 500 Hz (van Boxtel, 2001). After filtering the data, apply a nonoverlapping 100 ms window RMS moving average filter to the filtered data to provide data informing muscle tone and activation about the head of the animal. If the rectified analysis approach is desired, simply take the mean-removed absolute value of the raw signal. Then apply a similar 100 ms window moving average filter onto the rectified data allowing for indirect evidence of muscle tone and activity. As shown in Fig. 5, nuchal EMG biopotentials and accelerometry data are highly correlated signals.

5. CONCLUSIONS

In this chapter, we highlight the uses of mice as a model organism for assessing the effects of anesthetic drugs with the ultimate goal of determining both molecular and neuronal sites of sedation and hypnosis. Given the explosive technical advances in recent years (Bucan & Abel, 2002; Rossant & McMahon, 1999; Singh, Schimenti, & Bolcun-Filas, 2015; Susaki, Ukai, & Ueda, 2017) and new found ability to manipulate the activity of discrete populations of neurons with optogenetic and chemogenetic approaches (Ressler, 2017), it is increasingly likely that anesthetic mechanisms research in mice will continue to provide novel insights into drug action.

REFERENCES

Baker, M. (2011). Animal models: Inside the minds of mice and men. *Nature*, *475*(7354), 123–128.

Behlke, L. M., Foster, R. A., Liu, J., Benke, D., Benham, R. S., Nathanson, A. J., et al. (2016). A pharmacogenetic 'restriction-of-function' approach reveals evidence for anxiolytic-like actions mediated by alpha5-containing GABAA receptors in mice. *Neuropsychopharmacology*, *41*(10), 2492–2501.

Bouet, V., Boulouard, M., Toutain, J., Divoux, D., Bernaudin, M., Schumann-Bard, P., et al. (2009). The adhesive removal test: A sensitive method to assess sensorimotor deficits in mice. *Nature Protocols*, *4*(10), 1560–1564.

Bruhn, J., Bouillon, T. W., Radulescu, L., Hoeft, A., Bertaccini, E., & Shafer, S. L. (2003). Correlation of approximate entropy, bispectral index, and spectral edge frequency 95 (SEF95) with clinical signs of "anesthetic depth" during coadministration of propofol and remifentanil. *Anesthesiology*, *98*(3), 621–627.

Bucan, M., & Abel, T. (2002). The mouse: Genetics meets behaviour. *Nature Reviews Genetics*, *3*(2), 114–123.

Campagna, J. A., Miller, K. W., & Forman, S. A. (2003). Mechanisms of actions of inhaled anesthetics. *The New England Journal of Medicine*, *348*(21), 2110–2124.

Caro, A. C., Hankenson, F. C., & Marx, J. O. (2013). Comparison of thermoregulatory devices used during anesthesia of C57BL/6 mice and correlations between body temperature and physiologic parameters. *Journal of the American Association for Laboratory Animal Science*, *52*(5), 577–583.

Chen, X., Shu, S., & Bayliss, D. A. (2009). HCN1 channel subunits are a molecular substrate for hypnotic actions of ketamine. *The Journal of Neuroscience*, *29*(3), 600–609.

De Luca, C. J., Gilmore, L. D., Kuznetsov, M., & Roy, S. H. (2010). Filtering the surface EMG signal: Movement artifact and baseline noise contamination. *Journal of Biomechanics*, *43*(8), 1573–1579.

Deussing, J. M. (2013). Targeted mutagenesis tools for modelling psychiatric disorders. *Cell and Tissue Research*, *354*(1), 9–25.

Drexler, B., Antkowiak, B., Engin, E., & Rudolph, U. (2011). Identification and characterization of anesthetic targets by mouse molecular genetics approaches. *Canadian Journal of Anaesthesia*, *58*(2), 178–190.

Eger, E. I., 2nd, & Saidman, L. J. (2005). Illustrations of inhaled anesthetic uptake, including intertissue diffusion to and from fat. *Anesthesia and Analgesia, 100*(4), 1020–1033.
Fisher, S. P., Godinho, S. I., Pothecary, C. A., Hankins, M. W., Foster, R. G., & Peirson, S. N. (2012). Rapid assessment of sleep-wake behavior in mice. *Journal of Biological Rhythms, 27*(1), 48–58.
Franks, N. P. (2006). Molecular targets underlying general anaesthesia. *British Journal of Pharmacology, 147*(Suppl. 1), S72–S81.
Franks, N. P. (2008). General anaesthesia: From molecular targets to neuronal pathways of sleep and arousal. *Nature Reviews Neuroscience, 9*(5), 370–386.
Godfrey, A., Conway, R., Meagher, D., & O'Laighin, G. (2008). Direct measurement of human movement by accelerometry. *Medical Engineering & Physics, 30*(10), 1364–1386.
Hemmings, H. C., Jr., Akabas, M. H., Goldstein, P. A., Trudell, J. R., Orser, B. A., & Harrison, N. L. (2005). Emerging molecular mechanisms of general anesthetic action. *Trends in Pharmacological Sciences, 26*(10), 503–510.
Heurteaux, C., Guy, N., Laigle, C., Blondeau, N., Duprat, F., Mazzuca, M., et al. (2004). TREK-1, a K+ channel involved in neuroprotection and general anesthesia. *The EMBO Journal, 23*(13), 2684–2695.
Hjorth, B. (1991). Principles for transformation of scalp EEG from potential field into source distribution. *Journal of Clinical Neurophysiology, 8*(4), 391–396.
Hu, F. Y., Hanna, G. M., Han, W., Mardini, F., Thomas, S. A., Wyner, A. J., et al. (2012). Hypnotic hypersensitivity to volatile anesthetics and dexmedetomidine in dopamine beta-hydroxylase knockout mice. *Anesthesiology, 117*(5), 1006–1017.
Hwang, E., Kim, S., Shin, H. S., & Choi, J. H. (2010). The forced walking test: A novel test for pinpointing the anesthetic-induced transition in consciousness in mouse. *Journal of Neuroscience Methods, 188*(1), 14–23.
Jurd, R., Arras, M., Lambert, S., Drexler, B., Siegwart, R., Crestani, F., et al. (2003). General anesthetic actions in vivo strongly attenuated by a point mutation in the GABA(A) receptor beta3 subunit. *The FASEB Journal, 17*(2), 250–252.
Kafkafi, N., Pagis, M., Lipkind, D., Mayo, C. L., Bemjamini, Y., Golani, I., et al. (2003). Darting behavior: A quantitative movement pattern designed for discrimination and replicability in mouse locomotor behavior. *Behavioural Brain Research, 142*(1–2), 193–205.
Kelz, M. B., Sun, Y., Chen, J., Cheng Meng, Q., Moore, J. T., Veasey, S. C., et al. (2008). An essential role for orexins in emergence from general anesthesia. *Proceedings of the National Academy of Sciences of the United States of America, 105*(4), 1309–1314.
Lakhlani, P. P., MacMillan, L. B., Guo, T. Z., McCool, B. A., Lovinger, D. M., Maze, M., et al. (1997). Substitution of a mutant alpha2a-adrenergic receptor via "hit and run" gene targeting reveals the role of this subtype in sedative, analgesic, and anesthetic-sparing responses in vivo. *Proceedings of the National Academy of Sciences of the United States of America, 94*(18), 9950–9955.
Liles, J. T., Baber, S. R., Deng, W., Porter, J. R., Corll, C., Murthy, S. N., et al. (2007). Pressor responses to ephedrine are not impaired in dopamine beta-hydroxylase knockout mice. *British Journal of Pharmacology, 150*(1), 29–36.
Mashour, G. A., Orser, B. A., & Avidan, M. S. (2011). Intraoperative awareness: From neurobiology to clinical practice. *Anesthesiology, 114*(5), 1218–1233.
McShane, B. B., Galante, R. J., Biber, M., Jensen, S. T., Wyner, A. J., & Pack, A. I. (2012). Assessing REM sleep in mice using video data. *Sleep, 35*(3), 433–442.
McShane, B. B., Galante, R. J., Jensen, S. T., Naidoo, N., Pack, A. I., & Wyner, A. (2010). Characterization of the bout durations of sleep and wakefulness. *Journal of Neuroscience Methods, 193*(2), 321–333.
Nelson, L. E., Guo, T. Z., Lu, J., Saper, C. B., Franks, N. P., & Maze, M. (2002). The sedative component of anesthesia is mediated by GABA(A) receptors in an endogenous sleep pathway. *Nature Neuroscience, 5*(10), 979–984.

Noldus, L. P., Spink, A. J., & Tegelenbosch, R. A. (2001). Ethovision: A versatile video tracking system for automation of behavioral experiments. *Behavior Research Methods, Instruments, & Computers, 33*(3), 398–414.

Oishi, Y., Takata, Y., Taguchi, Y., Kohtoh, S., Urade, Y., & Lazarus, M. (2016). Polygraphic recording procedure for measuring sleep in mice. *Journal of Visualized Experiments*(107), e53678.

Ou-Yang, T. H., Tsai, M. L., Yen, C. T., & Lin, T. T. (2011). An infrared range camera-based approach for three-dimensional locomotion tracking and pose reconstruction in a rodent. *Journal of Neuroscience Methods, 201*(1), 116–123.

Pack, A. I., Galante, R. J., Maislin, G., Cater, J., Metaxas, D., Lu, S., et al. (2007). Novel method for high-throughput phenotyping of sleep in mice. *Physiological Genomics, 28*(2), 232–238.

Peters, L. L., Robledo, R. F., Bult, C. J., Churchill, G. A., Paigen, B. J., & Svenson, K. L. (2007). The mouse as a model for human biology: A resource guide for complex trait analysis. *Nature Reviews Genetics, 8*(1), 58–69.

Proekt, A. (2018). Brief introduction to electroencephalography. *Methods in Enzymology, 603* (in press).

Proekt, A., Banavar, J. R., Maritan, A., & Pfaff, D. W. (2012). Scale invariance in the dynamics of spontaneous behavior. *Proceedings of the National Academy of Sciences of the United States of America, 109*(26), 10564–10569.

Ressler, K. J. (2017). Neurocircuits to behavior: The new revolution. *Harvard Review of Psychiatry, 25*(2), 47–49.

Reynolds, D. S., Rosahl, T. W., Cirone, J., O'Meara, G. F., Haythornthwaite, A., Newman, R. J., et al. (2003). Sedation and anesthesia mediated by distinct GABA(A) receptor isoforms. *The Journal of Neuroscience, 23*(24), 8608–8617.

Rossant, J., & McMahon, A. (1999). "Cre"-ating mouse mutants-a meeting review on conditional mouse genetics. *Genes & Development, 13*(2), 142–145.

Rudolph, U., & Antkowiak, B. (2004). Molecular and neuronal substrates for general anaesthetics. *Nature Reviews Neuroscience, 5*(9), 709–720.

Singh, P., Schimenti, J. C., & Bolcun-Filas, E. (2015). A mouse geneticist's practical guide to CRISPR applications. *Genetics, 199*(1), 1–15.

Sonner, J. M. (2002). Issues in the design and interpretation of minimum alveolar anesthetic concentration (MAC) studies. *Anesthesia and Analgesia, 95*(3), 609–614.

Sonner, J. M., Antognini, J. F., Dutton, R. C., Flood, P., Gray, A. T., Harris, R. A., et al. (2003). Inhaled anesthetics and immobility: Mechanisms, mysteries, and minimum alveolar anesthetic concentration. *Anesthesia and Analgesia, 97*(3), 718–740.

Steele, B. G., Holt, L., Belza, B., Ferris, S., Lakshminaryan, S., & Buchner, D. M. (2000). Quantitating physical activity in COPD using a triaxial accelerometer. *Chest, 117*(5), 1359–1367.

Sun, Y., Chen, J., Pruckmayr, G., Baumgardner, J. E., Eckmann, D. M., Eckenhoff, R. G., et al. (2006). High throughput modular chambers for rapid evaluation of anesthetic sensitivity. *BMC Anesthesiology, 6*(1), 13.

Susaki, E. A., Ukai, H., & Ueda, H. R. (2017). Next-generation mammalian genetics toward organism-level systems biology. *NPJ Systems Biology and Applications, 3*, 15.

Swoap, S. J., Weinshenker, D., Palmiter, R. D., & Garber, G. (2004). Dbh(−/−) mice are hypotensive, have altered circadian rhythms, and have abnormal responses to dieting and stress. *American Journal of Physiology. Regulatory, Integrative and Comparative Physiology, 286*(1), R108–R113.

Thomas, S. A., Marck, B. T., Palmiter, R. D., & Matsumoto, A. M. (1998). Restoration of norepinephrine and reversal of phenotypes in mice lacking dopamine beta-hydroxylase. *Journal of Neurochemistry, 70*(6), 2468–2476.

Upton, R. N., Ludbrook, G. L., Grant, C., & Martinez, A. M. (1999). Cardiac output is a determinant of the initial concentrations of propofol after short-infusion administration. *Anesthesia and Analgesia, 89*(3), 545–552.

van Boxtel, A. (2001). Optimal signal bandwidth for the recording of surface EMG activity of facial, jaw, oral, and neck muscles. *Psychophysiology*, *38*(1), 22–34.

Wasilczuk, A. Z., Proekt, A., Kelz, M. B., & McKinstry-Wu, A. R. (2016). High-density electroencephalographic acquisition in a rodent model using low-cost and open-source resources. *Journal of Visualized Experiments*, (117), e54908.

Yang, C. C., & Hsu, Y. L. (2010). A review of accelerometry-based wearable motion detectors for physical activity monitoring. *Sensors (Basel)*, *10*(8), 7772–7788.

Zhang, Z., Ferretti, V., Guntan, I., Moro, A., Steinberg, E. A., Ye, Z., et al. (2015). Neuronal ensembles sufficient for recovery sleep and the sedative actions of alpha2 adrenergic agonists. *Nature Neuroscience*, *18*(4), 553–561.

Zumwalt, A. C., Hamrick, M., & Schmitt, D. (2006). Force plate for measuring the ground reaction forces in small animal locomotion. *Journal of Biomechanics*, *39*(15), 2877–2881.

SECTION III

Photolabeling

CHAPTER TWELVE

Identification of General Anesthetic Target Protein-Binding Sites by Photoaffinity Labeling and Mass Spectrometry

Kellie A. Woll*, William P. Dailey†, Roderic G. Eckenhoff*,[1]
*University of Pennsylvania Perelman School of Medicine, Philadelphia, PA, United States
†University of Pennsylvania School of Arts and Sciences, Philadelphia, PA, United States
[1]Corresponding author: e-mail address: roderic.eckenhoff@uphs.upenn.edu

Contents

Abstract

General anesthetics are unique in that they represent a diverse range of chemical structures. Therefore, it is not surprising that the desired and undesired molecular targets, and binding sites therein, are as equally diverse and unique. Photoaffinity labeling has proven to be a valuable strategy for the identification of anesthetic molecular targets, as well as binding sites within those targets. In combination with the advances in mass spectrometry-based proteomics, along with the ability to comprehensively map posttranslational modifications, the method is likely to undergo continued improvement. Here, we provide the fundamentals for the design and development of an anesthetic photolabel. We also outline a protocol for the identification of photolabeled residues by mass spectrometry. The major steps include the photolabeling experiment, sample preparation, high-resolution mass spectrometry,

Methods in Enzymology, Volume 602
ISSN 0076-6879
https://doi.org/10.1016/bs.mie.2018.01.009

and data analysis. The protocol can be used as a foundation for further optimization for the specific protein of interest and conditions of an experiment. The use of photoaffinity labeling adds an advantageous alternative and/or complementary approach to increase understanding of anesthetic molecular mechanisms.

1. INTRODUCTION

General anesthetics are among the most chemically diverse class of drugs used within clinics ranging from steroidal agents like allopregnanolone to ethers like isoflurane. Therefore, it comes as little surprise that these agents affect the function of multiple protein targets (Weiser, Woll, Dailey, & Eckenhoff, 2014) given that there is no distinct pharmacophore, and thus no unitary molecular target. Photoaffinity labeling is a vital tool for the direct identification of protein-binding sites for small molecules within proteins (Das, 2011; Dubinsky, Krom, & Meijler, 2012). In turn, characterization of these binding sites has provided a significant amount of insight into the molecular mechanisms of general anesthetics (Weiser et al., 2014; Woll, Dailey, Brannigan, & Eckenhoff, 2016). Photoaffinity labeling necessarily starts with the integration of an otherwise inert chemical group into the anesthetic chemical structure that, under specific irradiation conditions, generally near-ultraviolet (near-UV), undergoes activation to generate a highly reactive intermediate (Woll, Dailey, et al., 2016). This transient photogenerated chemical species immediately covalently inserts into a nearby molecule, resulting in an irreversible chemical modification.

This chemical modification can be detected by different methods, primarily Edman degradation or mass spectrometry (MS). Edman degradation (radiomicrosequencing) involves the protease digestion of a protein photolabeled with a photolabel that contains a radioactive isotope. The resulting peptides are purified and the N-terminal residues are labeled and are chemically cleaved via the sequential Edman reaction. These labeled amino acids are separated into two pools: one for the identification of the residue and the second is subjected to scintillation counting to determine the presence of the radioactive isotope. The correlation between the identified residue and a spike in radioactivity indicates a photolabeled residue (Woll, Dailey, et al., 2016). The identification of a photolabeled residue by MS, the method described within this chapter, relies on a change in the mass of a peptide as a consequence of the covalent insertion (Woll, Dailey, et al., 2016). As a result, using MS does not require a radioactive photolabel, a distinct advantage given the high cost of isotopic labeling. In this chapter we address the basic photolabel design and photolabeling experiment using MS for the identification of a photolabeled site.

2. PHOTOLABEL DESIGN AND SYNTHESIS

Design of an effective photolabel analog requires that it mimics the properties of the original drug as closely as possible, and once the analog is photoactivated, it should react indiscriminately with nearby residues in or near the binding site (Sumranjit & Chung, 2013). Introduction of the diazirine group into anesthetic analogs is a proven effective strategy that generally satisfies these requirements (Woll, Dailey, et al., 2016). The diazirine group is a relatively nonpolar and nonbasic functional group and, depending on neighboring atoms and bonds, is kinetically stable under physiological conditions (Schmitz, 1987). Light activation generates a singlet carbene by photoextrusion of dinitrogen. The resulting carbene is an extremely reactive intermediate and typically reacts by insertion into various nearby C—H, O—H, and N—H bonds. However, if there are C—H bonds adjacent to the carbene center, these preferentially react by intramolecular C—H insertion resulting in the generation of an alkene (Fig. 1). This undesired reaction limits the design of potential analogs to those substituted with groups resistant to migration such as fluorine or aromatic rings (Brahms & Dailey, 1996).

If the anesthetic of interest contains an aromatic group, then substitution of one of the ring hydrogen atoms by the trifluoromethyldiazirine group can be an effective strategy for preparation of a useful photolabel analog. Synthesis of the aryltrifluoromethyldiazirine can be accomplished from the corresponding trifluoromethylketone. Depending on the other

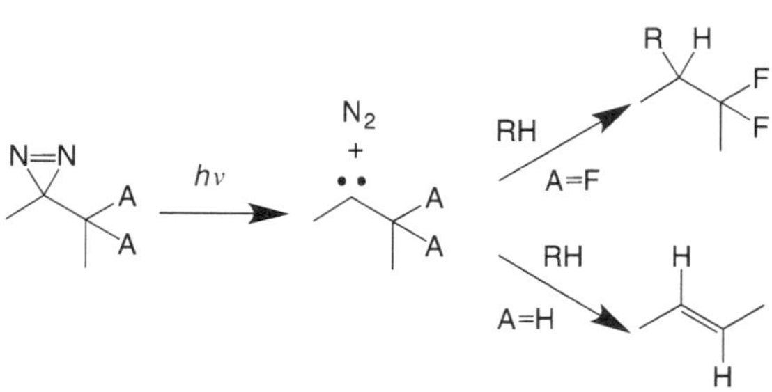

Fig. 1 Photochemical reaction of a diazirine produces a carbene intermediate which can undergo intramolecular rearrangement to an alkene if there are hydrogen atoms (A=H) adjacent to the carbene center. The desired intramolecular reaction can occur if there are groups resistant to insertion such as fluorine (A=F) or aromatic rings.

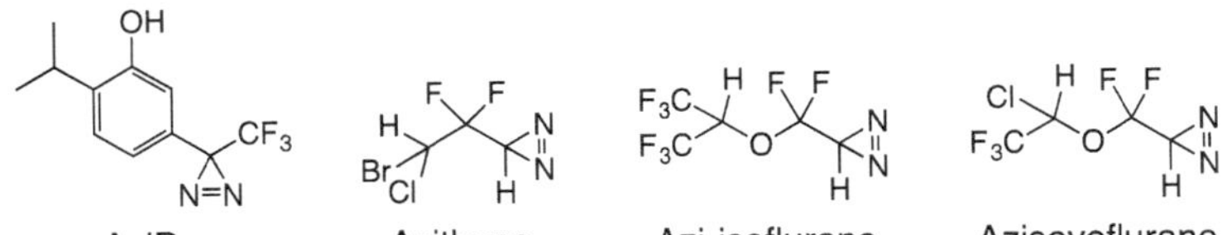

Fig. 2 Some successful anesthetic photolabels include AziPm (2-isopropyl-5-(3-(trifluoromethyl)-3*H*-diazirin-3-yl)phenol), azithane (3-(2-bromo-2-chloro-1,1-difluoroethyl)-3*H*-diazirine), azi-isoflurane (3-(1-chloro-2,2,2-trifluoroethoxy)difluoromethyl)-3*H*-diazirine), and azisevoflurane (3-(difluoro(1,1,1,3,3,3-hexafluoropropan-2-yl)oxy)methyl)-3*H*-diazirine).

functional groups present in the final molecule, protecting group strategies may be required. The synthesis of AziPm is an example of such a synthesis (Hall et al., 2010) (Fig. 2). For anesthetic molecules containing only alkane functional groups, introduction of fluorine atoms adjacent to a terminal diazirine group has proven to be an effective strategy (Das, 2011). These difluoro-substituted alkyl diazirines can be prepared from the corresponding difluoro-substituted aldehydes. Unfortunately, there is no general synthetic method for preparation of difluoro-aldehydes so each molecule requires a unique synthetic approach. The syntheses of azithane (Eckenhoff, Knoll, Greenblatt, & Dailey, 2002), azi-isoflurane (Eckenhoff et al., 2010), and azisevoflurane (Woll et al., 2017) are further examples of such photolabels (Fig. 2).

3. PHYSICOCHEMICAL CHARACTERIZATION AND BIOLOGICAL VALIDATION

After synthesis and purification (an often arduous task) the generated photolabel requires basic physicochemical characterization. This includes determination of the extinction coefficient as well as aqueous solubility. Both of these measurements can be taken from the distinct absorbance peak provided by the photoreactive group using a UV–vis spectrophotometer. Another important physicochemical characteristic to address is the photoactivation half-life, which can be determined by the decrease in the photoreactive absorbance peak over the course of light exposure. The half-life of the photoreactive group will be dependent on the photolabel structure as well as the intensity of the light source (Bayley, 1983; Das, 2011); therefore when determining the half-life, it is valuable to maintain the same light source or have a similar intensity as the source that will be used during future photolabeling experiments.

To allow for photoreactivity, modifications to the molecular structure are necessary, potentially changing its activity from the parent clinical anesthetic. As a result, multiple avenues for validation are required in order to verify that the photolabel acts as a suitable analog for the clinical anesthetic (Woll, Dailey, et al., 2016). Many of these methods are included within this volume, including protein binding (e.g., isothermal titration calorimetry), functional assays (e.g., electrophysiology), and in vivo assays (e.g., tadpole immobility assay), we therefore refer the reader to these detailed protocols. It is important to determine whether the photolabel analog shares similar functional activity as the clinical anesthetic within the specific protein of interest to indicate a common binding site, and therefore a shared mechanism of action.

Prior to determining the physicochemical characteristics and biological validation, the researcher needs to be aware of a couple potential pitfalls. First, the synthesized reagent needs to be pure. Contaminants, particularly those containing the photoreactive group, will likely skew the physicochemical characterization. Within biological evaluation, the photolabel occasionally demonstrates a slight change in potency compared to the clinical anesthetic (Bayley, 1983; Eckenhoff et al., 2010; Hall et al., 2010; Woll et al., 2017; Zhong, Rusch, & Forman, 2008), and it is important to consider these changes for concentrations used in downstream photolabeling experiments. Second, the stability of the reagent needs to be considered, both under ambient conditions and the conditions present in the functional assay. For example, some validation assays are fluorescence based, and these might be hindered by the photoreactive group absorbance (inner-filter effect). Therefore, it is important that the functional assay(s) are evaluated for prevailing chemical and enzymatic conditions that might be affected by the photolabel (or clinical anesthetic).

4. PHOTOAFFINITY LABELING

The amount of time for irradiation is dependent on the specific photolabel as well as the intensity of the light source (Bayley, 1983). For the optimal light exposure our lab uses quartz cuvettes with a 200- to 2500-nm spectral range and a 1-mm pathlength. Glass cuvettes and/or the composition of the sample (i.e., detergent, salts, etc.) may also have an absorbance at the activating wavelength therefore reducing photolabeling efficiency. By using quartz and a lower pathlength these limitations are reduced. Finally, using a light source that has a broad light output (e.g., Rayonet RPR-3500 lamp) the researcher should be aware that shorter UV wavelengths could over time deteriorate the protein sample. As such, we find it valuable to insert our quartz cuvette within an additional outside chamber that has a low-pass filter (295 nm low-pass glass filter) to prevent protein degradation. The below protocol includes a simple method for photolabeling a protein of interest.

a. Add the appropriate volume of concentrated photolabel stock with or without additional ligands to the protein(s) of interest. The photolabel should be sufficiently solubilized within the sample. If using a solubilizing agent, or vehicle, to introduce the photolabel into the protein sample, insure that the least amount of vehicle is used and it is compatible with the protein of interest/photolabel.

b. Incubate sample in the dark for appropriate timescale for equilibration (generally ~5–10 min). Introduce the sample into the quartz cuvette and then the secondary chamber with the low-pass glass filter.

c. Irradiate sample with the appropriate wavelength and time for sufficient photolabel photoactivation. Depending on duration, the sample may be cooled to prevent light-induced overheating.

d. Remove protein sample and continue the isolation and digestion protocol appropriate for the protein of interest. This can include delipidation, concentration, and/or precipitation; the following is a procedure for protein precipitation with acetone.

  i. Aliquot acetone, four volumes times that of protein samples, into acetone compatible tubes and cool tubes to −20°C.

  ii. Add photolabeled protein sample to the chilled acetone, vortex vigorously, and store at −20°C for 3 h to overnight.

  iii. Centrifuge at 4°C for 20 min at 14,000–16,000 × *g*.

  iv. Carefully decant supernatant and gently wash pellet with 300 μL of chilled acetone.

  v. Air-dry protein pellet for ~10 min, being careful not to over-dry the pellet.

5. MASS SPECTROMETRIC ANALYSIS

Mass spectrometry (MS) is centered on the measurement of mass; more specifically, the instrument determines the mass-to-charge (m/z) ratio of molecules (peptides) that become charged on entry into the spectrometer. Within proteomics, MS provides the primary structure of proteins and the modifications therein. The advancements in instrumentation and bioinformatics over the past few decades have increased the availability and the user friendliness of this means of analysis. As such, MS has emerged as the method-of-choice for the identification of posttranslational modifications (PTMs), including photolabeled residues. As a rule, anesthetic photoaffinity labeling has been applied using a "bottom-up" MS platform, and therefore will be the focus of this chapter. It should be noted that other platforms including "top-down," the MS analysis of intact proteins, and "middle-down," the MS analysis of intact proteins and long (~30–80 amino acid residue), exist (Han, Aslanian, & Yates, 2008), and do have potential for further development and usefulness in specific applications within photoaffinity labeling studies. Bottom-up MS involves the digestion of photolabeled proteins into small peptides (~5–17 amino acid residue) that subsequently undergo separation by online high-performance liquid chromatography followed by tandem MS (Han et al., 2008). The peptides are fragmented by collision-induced dissociation (CID) or high-energy dissociation (HCD), both of which are generally available in modern commercial instruments. The spectra obtained from these fragments are then used to determine the initial peptide sequences and modifications with the application of database search engines and additional software.

The bottom-up platform is highly sensitive and determines the amino acid sequences and PTMs therein. However, one major drawback is that bottom-up MS cannot definitively determine the concurrence of photolabeling within an individual protein. There are, usually, three major steps for MS analysis of a photolabeled sample: sample preparation, ionization and detection of the sample, and, finally, data analysis. These steps are highly dependent on the characteristics of the protein to be photolabeled and will likely require gradual optimization. In this section we describe the general protocol(s) for each mentioned step that we apply for the MS analysis of a photolabeled sample as well as recommendations. A researcher can use these workflows as a starting point for further development.

5.1 Sample Preparation

The preparation of samples for MS analysis for the identification of photolabeled residues will be dependent on the composition of the sample (e.g., salts, lipids, detergents, etc.) as well as the protein of interest. Various alterations to the method can be applied including different and/or multiple proteases, digestion times, and/or conditions. In our laboratory we generally apply multiple replicates for deep sequence coverage using in-solution and/or in-gel with trypsin digest or trypsin and chymotrypsin sequential digests. Additionally we include a protease surfactant enhancer to optimize the digestion of hydrophobic proteins. The following are procedures we use that can serve as the foundation for further optimization.

5.1.1 In-Gel Digestion

a. Suspend dried protein pellet (Section 4) with 1% SDS, 1% Triton X-100 in 50 m*M* Tris–HCl (pH 8).

b. Prepare 5 × SDS-loading buffer containing 0.6 *M* dithiothreitol (DTT), add appropriate volume to the solubilized protein pellet, and vortex vigorously.

c. Incubate the samples in the dark for 45 min to 1 h at room temperature then run the samples, on the SDS-PAGE gel.

d. Once the run is complete, fix and stain the gel using Coomassie Blue G-250 stain solution composed of 10% methanol (MeOH) and 10% acetic acid (AcOH) (v/v%) containing 0.5% Coomassie Blue G-250 (g/v%) in water for 5–10 min. Destain the gel overnight using a destain solution composed of 10% MeOH and 10% AcOH (v/v%) in water.

e. Excise the protein band of interest. Cut the gel slices into ~1-mm^3 pieces and transfer to a 1.5-mL microcentrifuge tube. The following procedure can accommodate gel slices up to 2 × 6 × 1 mm^3. If bands are larger, divide slices as needed.

The following procedure is for the preparation of each excised protein gel band for MS.

f. *Destain*: Add 200 μL 50% methanol (MeOH)/50 m*M* ammonium bicarbonate, pH 8 (AB) and incubate for 1 min with intermittent vortexing and decant solution. Repeat this step until the majority of stain has been largely removed.

g. *Dehydrate*: Add 200 μL 50% acetonitrile (ACN) and 50 m*M* AB, and incubate for 5 min with intermittent vortexing and decant solution. Add 200 μL of 100% ACN and incubate for 30 s and decant solution then speed vac until dryness.

h. *Reduce*: Suspend gel band in 100 μL 25 m*M* DTT in 50 m*M* AB and equilibrate for 10 min at ambient temperature. Then incubate the sample for 30 min at 56°C. Decant the DTT solution.

i. *Alkylate*: Add 100 μL 55 m*M* iodoacetamide (IAA) in 50 m*M* AB and incubate at ambient temperature in dark for 45 min. Decant the IAA solution and wash the bands with 400 μL of water.

j. *Dehydrate*: Same as above.

k. Suspend the dried gel pieces in 30 μL AB containing 0.015% ProteaseMax™ Surfactant Enhancer (Promega) and mass spectrometry sequencing grade trypsin at a 100:1 to 20:1 (w:w; protein:protease) ratio. Incubate the sample at ambient temperature for 10 min.

l. Dilute sample with 50 m*M* AB containing 0.015% ProteaseMax™ Surfactant Enhancer to 100 μL.

m. Incubate at 37°C for 3–18 h.

n. Gently centrifuge (<1000 × *g*) to collect condensation for 30 s.

o. If only utilizing a trypsin digest continue step (p), if undergoing a sequential chymotrypsin digest dilute sample with 100 μL of 175 m*M* AB containing 0.02% ProteaseMax™ Surfactant Enhancer and mass spectrometry sequencing grade chymotrypsin at a 100:1 to 20:1 protein:protease (w:w) ratio. Incubate at 37°C for 3–18 h, then gently centrifuge (<1000 × *g*) to collect condensation for 30 s.

p. *Peptide extraction (1)*: Remove peptide digest solution and place in new 1.5-mL tube. Suspend gel pieces in 100 μL of 35%:5%:70% ACN:AcOH:water (v%:v%:v%) and sonicate for 20 min. Remove first peptide extraction solution.

q. *Peptide extraction (2)*: Suspend gel pieces in 100 μL of 75%:5%:20% ACN:AcOH:water (v%:v%:v%) and sonicate for 20 min. Remove second peptide extraction solution.

r. Pool the initial peptide digest with the peptide extraction solutions. Snap–freeze samples with dry ice or liquid nitrogen and speed vac to dryness or until volume is <5 μL. Resuspend peptide digest in 100 μL (or until any remaining peptide extraction solution is <3% (v/v%)) 0.5% AcOH (v/v%) and sonicate for 20 min.

5.1.2 In-Solution Digestion

a. Solubilize protein pellet using 20 μL AB containing 0.2% ProteaseMax™ surfactant enhancer. Once the protein pellet is largely solubilized, dilute the sample with 73 μL AB.

b. *Reduce*: Add 1 μL of freshly prepared 0.5 *M* DTT and incubate at 56°C for 30 min (final concentration 5.3 m*M* DTT). Then allow to equilibrate to ambient temperature.

c. *Alkylate*: Add 3 μL of freshly prepared 0.55 *M* IAA and incubate in the dark at ambient temperature for 45 min (final concentration 17 m*M* IAA).

d. Add 2 μL of 1% ProteaseMax™ Surfactant Enhancer (final concentration 0.02% (v/v%) and 1 μL of mass spectrometry sequencing grade trypsin solution containing sufficient amount of protease to reach a 100:1–20:1 protein:protease (w:w) ratio.

e. Incubate at 37°C for 3–18 h, then gently centrifuge (<1000 × *g*) to collect condensation for 30 s.

f. If only utilizing a trypsin digest continue C18 stage-tip preparation. If undergoing a sequential chymotrypsin digest dilute sample with 100 μL of 175 m*M* AB containing 0.02% ProteaseMax™ Surfactant Enhancer and mass spectrometry sequencing grade chymotrypsin at a 100:1–20:1 protein:protease (w:w) ratio. Incubate at 37°C for 3–18 h, then gently centrifuge (<1000 × *g*) to collect condensation for 30 s.

All our samples are desalted prior to LC-MS/MS using C18 stage-tips that are made in-house. Desalting is an important step in sample preparation as salts that ionize during electrospray suppress the signal from peptides. Additionally, salts may form ionic adducts on peptides that result in an inaccurate mass as well as a reduced signal intensity from the nonsalt adducted peptide. Below is a protocol for the in-house preparation of C18 stage-tips; however, there are also commercially available forms.

a. Acidify all peptide digests to a pH <2 using AcOH and pH strips. Centrifuge samples for 20 min at 16,000 × *g* to remove insoluble debris.

b. Cut the last 0.5 cm from of a P200 tip to make the opening larger. Use this modified pipette tip to punch out a piece of C18 material from a solid phase extraction disk. Push the miniature C18 disk, which should be in the enlarged opening of the modified pipette tip, into a new P200 tip using a fused silica capillary. Gently pack the disk at the bottom of the P200 tip. Pack two miniature C18 disks in the same P200 tip if the peptide contains >25 μg of peptide, ensuring that there is no space between disks.

c. Hold the stage-tip in place within a 1.5-mL microcentrifuge tube using a centrifuge adaptor.

d. Place tubes and tips in centrifuge with slow rotation (<900 × *g*); the following solvents used will normally pass through the stage-tip <2 min. Ensure that all the solvent passes through the resin before progressing to the next step.

e. Flush and activate the resin by adding 100 μL MeOH.

f. Equilibrate the stage-tip by adding 125 μL 0.5% AcOH (v/v%) by slow centrifugation.

g. Load 100 μL of sample into the stage-tip by slow centrifugation. Repeat this step if sample volume is >100 μL until the entire sample is loaded.

h. Wash the sample with 125 μL 0.5% AcOH (v/v%) by slow centrifugation. Perform a second wash with 100 μL 0.5% AcOH by slow centrifugation.
i. Transfer the centrifuge adaptor and stage-tip to a new 1.5-mL tube and elute the sample by loading 25 μL per disk of 75% ACN and 5% AcOH (v/v%) in water by slow centrifugation.
j. Snap–freeze with dry ice or liquid nitrogen and dry sample by speed vac. Note: This is a good stopping point where the sample can be stored at −80°C.
k. Resuspend the sample in 0.1% formic acid prior to MS analysis.

5.2 Liquid Chromatography-Tandem Mass Spectrometry (LC-MS/MS)

The LC-MS/MS analysis of the peptide digests can be done in a number of ways depending on the availability of the HPLC and the MS. We suggest using a reverse-phase nano-HPLC that is coupled to an ultrahigh resolution MS. The optimal rpHPLC gradient for the separation is more or less dependent on the hydrophobicity of the protein. For the following we generally apply nanoelectrospray ionization (nano-ESI) on a Hybrid LTQ-Orbitrap MS (Thermo Orbitrap Elite™) for the analysis of peptide digests. We acquire our data using the data-dependent acquisition (DDA) mode where a high-resolution full MS1 scan is acquired that is followed by a low-resolution MS2 acquisition of the most abundant peaks from the full MS1 scan. This is in contrast to the data-independent acquisition (DIA) where a high-resolution full MS1 scan is followed by series of high-resolution MS2 scans across a fixed MS1 scan m/z window. Ultimately DIA is not recommended for the identification of unknown peptides; therefore, DDA is traditionally used for most PTM identifications.

a. Fit the C18 HPLC column (inner diameter: 75 μm, length: 10–20 cm) purchased commercially (or packed in-house).
b. Load peptides into autosampler.
c. *Program rpHPLC gradient*: We have found that a 100-min elution with linear gradients of ACN in 0.1% formic acid in water (v/v%) starting from 2% to 40% ACN (85 min), then 40% to 85% (5 min), and finally a washout of 85% ACN (10 min) provides a sufficient separation of peptides for both soluble and transmembrane proteins.
d. *Program the MS method*: We apply DDA with a dynamic exclusion every 45 s. For every 3 s cycle, we collect one full MS1 scan using the Orbitrap

with a scan range of 350–1500 m/z, at resolution of 60 K, and a maximum injection time of 50 ms and an automatic gain control of 500,000. Using the ion trap, the MS2 scans are from the most intense parent ions from the MS1 scan; these are filtered with a 2–5 charge and an isolation window of 1.5 m/z in quadruple isolation mode. CID is used to fragment the ions with a collision energy of 35%. We apply the normal scan range mode and rapid ion trap scan rate for the ion trap detection and set the automatic gain control to 10,000 with a maximal injection time of 100 ms.

5.3 Data Analysis

The identification of the peptide, and potential modification(s) therein, can be done by subjecting the acquired peptide MS/MS against a protein sequence or protein sequence database using a database search engine. A custom database will need to be applied for proteins that incorporate additional affinity tags or unique sequences. For proteins derived from native tissue often Swiss-Prot protein sequence database can be used to obtain required sequences (can be obtained from http://www.uniprot.org). In either case it is essential that the sequences using within the search exactly represent those within the samples to prevent false identification. Different database search engines may also be applied; the main engines include SEQUEST and Mascot. Additionally, we find that the Proteome Discoverer (Thermo Scientific) application (or similar) further facilitates the analysis. Before initiating the search includes the protease(s) used in digestion and the appropriate mass toleration for the MS1 (~10–20 ppm) and MS2 (~0.02–0.8 Da) scans. The parameters appropriate for the MS used, and a 0.01 false discovery rate filter should be applied. Modifications added during the sample preparation should be included; for the procedures provided these include the static alkylation of cysteine (+57.0215 m/z; IAA alkylation) and the dynamic oxidation of methionine (+15.9949 m/z). Other modifications unique to the protein of interest should also be included as dynamic or static modifications accordingly. At this point the photolabel modification should be treated as a dynamic modification of any residue using the expected gain in m/z.

It is optimal for the high confidence coverage (e.g., ~80%) of a purified protein to help insure that most if not all photolabeled residues were identified. If sufficient coverage is not achieved, alternative sample preparation prior to MS may be required. These can include the use of different protease(s) and protease digestion times as well as the addition of offline separation such as strong cation exchange (Villen & Gygi, 2008). The peptide matches with photolabel modification(s) should be high confidence and further examined for sufficient dispersal of b and y fragment ions and using scores provided by Mascot (e.g., ion score) or SEQUEST (e.g., XCorr). Peptide matches that have a high number of dynamic modifications should be heavily scrutinized. Each additional dynamic modification included in a search increases the chance for random peptide matches. We also suggest conducting a search against a nonphotolabeled protein sample and applying the photolabel modification as a dynamic modification to determine if a peptide match has the potential for a false positive. Finally it has been shown that a photolabel modification may result in a peptide that cannot be successfully identified by MS analysis (He et al., 2016) resulting in a "gap" in protein coverage. There is a potential for this occurrence for photolabeled peptides as well; therefore, it can beneficial to run separate experiments with nonphotolabeled samples to compare protein coverage.

6. ADDITIONAL METHODOLOGIES AND CONSIDERATIONS

Recent developments in "click" chemistry coupled with photoaffinity labeling have allowed identification of specific biological targets (Lapinsky, 2012). General design requirements of click-active photolabel compounds involve the same analysis as that for a regular photolabel compound, but with the added requirement that a terminal alkyne group also be present in the molecule, and importantly, that the alkyne group is far enough away from the future-generated carbene so that the two groups will not react with each other. For anesthetics containing alkyl groups like propofol, substitution of a terminal alkyne group for one of the alkyl groups is an obvious approach. A successful example of this design is illustrated in Fig. 3 for a propofol analog, AziPm-click (Woll, Murlidaran, et al., 2016).

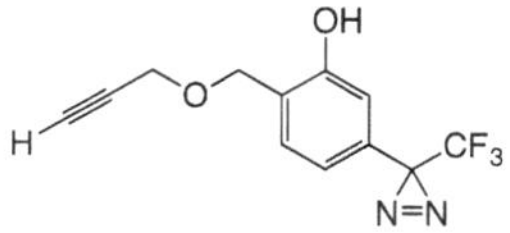

AziPm-click

Fig. 3 A successful click-enabled anesthetic photolabel is AziPm-click (2-((prop-2-yn-1-yloxy)methyl)-5-(3-(trifluoromethyl)-3*H*-diazirin-3-yl)phenol).

A vital area for determining the specific binding of a clinical anesthetic within a photolabeled cavity is the capability of the clinical anesthetic to compete with the photolabel for the binding site when under equilibrium. This competition confers protection from photolabeling when the sample is irradiated with near-UV light. Previously, the inhibition of photoincorporation has been observed by a decrease in photoincorporation of radiolabeled photolabels in proteins and/or amino acid residues by scintillation counting and/or Edman degradation (Chiara et al., 2012; Hamouda, Stewart, Husain, & Cohen, 2011; Jayakar et al., 2013; Weiser, Kelz, & Eckenhoff, 2013). Both of these methods involve known quantities of radioactivity and protein and therefore allows for quantification. A variety of different MS strategies can also be applied to allow for effective quantification of the clinical anesthetic protection from photolabeling. Among these include the use of heavy isotope labeling, including Stable Isotope Labeling of Amino acids in Cell culture and Tandem-Mass-Tag labeling, and/or label-free methods such as spectral counting (Li et al., 2012; Park et al., 2012). A consistent observation during these protection experiments is that magnitudes higher concentrations of the clinical anesthetic are required in order to effectively inhibit photoincorporation of the photolabel (Jayakar et al., 2014; Woll et al., 2017). This is likely a result of the reversible- (clinical anesthetic) vs irreversible (photolabel)-binding events that occur during the nonequilibrium system (i.e., during near-UV irradiation). These interactions events are further discussed in a previous review (Woll, Dailey, et al., 2016).

7. CONCLUSIONS

The widely used photoaffinity labeling strategy has provided crucial information of the binding protein targets and sites for general anesthetics. In this chapter, we provided the basics behind the successful design and development of an anesthetic photolabel as well as the workflow and protocol for the identification of photolabeled residues within a protein using MS. Photolabels have been developed for a range of anesthetics including both the intravenous and volatile agents (Woll, Dailey, et al., 2016) and their employment will provide further insight into anesthetic molecular mechanisms. In contrast to other methodologies, photoaffinity labeling allows the direct and unbiased identification of a protein-binding site. An important limitation is the required change in the chemical structure of the anesthetic in order to allow for photoreactivity. Through appropriate controls and safeguards, photoaffinity labeling represents an effective alternative and/or complement to the identification of anesthetic-binding sites and targets.

ACKNOWLEDGMENT

This work was supported by National Institutes of Health (GM055876, GM107117).

REFERENCES

Bayley, H. (1983). Photogenerated reagents in biochemistry and molecular biology. *Laboratory techniques in biochemistry and molecular biology, Vol. 12*, Elsevier. https://doi.org/10.1016/S0075-7535(08)70436-4.

Brahms, D. L. S., & Dailey, W. P. (1996). Fluorinated carbenes. *Chemical Reviews, 96*(5), 1585–1632.

Chiara, D. C., Dostalova, Z., Jayakar, S. S., Zhou, X. J., Miller, K. W., & Cohen, J. B. (2012). Mapping general anesthetic binding site(s) in human alpha 1 beta 3 gamma-aminobutyric acid type a receptors with H-3 TDBzl-etomidate, a photoreactive etomidate. *Biochemistry, 51*(4), 836–847. https://doi.org/10.1021/bi201772m.

Das, J. (2011). Aliphatic diazirines as photoaffinity probes for proteins: Recent developments. *Chemical Reviews, 111*(8), 4405–4417. https://doi.org/10.1021/cr1002722.

Dubinsky, L., Krom, B. P., & Meijler, M. M. (2012). Diazirine based photoaffinity labeling. *Bioorganic & Medicinal Chemistry, 20*(2), 554–570. https://doi.org/10.1016/j.bmc.2011.06.066.

Eckenhoff, R. G., Knoll, F. J., Greenblatt, E. P., & Dailey, W. P. (2002). Halogenated diazirines as photolabel mimics of the inhaled haloalkane anesthetics. *Journal of Medicinal Chemistry, 45*(9), 1879–1886. https://doi.org/10.1021/jm0104926.

Eckenhoff, R. G., Xi, J., Shimaoka, M., Bhattacharji, A., Covarrubias, M., & Dailey, W. P. (2010). Azi-isoflurane, a photolabel analog of the commonly used inhaled general anesthetic isoflurane. *ACS Chemical Neuroscience, 1*(2), 139–145. https://doi.org/10.1021/cn900014m.

Hall, M. A., Xi, J., Lor, C., Dai, S., Pearce, R., Dailey, W. P., et al. (2010). m-Azipropofol (AziPm) a photoactive analogue of the intravenous general anesthetic propofol. *Journal of Medicinal Chemistry, 53*(15), 5667–5675.

Hamouda, A. K., Stewart, D. S., Husain, S. S., & Cohen, J. B. (2011). Multiple transmembrane binding sites for p-trifluoromethyldiazirinyl-etomidate, a photoreactive torpedo nicotinic acetylcholine receptor allosteric inhibitor. *The Journal of Biological Chemistry, 286*(23), 20466–20477. https://doi.org/10.1074/jbc.M111.219071.

Han, X., Aslanian, A., & Yates, J. R., III. (2008). Mass spectrometry for proteomics. *Current Opinion in Chemical Biology, 12*(5), 483–490. https://doi.org/10.1016/j.cbpa.2008.07.024.

He, C., Sidoli, S., Warneford-Thomson, R., Tatomer, D. C., Wilusz, J. E., Garcia, B. A., et al. (2016). High-resolution mapping of RNA-binding regions in the nuclear proteome of embryonic stem cells. *Molecular Cell, 64*(2), 416–430. https://doi.org/10.1016/j.molcel.2016.09.034.

Jayakar, S. S., Zhou, X., Chiara, D. C., Dostalova, Z., Savechenkov, P. Y., Bruzik, K. S., et al. (2014). Multiple propofol binding sites in a gamma-aminobutyric acid type a receptor (GABAAR) identified using a photoreactive propofol analog. *The Journal of Biological Chemistry, 1*(289), 27456–27468.

Jayakar, S. S., Zhou, X. J., Dostalova, Z., Dailey, W. P., Miller, K. W., Eckenhoff, R. G., et al. (2013). Identifying the propofol binding site(S) in heterologously expressed human alpha1 beta3 gaba-A receptors using a photoreactive propofol analog. *Biophysical Journal, 104*(2), 637A.

Lapinsky, D. J. (2012). Tandem photoaffinity labeling-bioorthogonal conjugation in medicinal chemistry. *Bioorganic & Medicinal Chemistry, 20*(21), 6237–6247. https://doi.org/10.1016/j.bmc.2012.09.010.

Li, Z., Adams, R. M., Chourey, K., Hurst, G. B., Hettich, R. L., & Pan, C. (2012). Systematic comparison of label-free, metabolic labeling, and isobaric chemical labeling for quantitative proteomics on LTQ Orbitrap Velos. *Journal of Proteome Research*, *11*(3), 1582–1590.

Park, S.-S., Wu, W. W., Zhou, Y., Shen, R.-F., Martin, B., & Maudsley, S. (2012). Effective correction of experimental errors in quantitative proteomics using stable isotope labeling by amino acids in cell culture (SILAC). *Journal of Proteomics*, *75*(12), 3720–3732. https://doi.org/10.1016/j.jprot.2012.04.035.

Schmitz, E. (1987). Synthesis and reactions of diazirines. *Chemistry of diazirines: Vol. 1* (1st ed., pp. 57–98). Boca Raton, FL: CRC Press.

Sumranjit, J., & Chung, S. J. (2013). Recent advances in target characterization and identification by photoaffinity probes. *Molecules (Basel, Switzerland)*, *18*(9), 10425–10451. https://doi.org/10.3390/molecules180910425.

Villen, J., & Gygi, S. P. (2008). The SCX/IMAC enrichment approach for global phosphorylation analysis by mass spectrometry. *Nature Protocols*, *3*(10), 1630–1638.

Weiser, B. P., Kelz, M. B., & Eckenhoff, R. G. (2013). In vivo activation of azipropofol prolongs anesthesia and reveals synaptic targets. *The Journal of Biological Chemistry*, *288*(2), 1279–1285.

Weiser, B. P., Woll, K. A., Dailey, W. P., & Eckenhoff, R. G. (2014). Mechanisms revealed through general anesthetic photolabeling. *Current Anesthesiology Reports*, *4*(1), 57–66.

Woll, K. A., Dailey, W. P., Brannigan, G., & Eckenhoff, R. G. (2016). Shedding light on anesthetic mechanisms: Application of photoaffinity ligands. *Anesthesia and Analgesia*, *123*(5), 1253–1262. https://doi.org/10.1213/ANE.0000000000001365.

Woll, K. A., Murlidaran, S., Pinch, B. J., Henin, J., Wang, X., Salari, R., et al. (2016). A novel bifunctional alkylphenol anesthetic allows characterization of gamma-aminobutyric acid, type A (GABAA), receptor subunit binding selectivity in synaptosomes. *The Journal of Biological Chemistry*, *291*(39), 20473–20486. https://doi.org/10.1074/jbc.M116.736975.

Woll, K. A., Peng, W., Liang, Q., Zhi, L., Jacobs, J. A., Maciunas, L., et al. (2017). Photoaffinity ligand for the inhalational anesthetic sevoflurane allows mechanistic insight into potassium channel modulation. *ACS Chemical Biology*, *12*(5), 1353–1362. https://doi.org/10.1021/acschembio.7b00222.

Zhong, H. J., Rusch, D., & Forman, S. A. (2008). Photo-activated azi-etomidate, a general anesthetic photolabel, irreversibly enhances gating and desensitization of gamma-aminobutyric acid type A receptors. *Anesthesiology*, *108*(1), 103–112.

SECTION IV

Xenon

CHAPTER THIRTEEN

Xenon–Protein Interactions: Characterization by X-Ray Crystallography and Hyper-CEST NMR

Benjamin W. Roose, Serge D. Zemerov, Ivan J. Dmochowski[1]
University of Pennsylvania, Philadelphia, PA, United States
[1]Corresponding author: e-mail address: ivandmo@sas.upenn.edu

Contents

Abstract

The physiological activity of xenon has long been recognized, though the exact nature of its interactions with biomolecules remains poorly understood. Xe is an inert noble gas, but can act as a general anesthetic, most likely by binding internal hydrophobic cavities within proteins. Understanding Xe–protein interactions, therefore, can provide crucial insight regarding the mechanism of Xe anesthesia and potentially other general anesthetic agents. Historically, Xe–protein interactions have been studied primarily through X-ray crystallography and nuclear magnetic resonance (NMR). In this chapter, we first describe our methods for preparing Xe derivatives of protein crystals and identifying Xe-binding sites. Second, we detail our procedure for ^{129}Xe hyper-CEST NMR spectroscopy, a versatile NMR technique well suited for characterizing the weak, transient nature of Xe–protein interactions.

Methods in Enzymology, Volume 602
ISSN 0076-6879
https://doi.org/10.1016/bs.mie.2018.02.005

ABBREVIATIONS

Bla TEM-1 β-lactamase
CEST chemical exchange saturation transfer
$GABA_A$ γ-aminobutyric acid type-A
HSQC heteronuclear single-quantum correlation
LTP1 lipid transfer protein 1
MAC median alveolar concentration
MBP maltose binding protein
MRI magnetic resonance imaging
NMDA *N*-methyl-D-aspartate
NMR nuclear magnetic resonance
NOESY nuclear Overhauser effect spectroscopy
PBS phosphate-buffered saline
PDB protein data bank
SEOP spin exchange optical pumping

1. INTRODUCTION

Xenon, first discovered by British chemists Sir William Ramsay and Morris Travers in 1898, possesses a broad range of beneficial biological properties (Winkler, Thornton, Farjot, & Katz, 2016). Xe is clinically valuable, having been shown to act as a general anesthetic (Aziz, 2001), analgesic (Giacalone, Abramo, Giunta, & Forfori, 2013), and neuroprotective agent (David, Haelewyn, Risso, Colloc'h, & Abraini, 2010). In structural biology, Xe–lipid (Weinrich & Worcester, 2013; Yamamoto et al., 2012) and Xe–protein (Abraini et al., 2014; Colloc'h et al., 2007) interactions have been studied as a means of elucidating the mechanisms of general anesthesia, an enduring area of inquiry. Xe, unlike other general anesthetics such as nitrous oxide (N_2O), propofol, or the halocarbons (e.g., sevoflurane), is a monoatomic species. Being a noble gas, Xe has a filled outer electron shell and is generally unreactive, although its inertness can be overcome under extreme conditions to form inorganic (Bartlett, 1962) and organic (Frohn & Bardin, 2001) compounds. Xenon has more than 50 isotopes of which 9 are stable. ^{133}Xe, a byproduct of nuclear fission, serves as a marker for monitoring compliance with nuclear weapons testing bans (Bowyer et al., 2002). ^{135}Xe is a robust neutron absorber used to control fission in nuclear reactors. The nuclei of two stable isotopes, ^{129}Xe and ^{131}Xe, have nuclear spins of 1/2 and 3/2, respectively, and are suitable as nuclear magnetic resonance (NMR)/magnetic resonance imaging (MRI) probes.

Xe is the largest of the stable noble gases, with a van der Waals diameter of 4.3 Å. Its large atomic volume confers a large degree of polarizability—4.0 $Å^3$ for Xe, compared to just 0.2 $Å^3$ for He and 1.5 $Å^3$ for water (Rumble, 2017). This polarizability allows Xe to have interactions with solutes, and thus Xe can act as an inert solvent capable of dissolving biological and organic molecules (Rentzepis & Douglass, 1981). Conversely, xenon's polarizability makes it more soluble in biologically relevant fluids than the other noble gases (Table 1) (Weathersby & Homer, 1980).

The anesthetic properties of the inert gases were first noted in 1939 by US Navy physician Albert Behnke while investigating the cause of "drunkenness" presented by deep-sea divers. In 1946, Lawrence et al. predicted that Xe should also possess anesthetic properties based on its lipophilicity relative to the inert gases (Ar, N_2), whose narcotic effects had already been recognized (Lawrence, Loomis, Tobias, & Turpin, 1946). Lawrence based his reasoning on the Meyer–Overton hypothesis—the near-perfect linear correlation ($r=0.99$) between the partial pressure of inhaled anesthetic required to achieve general anesthesia (a.k.a. the median alveolar concentration or MAC) and its olive-oil gas partition coefficient (Campagna, Miller, & Forman, 2003; Meyer, 1899; Overton, 1901). Indeed, Lawrence and coworkers were successful in demonstrating the anesthetic properties of Xe in mice, and 5 years later Xe was first used as an anesthetic in a clinical setting (Cullen & Gross, 1951).

Unlike most general anesthetics, which heighten the activity of inhibitory $GABA_A$ (γ-aminobutyric acid type-A) receptors, xenon's anesthetic and analgesic properties arise from its noncompetitive inhibition of excitatory *N*-methyl-D-aspartate (NMDA) receptors (Franks, Dickinson, de Sousa, Hall, & Lieb, 1998). There is renewed interest in using Xe as an anesthetic, largely due to its nontoxicity, circulatory stability, favorable perfusion rates, and potency (Xe has a MAC of around 71%) (Arch & Harper, 2011;

Table 1 Solubility of Noble Gases (mL/mL) at 37°C

Gas	Water	Plasma	Blood
He	0.010	0.009	0.008
Ne	0.011	N/A	0.009
Ar	0.030	0.028	0.030
Kr	0.050	0.051	0.060
Xe	0.083	0.094	0.146

Marx, Schmidt, Schirmer, & Reinelt, 2000). Xe was approved for use as an anesthetic in Russia in 2000, Germany in 2005, and ultimately all of Europe in 2007 (Esencan et al., 2013). In 2007, Air Liquide introduced LENOXe™, Europe's first Xe-based anesthetic. In addition to its efficacy and safety, Xe is also an environmentally friendly alternative to halocarbon anesthetics such as sevoflurane, desflurane, and isoflurane. The primary disadvantage of xenon is its high price, roughly $10/L, which stems from the relative scarcity of atmospheric Xe (only 87 ppb in dry air) (Neice & Zornow, 2016). Currently, Xe is prepared on an industrial scale as a byproduct of liquid oxygen and nitrogen production by fractional distillation of liquefied air. Increased Xe gas production, coupled with improved Xe-recovery methods such as the Felix Dual™ closed-circuit respirator (Air Liquide Medical Systems, France), have the potential to lower the price of Xe enough to make it marketable as a clinical anesthetic (Dingley et al., 2001; Sanders, 2003; Stoppe et al., 2013).

Xe is amenable to mechanistic study because of its chemical inertness, structural simplicity (just a lone atom!), and because its interactions with proteins can be characterized by a variety of techniques such as X-ray crystallography, hyperpolarized ^{129}Xe NMR and other NMR methods, manometry, computational chemistry, and activity assays. Xenon is spherically symmetric with zero net charge, limiting its interactions to those arising from charge-induced, dipole-induced, and London (dispersion) forces. Xe is observed to occupy a wide range of sites in proteins, though typically it is found in small hydrophobic cavities, predominantly lined with short hydrophobic residues such as leucine, isoleucine, valine, and alanine (Prangé et al., 1998). These buried hydrophobic cavities are usually devoid of water (Wolfenden & Radzicka, 1994). Although the atomic packing density of proteins is comparable to the packing efficiency observed in small organic crystals (Richards, 1974), internal void spaces (a.k.a. cavities) are nearly always present in proteins larger than 100 residues (Hubbard, Gross, & Argos, 1994). Xenon interacts with proteins via noncovalent, low-energy van der Waals forces and is driven by the weak but favorable enthalpies of absorption (Ewing & Maestas, 1970; Tilton et al., 1986). Xe binding typically does not perturb protein structure, making it useful for overcoming the phase problem in X-ray crystallography (Mueller-Dieckmann et al., 2004; Quillin & Matthews, 2002, 2003; Schiltz, Fourme, & Prangé, 2003; Schiltz et al., 1997; Vitali, Robbins, Almo, & Tilton, 1991). Xe binding, however, can affect protein function by restricting the number of available conformational states. This

Xe-mediated restriction is observed in the loss of rotational degrees of freedom of protein-bound waters (Schoenborn, Featherstone, Vogelhut, & Süsskind, 1964). More recently, it has been proposed that xenon can reduce the biological function of proteins through the formation of nanobubbles at hydrophobic sites on or near the protein surface (Zhang et al., 2017).

Myoglobin is the protein whose interactions with Xe have been most extensively characterized, and it serves as a useful case study for understanding protein–Xe interactions in general. Sperm whale myoglobin (Schoenborn, 1965) and horse hemoglobin (Schoenborn, 1965) were the first proteins whose structures were solved in the presence of Xe. At a Xe pressure of 2.5 bar, a single Xe atom was observed to bind inside a preexisting void space in the interior of myoglobin. It was later shown that at 30 bar Xe pressure, four additional Xe atoms bind to myoglobin, though at low (<0.3) occupancies (Fig. 1) (Abraini et al., 2014). Computations based on crystallography coordinates determined favorable enthalpic contributions of 0.6–4.2 kcal/mol for Xe binding to myoglobin (Tilton et al., 1986). The ^{129}Xe NMR spectrum of myoglobin in solution shows a single broadened peak near the Xe@solvent frequency, indicating that the population of myoglobin-bound xenon is small relative to the pool of dissolved

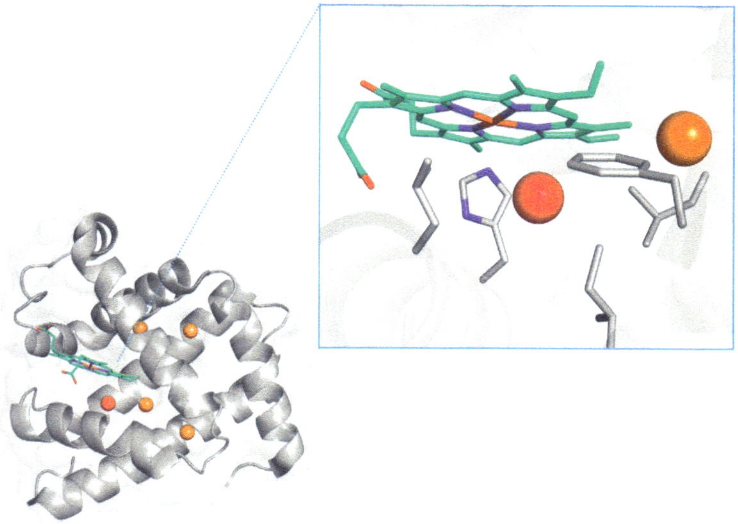

Fig. 1 Crystal structure of sperm whale myoglobin under 30 bar Xe pressure (PDB ID 4NXA). The Xe proximal to the heme has occupancy of 0.8, whereas the other four Xe have occupancies less than 0.3. (*Inset*) Close-up of the major Xe-binding site of myoglobin.

xenon, and that exchange between the two Xe pools is fast on the NMR timescale (Miller et al., 1981). Increasing the concentration of myoglobin in solution moves the chemical shift of the lone NMR peak either upfield or downfield depending on the oxidation state and spin state of the heme Fe (Rubin, Lee, Ruiz, Pines, & Wemmer, 2002; Tilton & Kuntz, 1982). In silico analysis of Xe–myoglobin interactions suggest that Xe exchange is controlled by complex gating processes in which structural fluctuations within the protein control the rate of Xe access to the buried interior binding cavity (Teeter, 2004; Tilton, Singh, Kuntz, & Kollman, 1988). Finally, although Xe binding minimally perturbs myoglobin structure, Xe (as well as nitrogen and cyclopropane) binding to myoglobin alters its function, specifically its binding affinity for carbon monoxide (Settle, 1973). It has been proposed that such functional inhibition may result from xenon reorienting the proximal histidine (H93), xenon disrupting the side chains that control ligand entry, and/or xenon rigidifying the protein (Tilton, Kuntz, & Petsko, 1984).

2. X-RAY CRYSTALLOGRAPHY

2.1 Background

One prevailing hypothesis is that general anesthetics act on proteins by heightening the activity of inhibitory receptors and/or tempering the activity of excitatory receptors (Campagna et al., 2003; Colloc'h et al., 2007). The inhaled general anesthetics bind weakly and promiscuously to occupy internal cavities contained within numerous protein targets. These cavities, rather than merely being packing defects, confer a degree of flexibility necessary for protein function, thus molecules bound within these sites rigidify the protein and subsequently limit their activity (Hubbard & Argos, 1996; Hubbard et al., 1994; Vallone & Brunori, 2004. This theory can be neatly described as "calm molecules, calm minds" (Eckenhoff, 2001). Though structural studies with receptor protein complexes have proven challenging, experiments with simpler model proteins using Xe have been recognized as useful for understanding the binding modes of general anesthetics and the resulting molecular changes of the target proteins (Colloc'h et al., 2007).

Urate oxidase, a globular protein possessing large hydrophobic cavities, has frequently been used as a model system for understanding anesthetic mechanisms (Abraini et al., 2014; Colloc'h et al., 2007; Marassio et al., 2011). Xe binding to urate oxidase was discovered in 1997 when 8 bar

(0.8 MPa) Xe was employed as a heavy atom for phasing (Colloc'h et al., 1997). Structural studies with Xe were repeated to improve the resolution to 1.75 Å, where it was observed that at 2 MPa Xe binds with 0.9 occupancy in a cavity near the active site (Fig. 2) (Colloc'h et al., 2007). As is typical, Xe binds to a preexisting cavity lined with hydrophobic residues and devoid of solvent in the native structure. The key conclusion drawn was that Xe binding expands the volume of the cavity by 23.5%, an unexpected result given the compression from the high Xe pressure. This volume expansion, in conjunction with the Xe-binding site being near the active site of urate oxidase, supports the hypothesis that gaseous anesthetics function by expanding the volume of internal protein cavities, thereby leading to conformational rigidity and (reversible) disruption of protein (and subsequently cellular) function.

Comparisons of crystallographic studies performed with N_2O and Xe show that both gaseous anesthetics typically bind at the same cavities and lead to volume expansion, suggesting that xenon's interactions with proteins are generally indicative of gaseous anesthetics. Structural studies with Xe and N_2O, however, have also revealed subtle differences about their respective binding to protein. Abraini et al. analyzed crystals of urate oxidase, lysozyme, myoglobin, and neuroglobin under varying pressures of Xe and N_2O and observed that Xe, the more polarizable of the two species, prefers very hydrophobic environments, whereas N_2O prefers mildly hydrophobic

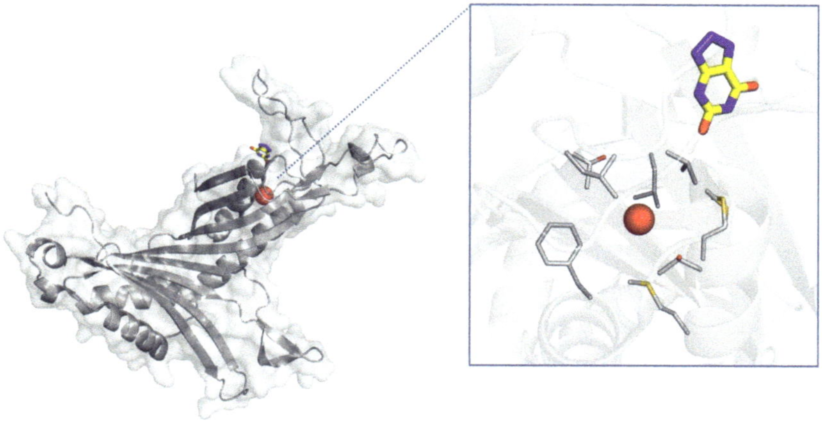

Fig. 2 Crystal structure of urate oxidase complexed with the inhibitor 8-azaxanthine under 2.0 MPa Xe pressure (PDB ID 2IC0). Xe and 8-azaxanthine are shown bound to a single subunit of the urate oxidase homotetramer. (*Inset*) Close-up of the Xe-binding pocket, with surrounding residues shown.

environments (Abraini et al., 2014). Additionally, it was found that Xe or N_2O occupancies were primarily correlated to cavity volume, and not hydrophobicity. In short, while structural studies with Xe provide useful leads for understanding anesthetic mechanisms in general, caution should be taken to recognize the differences in binding site preference and binding affinities of different inhaled anesthetics.

The process of preparing Xe derivatives of crystallized protein is fairly straightforward and well-established, with numerous methodology papers having been published detailing the methodology of Xe derivatization and follow-up data analysis (Sauer, Schmidt, & Kratky, 1997; Schiltz, Prangé, & Fourme, 1994; Soltis, Stowell, Wiener, Phillips, & Rees, 1997; Stowell et al., 1996). X-ray diffraction data from Xe derivatives can be collected at room temperature or at 100 K under a cryostream of liquid nitrogen. In the first method, the native protein crystal is transferred to a sealed capillary tube connected to a reservoir of pressurized Xe, and X-ray diffraction data are collected from the protein crystal while under a set Xe pressure (Mizuno, Makino, & Kumasaka, 2013; Schiltz et al., 2003, 1994). Although more specialized equipment is needed, the main advantage of this method is the ability to finely adjust and maintain the Xe pressure experienced by the protein crystal. This approach is preferred for quantitative structural studies in which Xe occupancies, B-factors, etc., are calculated as a function of Xe pressure (Abraini et al., 2014; Colloc'h, Marassio, & Prangé, 2008). There are several papers detailing the methodology of this Xe derivatization technique (Schiltz et al., 2003, 1994; Stowell et al., 1996).

In the second method, the native protein crystal is pressurized with Xe and then immediately flash frozen in liquid nitrogen, effectively immobilizing Xe bound to the protein. Though this method is somewhat imprecise in regard to determining the actual Xe gas pressure within the crystal (due to some Xe escaping between depressurization and freezing), the advantage of this method is the ease of the Xe derivatization process as well as the option to use standard cryoloops, goniometers, and synchrotron equipment. The cryocrystallography method is well suited for preparing Xe derivatives for phasing purposes as well as identifying Xe-binding sites within proteins. Our laboratory follows this method using a Xenon Chamber™ (Hampton Research) to study Xe binding to proteins and its implications for Xe NMR/MRI biosensing applications.

2.2 Freeze-Trapping Xe Derivatives

2.2.1 Equipments

- Optical microscope for looping protein crystals
- Cryocrystallography tools (e.g., tongs, magnetic transfer wand)
- CrystalCap™ with a mounted CryoLoop (Hampton Research)
- CrystalCap™ Vial with stand (Hampton Research)
- 1000 mL liquid nitrogen dewar
- Xe derivatization apparatus such as the Xenon Chamber™ (Hampton Research, Alisa Viejo, CA, USA) (Fig. 3A), XCell (Oxford Cryosystems, Oxford, UK), or Cryo-Xe-Siter (Rigaku, The Woodlands, TX, USA)
- Mini-Vial™ with Wick (Hampton Research)

2.2.2 Reagents

- Compressed Xe gas
- Liquid nitrogen

2.2.3 Procedure

1. Crystallize the protein to be derivatized following standard techniques (e.g., vapor diffusion).

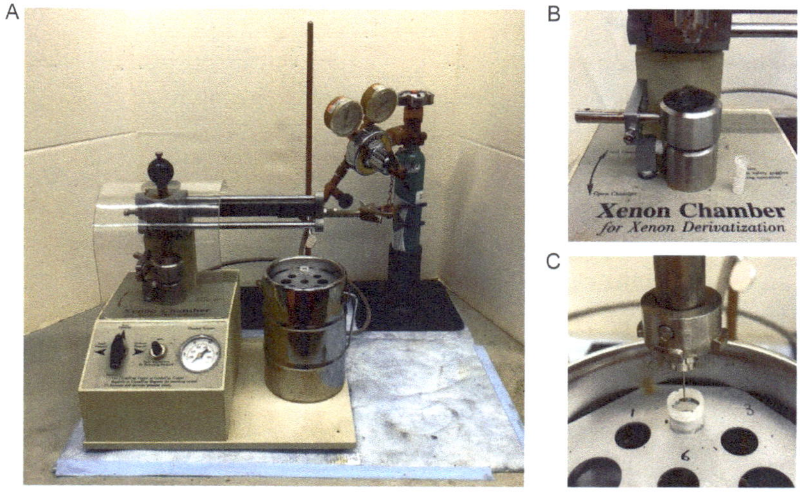

Fig. 3 (A) Xenon derivatization chamber manufactured by Hampton Research; (B) Xe pressurization chamber with a Mini-Vial™ with Wick on the side; and (C) CrystalCap™ aligned with a CrystalCap™ Vial inside a 1000 mL dewar.

2. Transfer the protein crystal to a cryoprotectant solution or slowly introduce cryoprotectant to the mother liquor containing the protein crystal.
3. Collect the cryoprotected crystal in a CryoLoop attached to a CrystalCap™.
4. Place the CrystalCap™ onto the magnetic base of the derivatization chamber plunger.
5. Insert the CrystalCap™ into the Xe pressurization chamber containing a Mini-Vial™ with Wick saturated with cryoprotectant solution (Fig. 3B). The Mini-Vial™ with Wick prevents the CryoLoop from drying out during the Xe derivatization process.
6. Seal the chamber and pressurize with up to 4.1 MPa Xe. If Xe pressure within the chamber exceeds 4.1 MPa, the pressure will vent through the safety valve.
7. Incubate the protein crystal with Xe for 15–60 min.
8. Slowly release Xe from the chamber so as not to disturb the protein crystal.
9. Immediately freeze the derivatized protein crystal by transferring the CrystalCap™ to a vial submerged in liquid nitrogen within the 1000 mL dewar (Fig. 3C). Shock freezing the crystal will immobilize Xe bound to the protein as well as Xe dissolved in the mother liquor.

2.2.4 Notes

1. The Xe pressure required to derivatize crystallized protein should be 5–10 times higher than the physiological concentration of Xe during anesthesia (~0.7 atm) (Abraini et al., 2014; Colloc'h et al., 2008; Miller, 2002).
2. Xe is soluble in water (~5 m*M* per atm) and quickly diffuses via solvent channels to binding sites within the protein crystal on a timescale of minutes (Schiltz et al., 1994). Other studies have shown that Xe binding to crystallized protein is typically complete after 10 min (Soltis et al., 1997).
3. After Xe depressurization, the diffusion of Xe from a tight binding site within a protein is roughly 5 min (Soltis et al., 1997). For a practiced Xe crystallographer, depressurizing and flash freezing the derivatized protein crystal takes only a few seconds, so loss of Xe from the protein crystal is minimal.

2.3 Diffraction Data Collection and Analysis

2.3.1 Software

- *iMosflm* (Battye, Kontogiannis, Johnson, Powell, & Leslie, 2011), *AIMLESS* (Evans, 2011), and *CCP4* suite of programs (Winn et al., 2011) are used for processing diffraction data.
- *Phaser* (McCoy et al., 2007) and *PHENIX* (Adams et al., 2010; Afonine et al., 2012) are used for phasing and model refinement.
- *COOT* (Emsley, Lohkamp, Scott, & Cowtan, 2010) enables visualizing electron density maps and manual model manipulation.

2.3.2 Procedure

1. Collect diffraction data at the native wavelength of the synchrotron beamline, or at the Cu K_α wavelength (1.54 Å) if using a home-source X-ray instrument. However, if one is using Xe as a heavy atom for de novo phasing, or if one aims to maximize anomalous scattering, diffraction data can be collected at longer wavelengths closer to the absorption edges of Xe: 2.29 Å (L_I), 2.43 Å (L_{II}), or 2.59 Å (L_{III}).
2. Generate isomorphous difference maps using $|F_{\text{xenon}}| - |F_{\text{native}}|$ as amplitudes and phase using the atomic coordinates of the native protein structure.
3. While viewing the protein model in *COOT*, add Xe atoms into the model where there are large spheres of positive electron density in the isomorphous difference map (Fig. 4A).

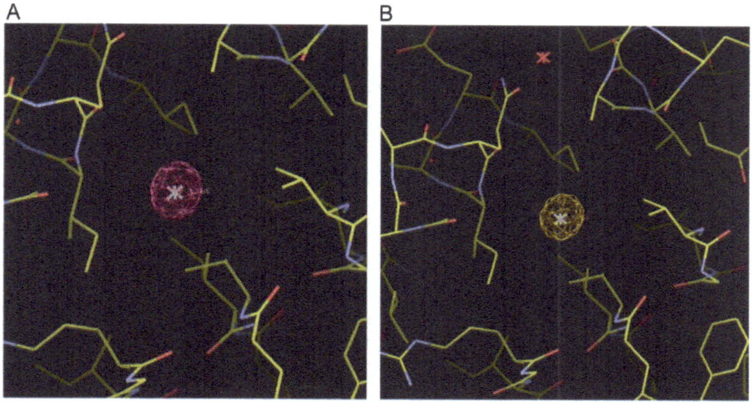

Fig. 4 Major Xe-binding site of TEM-1 β-lactamase derivatized with 1.2 MPa Xe (PDB ID 5HW1). (A) Isomorphous difference map contoured at 15σ and (B) anomalous map contoured at 5σ.

4. (Optional) Evaluate Xe binding by examining the anomalous maps (Fig. 4B).
5. Set the occupancies of the added Xe atoms to 0.5, and manually change the B-factors to roughly match those of the atoms from surrounding protein side chains.
6. As additional rounds of refinement of the Xe derivative structure are performed, the occupancies and B-factors of Xe within the protein model should converge to their final values.

2.3.3 Notes

1. There are many other software programs that offer the same functionality as the ones listed earlier. The ones described here are simply the ones used by our laboratory.
2. Diffraction data of Xe-derivatized crystals can be collected at the Cu K_α wavelength (1.54 Å) or the native wavelength of the synchrotron because Xe has appreciable anomalous dispersion even at wavelengths far from its absorption edges. Sperm whale metmyoglobin was successfully phased from diffraction data collected at the Cu K_α wavelength, where $f' = -0.783$, $f'' = 7.384$ (Vitali et al., 1991).
3. Anomalous scattering is typically not used to identify protein-bound Xe, as bound Xe is readily identified from Fourier isomorphous difference maps (vide infra). Because Xe binding is weak and generally not structurally perturbing, Xe derivatives are isomorphous to the native structure and thus the coordinates of bound Xe are readily identified from Fourier isomorphous difference maps (Prangé et al., 1998).
4. Crystallographic symmetry mates should be displayed when inspecting potential Xe-binding sites on or near the protein surface. Sometimes Xe binding arises due to crystal packing (e.g., lysozyme, CytB), and as such the physiological relevance of these sites should be treated with caution.

3. Xe NMR

3.1 Background

The Xe-129 nuclide has a spin of ½ and is readily detectable by NMR. The large polarizable electron cloud of ^{129}Xe gives it a ~200 ppm chemical shift range in common solvents and tissues (Miller et al., 1981). The magnetic signal of ^{129}Xe can be increased by several orders of magnitude by spin exchange optical pumping (SEOP), a method that uses circularly polarized light to excite the valence electrons of rubidium which then, in turn, collide with Xe in the gas state and polarizes the Xe nuclei (Barskiy et al., 2017; Walker & Happer, 1997). NMR studies of proteins with Xe have proven complementary to structural studies by X-ray crystallography. Whereas crystal structures provide a detailed but static view of Xe binding to specific sites in proteins, NMR can characterize specific and nonspecific Xe–protein interactions as well as determine Xe exchange rates and binding affinities. Xenon in solution exchanges rapidly between dissolved protein and solvent. Thus, the ^{129}Xe NMR spectrum of a protein solution reports a single peak whose chemical shift change relative to Xe in solvent alone (δ_{obs}) is an average of the chemical shifts of Xe at specific sites on or within the protein (δ_i) weighted by their respective binding constants (K_i), multiplied by protein concentration ($c_{protein}$) (Eq. 1) (Locci et al., 2001; Lowery et al., 2003).

$$\delta_{obs} = \left(\sum \delta_i K_i\right) c_{protein} \tag{1}$$

Dividing δ_{obs} by protein concentration gives a concentration-normalized chemical shift α (expressed in units of ppm/mM) that quantifies the sum of Xe–protein interactions. Xe NMR experiments with denatured proteins observed a linear dependence of the α value on protein size, with an average α of 0.005 ppm/mM per amino acid (Rubin et al., 2001). The α values of native proteins with known Xe-binding sites are typically between 1 and 3 ppm/mM (Corda, Era, Fais, & Casu, 2004; Lowery et al., 2005; Rubin et al., 2002). Conformational changes in proteins have been characterized by measuring changes of α in response to ligand-induced structural change (Lowery et al., 2005; Rubin, Spence, Pines, & Wemmer, 2001). Beyond being used to measure global Xe–protein interactions, Xe NMR has also been used to measure the kinetics and thermodynamics of Xe–protein interactions. Xenon resonance peak line width analysis at varying Xe and protein concentrations determined Xe exchange rates for proteins such as myoglobin (Tilton & Kuntz, 1982) and T4 lysozyme (Desvaux et al., 2005). Additionally, observing chemical shift change as a function of Xe pressure through nuclear Overhauser effect spectroscopy (NOESY) and heteronuclear single-quantum correlation (HSQC) experiments measured Xe-binding affinities to T4 lysozyme (Desvaux et al., 2005), maltose binding protein (MBP) (Rubin et al., 2002), and lipid transfer protein 1 (LTP1) (Dubois et al., 2004).

Though much information about Xe–protein interactions can be gleaned from direct detection NMR, a major shortcoming is the inability to resolve resonance peaks corresponding to Xe bound to protein.

Moreover, large amounts of protein (typically high-μ*M* to low-m*M*) are needed to observe any chemical shift changes and/or peak broadening. These limitations can be overcome through chemical exchange saturation transfer (CEST), an alternative contrast approach that takes advantage of the transient nature of Xe–protein interactions to heighten sensitivity (Schröder, 2013; van Zijl & Yadav, 2011). CEST used in combination with hyperpolarized ^{129}Xe, a technique known as hyper-CEST, works by applying a saturation pulse that selectively depolarizes Xe bound to solute (e.g., protein, macrocycle). Depolarized Xe rapidly exchanges into solvent, depolarizing the reservoir of bulk Xe which, in turn, generates MR contrast. Our lab has employed hyper-CEST to study Xe interactions with spores (Bai, Wang, Goulian, Driks, & Dmochowski, 2014) as well as the proteins TEM-1 β-lactamase (Bla) (Wang, Roose, Palovcak, Carnevale, & Dmochowski, 2016) and MBP (Roose, Zemerov, & Dmochowski, 2017). Representative hyper-CEST *z*-spectra for 80 μ*M* Bla and MBP in phosphate-buffered saline (PBS) are shown in Fig. 5. In both spectra, two peaks are resolved—one centered at 0 ppm, corresponding to Xe@H_2O, and the other downfield corresponding to Xe@protein. The degree of downfield shift of the Xe@protein peak provides information as to the binding site of Xe, and the broadness of the peak gives an indication of the exchange kinetics. Additionally, in the case of MBP, the presence or absence of the downfield peak is a function of the conformational state of the protein. In the absence of maltose, MBP adopts an open conformation that does not generate CEST contrast, whereas adding maltose transitions the protein into a closed state that reports a distinct downfield saturation response. Our lab has used hyper-CEST NMR to monitor Xe–protein interactions primarily for biosensing and molecular imaging applications, but this technique has the potential to shed light on the anesthetic mechanisms of Xe as well.

The following sections will describe the procedure for generating hyperpolarized (hp) ^{129}Xe using a gas mixture of 10% nitrogen, 88% helium, and 2% natural abundance xenon (Linde Group); as well as for running hyper-CEST NMR experiments using a 500 MHz Bruker BioDRX NMR spectrometer.

3.2 Hyperpolarization

Briefly, hyperpolarized (hp) Xe is generated via the SEOP method (Barskiy et al., 2017) using a home-built ^{129}Xe hyperpolarizer, based on the commercial model IGI.Xe.2000 by GE (Wang, Roose, Philbin, Doman, &

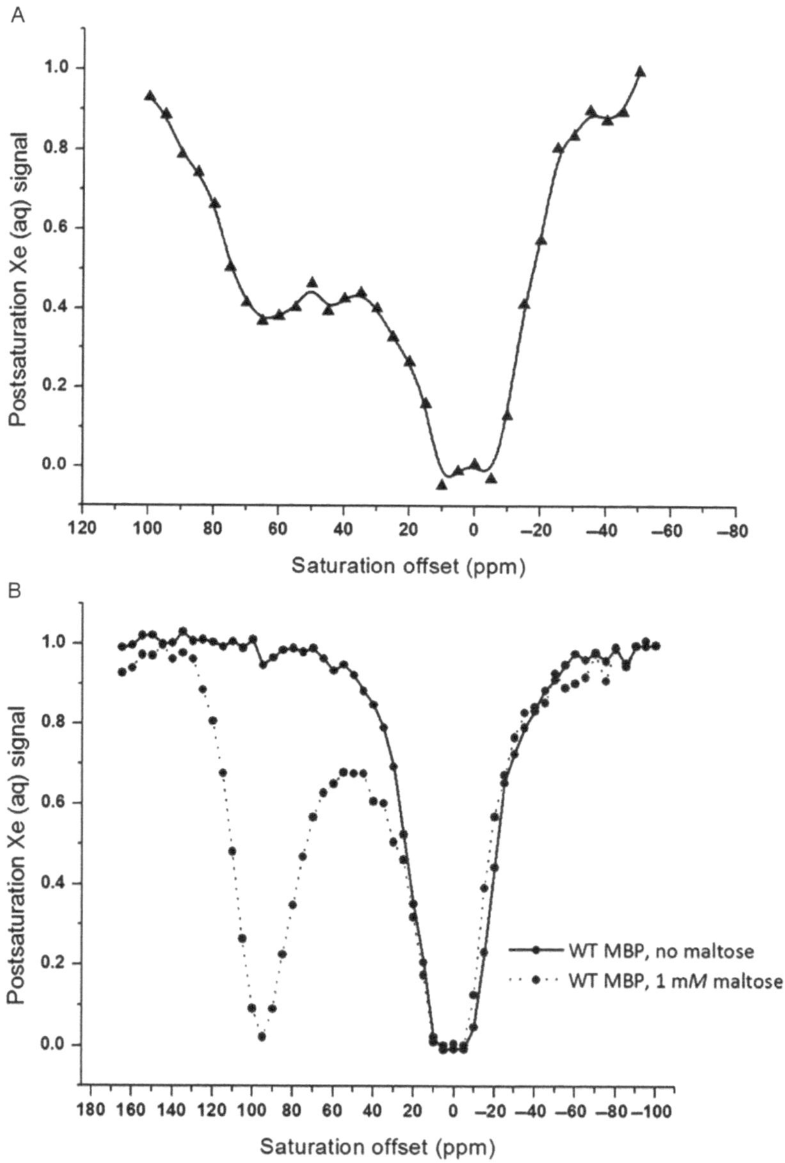

Fig. 5 Hyper-CEST *z*-spectra for Bla (A), and MBP with and without maltose (B). Spectra were obtained from 80 μ*M* protein in PBS at 300 K.

Dmochowski, 2016). A Shark 65W tunable ultra-narrowband laser (OptiGrate) set to 795 nm is used for optical pumping of Rb vapor contained in a glass cell. A water chiller (K–O concepts) set to 16°C supplies chilled water to the laser diodes. This setup is shown in Fig. 6.

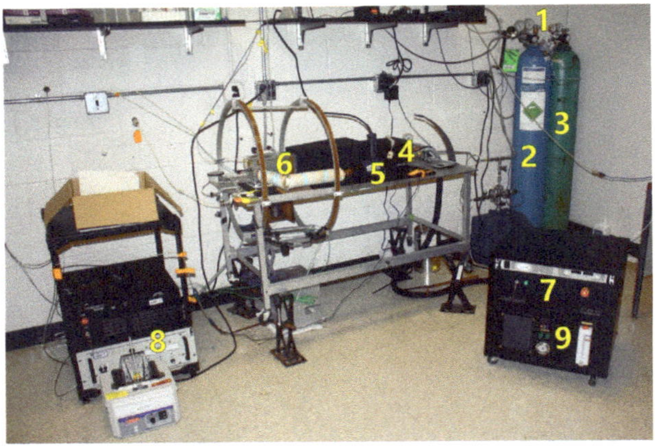

Fig. 6 Overview of a home-built ^{129}Xe hyperpolarizer. (1) Pressure gage A; (2) Xe gas mixture tank; (3) N_2 tank; (4) laser; (5) heat gun; (6) Rb cell; (7) laser control panel; (8) magnet (Helmholtz coil power supply); and (9) chiller.

A standard procedure for preparing the SEOP setup and generating hp. ^{129}Xe is available in the literature (Witte, Kunth, Döpfert, Rossella, & Schröder, 2012).

3.3 Hyper-CEST

1. Turn on the module supplying chilled water to the laser diodes. Allow chilled water to circulate for at least 10 min before turning on the laser electronics.
2. If the pressure in the tubing leading from the Xe gas tank to the Rb cell is below 40 psi, the tubing should be degassed using the vacuum. It is important to keep this line pressurized to 40–50 psi to avoid leaking of O_2 into the line. If the pressure is sufficient, allow Xe to enter the Rb cell.
3. Start heating the Rb cell to 180°C. For best signal to noise, it is advised to heat the cell for 2–3 h before collecting spectra, especially if the cell has not been operated for several days.
4. Turn on the laser electronics, make sure the wavelength is set to 794.7 nm, and slowly increase the current to the laser diode while monitoring the laser emission profile.
5. While the current is being increased, degas the tubing leading from the polarizer outlet to the valve regulating Xe flow into the NMR tube.
6. After the laser is set to full power, turn on the power supply to the Helmholtz coils.

7. If necessary, change the current NMR probe to a 10-mm BBO NMR probe.
8. Insert a 90% H_2O/10% D_2O sample into the magnet.
9. Lock on the D_2O signal and shim Z^1, Z^2, XZ^0, XZ^1, XZ^1, and YZ^1.
10. Run an acquisition and make sure the H_2O peak is symmetric and that the FWHM is ~7–13 Hz.
11. Remove the 90% H_2O/10% D_2O sample and insert the Xe in ethanol sample.
12. Manually tune the probe to the Xe frequency (138.12 MHz).
13. Run an acquisition and make sure that the Xe signal is visible.
14. Turn off the vacuum and make sure that the only open valves are those leading from the Rb cell outlet to the tubing leading from the polarizer outlet.
15. Open the outlet valve of the Rb cell.
16. For convenience, place a webcam in front of the flow rate monitor so that the flow rate can be clearly monitored in the NMR room while performing experiments.
17. Add 2.5 mL of ddH_2O to a 10 mm NMR tube and screw on the capillary cap onto the NMR tube. This cap should have the tubing leading from the polarizer as its inlet tubing and several long, thin capillaries for efficient bubbling of Xe as its outlet.
18. Pressurize the NMR tube with Xe. This can be done by rapidly opening and closing the valve leading from the polarizer outlet to the NMR tube several times.
19. Slowly, manually lower the NMR tube into the magnet until it touches a solid surface.
20. Open the NMR software and run a Xe stability test. To do this, apply a dSNOB-shaped saturation pulse at a frequency of 300 ppm and monitor the Xe(aq) signal. Run 20 acquisitions using a flow rate of approximately 0.7 SLM, making sure that this flow rate is stable throughout the acquisitions. Integrate the Xe(aq) peaks of all 20 experiments to determine whether or not the polarization is stable over time. If the intensities do not fluctuate by more than ± 0.1, move on to the next step. If they do, it may be necessary to heat the Rb cell for a longer period of time, or check to make sure that no leaks are present in the system. *Note*: The user-generated code to automatically run this stability test and process the data is available upon request.
21. Remove the NMR tube from the magnet and depressurize the tube.
22. After bubbling stops in the NMR tube, unscrew and remove the cap.

23. Prepare sample for hyper-CEST experiment. The sample volume should be 2.5 mL and 0.1% (v/v) pluronic may be added to mitigate foaming. For protein *z*-spectra, start with a protein sample concentration of 80 μ*M*.
24. Screw in the cap and pressurize the NMR tube as before.
25. To obtain a *z*-spectrum, apply a saturation train of 600 dSNOB-shaped pulses with bandwidth of 690 Hz at each frequency between the starting and ending chemical shift of interest. It is recommended to use a chemical shift step width of 5 ppm for preliminary data collection. Pulse length $t_{pulse} = 3.80$ ms, field strength $B_{1,max} = 77$ μT, saturation time $T_{sat} = 1.52$ s. The Xe(aq) signal will be displayed as the output. After data acquisition is complete, integrate all of the Xe(aq) peaks and plot the postsaturation Xe(aq) signal as a function of saturation frequency. Repeat 2 times, or as necessary until three sets of similar data are obtained. *Note*: The user-generated code to automatically run a *z*-spectrum and process the data is available upon request.
26. Depressurize the tube, and recover the sample for further analysis, if necessary.
27. In the polarizer room, turn off the heat, the laser electronics, the power supply to the Helmholtz coils, and the chiller module.
28. After the temperature of the cell reaches <30°C, close the inlet valve to the Rb cell, all valves between the Rb cell and the Xe tank, and finally the Xe tank.

4. SUMMARY AND CONCLUSIONS

The underlying physiological mechanisms of general anesthetics remain poorly understood despite their critical importance in modern medicine. Xenon, seen by some as the "anesthetic gas of the future" (Delhaye et al., 2010), is well positioned to serve as a proxy for unraveling some of the long-standing mysteries of general anesthesia due to its utility as a probe in X-ray crystallography and NMR experiments. More recently, the hyper-CEST approach to NMR contrast has allowed even weak, transient Xe–protein interactions to be characterized with high spectral resolution, far beyond what has been observed in direct detection NMR experiments. Xe hyper-CEST, complemented by structural information from X-ray crystallography, has the potential to characterize Xe interactions with receptor proteins and ion channels implicated in general anesthesia but heretofore undetectable by conventional methods.

REFERENCES

Abraini, J. H., Marassio, G., David, H. N., Vallone, B., Prangé, T., & Colloc'h, N. (2014). Crystallographic studies with xenon and nitrous oxide provide evidence for protein-dependent processes in the mechanisms of general anesthesia. *Anesthesiology, 121*(5), 1018–1027. https://doi.org/10.1097/ALN.0000000000000435.

Adams, P. D., Afonine, P. V., Bunkóczi, G., Chen, V. B., Davis, I. W., Echols, N., et al. (2010). PHENIX: A comprehensive python-based system for macromolecular structure solution. *Acta Crystallographica, Section D: Biological Crystallography, 66*(2), 213–221. https://doi.org/10.1107/S0907444909052925.

Afonine, P. V., Grosse-Kunstleve, R. W., Echols, N., Headd, J. J., Moriarty, N. W., Mustyakimov, M., et al. (2012). Towards automated crystallographic structure refinement with phenix.refine. *Acta Crystallographica, Section D: Biological Crystallography, 68*(4), 352–367. https://doi.org/10.1107/S0907444912001308.

Arch, A. M., & Harper, N. J. N. (2011). Xenon: An element of protection. *Trends in Anaesthesia and Critical Care, 1*(5–6), 238–242. https://doi.org/10.1016/j.tacc.2011.08.001.

Aziz, T. S. (2001). Xenon in anesthesia. *International Anesthesiology Clinics, 39*(2), 1–14. https://doi.org/10.1097/00004311-200104000-00003.

Bai, Y., Wang, Y., Goulian, M., Driks, A., & Dmochowski, I. J. (2014). Bacterial spore detection and analysis using hyperpolarized 129 Xe chemical exchange saturation transfer (hyper-CEST) NMR. *Chemical Science, 5*(8), 3197–3203. https://doi.org/10.1039/C4SC01190B.

Barskiy, D. A., Coffey, A. M., Nikolaou, P., Mikhaylov, D. M., Goodson, B. M., Branca, R. T., et al. (2017). NMR hyperpolarization techniques of gases. *Chemistry (Weinheim an Der Bergstrasse, Germany), 23*(4), 725–751. https://doi.org/10.1002/chem.201603884.

Bartlett, N. (1962). Xenon hexafluoroplatinate (V) $Xe^{+}[PtF_6]^{-}$. *Proceedings of the Chemical Society of London, (JUN)*, 218.

Battye, T. G. G., Kontogiannis, L., Johnson, O., Powell, H. R., & Leslie, A. G. W. (2011). iMOSFLM: A new graphical interface for diffraction-image processing with MOSFLM. *Acta Crystallographica, Section D: Biological Crystallography, 67*(4), 271–281. https://doi.org/10.1107/S0907444910048675.

Bowyer, T. W., Schlosser, C., Abel, K. H., Auer, M., Hayes, J. C., Heimbigner, T. R., et al. (2002). Detection and analysis of xenon isotopes for the comprehensive nuclear-test-ban treaty international monitoring system. *Journal of Environmental Radioactivity, 59*(2), 139–151. https://doi.org/10.1016/S0265-931X(01)00042-X.

Campagna, J. A., Miller, K. W., & Forman, S. A. (2003). Mechanisms of actions of inhaled anesthetics. *The New England Journal of Medicine, 348*(21), 2110–2124. https://doi.org/10.1056/NEJMra021261.

Colloc'h, N., el Hajji, M., Bachet, B., L'Hermite, G., Schiltz, M., Prangé, T., et al. (1997). Crystal structure of the protein drug urate oxidase-inhibitor complex at 2.05 A resolution. *Nature Structural Biology, 4*(11), 947–952. https://doi.org/10.1038/nsb1197-947.

Colloc'h, N., Marassio, G., & Prangé, T. (2008). Protein–noble gas interactions investigated by crystallography on three enzymes—Implication on anesthesia and neuroprotection mechanisms. In A. Chandrasekeran (Ed.), *Current trends in X-ray crystallography* (pp. 285–308). Rijeka: InTech.

Colloc'h, N., Sopkova-de Oliveira Santos, J., Retailleau, P., Vivarès, D., Bonneté, F., Langlois d'Estainto, B., et al. (2007). Protein crystallography under xenon and nitrous oxide pressure: Comparison with in vivo pharmacology studies and implications for the mechanism of inhaled anesthetic action. *Biophysical Journal, 92*(1), 217–224. https://doi.org/10.1529/biophysj.106.093807.

Corda, M., Era, B., Fais, A., & Casu, M. (2004). Structural investigation of pig metmyoglobin by 129Xe NMR spectroscopy. *Biochimica et Biophysica Acta, 1674*(2), 182–192. https://doi.org/10.1016/j.bbagen.2004.06.011.

Cullen, S. C., & Gross, E. G. (1951). The anesthetic properties of xenon in animals and human beings, with additional observations on krypton. *Science*, *113*(2942), 580–582.

David, H. N., Haelewyn, B., Risso, J.-J., Colloc'h, N., & Abraini, J. H. (2010). Xenon is an inhibitor of tissue-plasminogen activator: Adverse and beneficial effects in a rat model of thromboembolic stroke. *Journal of Cerebral Blood Flow and Metabolism: Official Journal of the International Society of Cerebral Blood Flow and Metabolism*, *30*(4), 718–728. https://doi.org/10.1038/jcbfm.2009.275.

Delhaye, O., Robin, E., Bazin, J.-E., Ripart, J., Lebuffe, G., & Vallet, B. (2010). Avantages décisifs, indications et limites de l'anesthésie au xénon. *Annales Françaises d'Anesthésie et de Réanimation*, *29*(9), 635–641. https://doi.org/10.1016/j.annfar.2010.04.006.

Desvaux, H., Dubois, L., Huber, G., Quillin, M. L., Berthault, P., & Matthews, B. W. (2005). Dynamics of xenon binding inside the hydrophobic cavity of pseudo-wild-type bacteriophage T4 lysozyme explored through xenon-based NMR spectroscopy. *Journal of the American Chemical Society*, *127*(33), 11676–11683. https://doi.org/10.1021/ja053074p.

Dingley, J., Findlay, G. P., Foëx, B. A., Mecklenburgh, J., Esmail, M., & Little, R. A. (2001). A closed xenon anesthesia delivery system. *Anesthesiology*, *94*(1), 173–176. https://doi.org/10.1097/00000542-200101000-00034.

Dubois, L., Da Silva, P., Landon, C., Huber, J. G., Ponchet, M., Vovelle, F., et al. (2004). Probing the hydrophobic cavity of lipid transfer protein from Nicotiana tabacum through xenon-based NMR spectroscopy. *Journal of the American Chemical Society*, *126*(48), 15,738–15,746. https://doi.org/10.1021/ja046195i.

Eckenhoff, R. G. (2001). Promiscuous ligands and attractive cavities: How do the inhaled anesthetics work? *Molecular Interventions*, *1*(5), 258–268. Retrieved from http://www.ncbi.nlm.nih.gov/pubmed/14993365.

Emsley, P., Lohkamp, B., Scott, W. G., & Cowtan, K. (2010). Features and development of Coot. *Acta Crystallographica, Section D: Biological Crystallography*, *66*, 486–501. http://doi.org/10.1107/S0907444910007493.

Esencan, E., Yuksel, S., Tosun, Y. B., Robinot, A., Solaroglu, I., & Zhang, J. H. (2013). XENON in medical area: Emphasis on neuroprotection in hypoxia and anesthesia. *Medical Gas Research*, *3*(1), 4. https://doi.org/10.1186/2045-9912-3-4.

Evans, P. R. (2011). An introduction to data reduction: Space-group determination, scaling and intensity statistics. *Acta Crystallographica, Section D: Biological Crystallography*, *67*(4), 282–292. https://doi.org/10.1107/S090744491003982X.

Ewing, G. J., & Maestas, S. (1970). The thermodynamics of absorption of xenon by myoglobin. *The Journal of Physical Chemistry*, *74*(11), 2341–2344. https://doi.org/10.1021/j100705a018.

Franks, N. P., Dickinson, R., de Sousa, S. L., Hall, A. C., & Lieb, W. R. (1998). How does xenon produce anaesthesia? *Nature*, *396*(6709), 324. https://doi.org/10.1038/24525.

Frohn, H.-J., & Bardin, V. V. (2001). Preparation and reactivity of compounds containing a carbon–xenon bond. *Organometallics*, *20*(23), 4750–4762. https://doi.org/10.1021/om010490j.

Giacalone, M., Abramo, A., Giunta, F., & Forfori, F. (2013). Xenon-related analgesia: A new target for pain treatment. *The Clinical Journal of Pain*, *29*(7), 639–643. https://doi.org/10.1097/AJP.0b013e31826b12f5.

Hubbard, S. J., & Argos, P. (1996). A functional role for protein cavities in domain: Domain motions. *Journal of Molecular Biology*, *261*(2), 289–300. https://doi.org/10.1006/jmbi.1996.0460.

Hubbard, S. J., Gross, K. H., & Argos, P. (1994). Intramolecular cavities in globular proteins. *Protein Engineering*, 7(5), 613–626. http://www.ncbi.nlm.nih.gov/pubmed/8073031.

Lawrence, J. H., Loomis, W. F., Tobias, C. A., & Turpin, F. H. (1946). Preliminary observations on the narcotic effect of xenon with a review of values for solubilities of gases in water and oils. *The Journal of Physiology*, *105*(3), 197–204.

Locci, E., Dehouck, Y., Casu, M., Saba, G., Lai, A., Luhmer, M., et al. (2001). Probing proteins in solution by 129Xe NMR spectroscopy. *Journal of Magnetic Resonance, 150*(2), 167–174. https://doi.org/10.1006/jmre.2001.2325.

Lowery, T. J., Doucleff, M., Ruiz, E. J., Rubin, S. M., Pines, A., & Wemmer, D. E. (2005). Distinguishing multiple chemotaxis Y protein conformations with laser-polarized 129Xe NMR. *Protein Science: A Publication of the Protein Society, 14*(4), 848–855. https://doi.org/10.1110/ps.041231005.

Lowery, T. J., Rubin, S. M., Ruiz, E. J., Spence, M. M., Winssinger, N., Schultz, P. G., et al. (2003). Applications of laser-polarized 129Xe to biomolecular assays. *Magnetic Resonance Imaging, 21*(10), 1235–1239. https://doi.org/10.1016/j.mri.2003.08.025.

Marassio, G., Prangé, T., David, H. N., Sopkova-de Oliveira Santos, J., Gabison, L., Delcroix, N., et al. (2011). Pressure-response analysis of anesthetic gases xenon and nitrous oxide on urate oxidase: A crystallographic study. *The FASEB Journal, 25*(7), 2266–2275. https://doi.org/10.1096/fj.11-183046.

Marx, T., Schmidt, M., Schirmer, U., & Reinelt, H. (2000). Xenon anaesthesia. *Journal of the Royal Society of Medicine, 93*(10), 513–517. https://doi.org/10.1177/014107680009301005.

McCoy, A. J., Grosse-Kunstleve, R. W., Adams, P. D., Winn, M. D., Storoni, L. C., & Read, R. J. (2007). Phaser crystallographic software. *Journal of Applied Crystallography, 40*(4), 658–674. https://doi.org/10.1107/S0021889807021206.

Meyer, H. (1899). Zur theorie der alkoholnarkose. *Naunyn-Schmiedeberg's Archives of Pharmacology, 42*(2), 109–118.

Miller, K. W. (2002). The nature of sites of general anaesthetic action. *British Journal of Anaesthesia, 89*(1), 17–31. https://doi.org/10.1093/bja/aef167.

Miller, K. W., Reo, N. V., Schoot Uiterkamp, A. J., Stengle, D. P., Stengle, T. R., & Williamson, K. L. (1981). Xenon NMR: Chemical shifts of a general anesthetic in common solvents, proteins, and membranes. *Proceedings of the National Academy of Sciences of the United States of America, 78*(8), 4946–4949. https://doi.org/10.1073/pnas.78.8.4946.

Mizuno, N., Makino, M., & Kumasaka, T. (2013). A convenient tool for gas derivatization using fine-needle capillary mounting for protein crystals. *Journal of Synchrotron Radiation, 20*(6), 999–1002. https://doi.org/10.1107/S0909049513021584.

Mueller-Dieckmann, C., Polentarutti, M., Djinovic Carugo, K., Panjikar, S., Tucker, P. A., & Weiss, M. S. (2004). On the routine use of soft X-rays in macromolecular crystallography. Part II. Data-collection wavelength and scaling models. *Acta Crystallographica, Section D: Biological Crystallography, 60*(1), 28–38. https://doi.org/10.1107/S0907444903020833.

Neice, A. E., & Zornow, M. H. (2016). Xenon anaesthesia for all, or only a select few? *Anaesthesia, 71*(11), 1267–1272. https://doi.org/10.1111/anae.13569.

Overton, E. (1901). *Studien uber die Narkose*. Jena: Fischer, p. 45.

Prangé, T., Schiltz, M., Pernot, L., Colloc'h, N., Longhi, S., Bourguet, W., et al. (1998). Exploring hydrophobic sites in proteins with xenon or krypton. *Proteins, 30*(1), 61–73. https://doi.org/10.1002/(SICI)1097-0134(19980101)30:1<61::AID-PROT6>3.0.CO;2-N.

Quillin, M. L., & Matthews, B. W. (2002). Generation of noble-gas binding sites for crystallographic phasing using site-directed mutagenesis. *Acta Crystallographica, Section D: Biological Crystallography, 58*(1), 97–103. https://doi.org/10.1107/S0907444901018145.

Quillin, M. L., & Matthews, B. W. (2003). Selling candles in a post-Edison world: Phasing with noble gases bound within engineered sites. *Acta Crystallographica, Section D: Biological Crystallography, 59*(11), 1930–1934. https://doi.org/10.1107/S0907444903018018.

Rentzepis, P. M., & Douglass, D. C. (1981). Xenon as a solvent. *Nature, 293*(5828), 165–166. https://doi.org/10.1038/293165a0.

Richards, F. M. (1974). The interpretation of protein structures: Total volume, group volume distributions and packing density. *Journal of Molecular Biology*, *82*(1), 1–14. https://doi.org/10.1016/0022-2836(74)90570-1.

Roose, B. W., Zemerov, S. D., & Dmochowski, I. J. (2017). Nanomolar small-molecule detection using a genetically encoded 129 Xe NMR contrast agent. *Chemical Science*, *8*(11), 7631–7636. https://doi.org/10.1039/C7SC03601A.

Rubin, S. M., Lee, S. Y., Ruiz, E. J., Pines, A., & Wemmer, D. E. (2002). Detection and characterization of xenon-binding sites in proteins by 129Xe NMR spectroscopy. *Journal of Molecular Biology*, *322*(2), 425–440. https://doi.org/10.1016/S0022-2836(02)00739-8.

Rubin, S. M., Spence, M. M., Dimitrov, I. E., Ruiz, E. J., Pines, A., & Wemmer, D. E. (2001). Detection of a conformational change in maltose binding protein by 129Xe NMR spectroscopy. *Journal of the American Chemical Society*, *123*(35), 8616–8617. https://doi.org/10.1021/ja0110325.

Rubin, S. M., Spence, M. M., Pines, A., & Wemmer, D. E. (2001). Characterization of the effects of nonspecific xenon–protein interactions on 129Xe chemical shifts in aqueous solution: Further development of xenon as a biomolecular probe. *Journal of Magnetic Resonance*, *152*(1), 79–86. https://doi.org/10.1006/jmre.2001.2389.

Rumble, J. R. (2017). *CRC handbook of chemistry and physics: A ready-reference book of chemical and physical data*. CRC Press, ISBN: 9781498784542.

Sanders, R. D. (2003). Xenon: No stranger to anaesthesia. *British Journal of Anaesthesia*, *91*(5), 709–717. https://doi.org/10.1093/bja/aeg232.

Sauer, O., Schmidt, A., & Kratky, C. (1997). Freeze-trapping isomorphous xenon derivatives of protein crystals. *Journal of Applied Crystallography*, *30*(4), 476–486. https://doi.org/10.1107/S0021889897000745.

Schiltz, M., Fourme, R., & Prangé, T. (2003). Use of noble gases xenon and krypton as heavy atoms in protein structure determination. *Methods in Enzymology*, *374*, 83–119. https://doi.org/10.1016/S0076-6879(03)74004-X.

Schiltz, M., Kvick, Å., Svensson, O. S., Shepard, W., de La Fortelle, E., Prangé, T., et al. (1997). Protein crystallography at ultra-short wavelengths: Feasibility study of anomalous-dispersion experiments at the xenon K-edge. *Journal of Synchrotron Radiation*, *4*(5), 287–297. https://doi.org/10.1107/S0909049597008571.

Schiltz, M., Prangé, T., & Fourme, R. (1994). On the preparation and X-ray data collection of isomorphous xenon derivatives. *Journal of Applied Crystallography*, *27*(6), 950–960. https://doi.org/10.1107/S0021889894005923.

Schoenborn, B. P. (1965). Binding of xenon to horse haemoglobin. *Nature*, *208*(5012), 760–762. https://doi.org/10.1038/208760a0.

Schoenborn, B. P., Featherstone, R. M., Vogelhut, P. O., & Süsskind, C. (1964). Influence of xenon on protein hydration as measured by a microwave absorption technique. *Nature*, *202*(4933), 695–696. https://doi.org/10.1038/202695a0.

Schröder, L. (2013). Xenon for NMR biosensing—Inert but alert. *Physica Medica*, *29*(1), 3–16. https://doi.org/10.1016/j.ejmp.2011.11.001.

Settle, W. (1973). Function of the myoglobin molecule as influenced by anesthetic molecules. In R. M. Featherstone (Ed.), *A guide to molecular pharmacology and toxicology* (pp. 477–493). New York, NY: Marcel Dekker.

Soltis, S. M., Stowell, M. H. B., Wiener, M. C., Phillips, G. N., & Rees, D. C. (1997). Successful flash-cooling of xenon-derivatized myoglobin crystals. *Journal of Applied Crystallography*, *30*(2), 190–194. https://doi.org/10.1107/S0021889896012939.

Stoppe, C., Rimek, A., Rossaint, R., Rex, S., Stevanovic, A., Schälte, G., et al. (2013). Xenon consumption during general surgery: A retrospective observational study. *Medical Gas Research*, *3*(1), 12. https://doi.org/10.1186/2045-9912-3-12.

Stowell, M. H. B., Soltis, S. M., Kisker, C., Peters, J. W., Schindelin, H., Rees, D. C., et al. (1996). A simple device for studying macromolecular crystals under moderate gas pressures (0.1–10 MPa). *Journal of Applied Crystallography*, *29*(5), 608–613. https://doi.org/10.1107/S0021889896004712.

Teeter, M. M. (2004). Myoglobin cavities provide interior ligand pathway. *Protein Science*, *13*(2), 313–318. https://doi.org/10.1110/ps.03334304.

Tilton, R. F., & Kuntz, I. D. (1982). Nuclear magnetic resonance studies of xenon-129 with myoglobin and hemoglobin. *Biochemistry*, *21*(26), 6850–6857. https://doi.org/10.1021/bi00269a035.

Tilton, R. F., Kuntz, I. D., & Petsko, G. A. (1984). Cavities in proteins: Structure of a metmyoglobin-xenon complex solved to 1.9 A. *Biochemistry*, *23*(13), 2849–2857. https://doi.org/10.1021/bi00308a002.

Tilton, R. F., Singh, U. C., Kuntz, I. D., & Kollman, P. A. (1988). Protein–ligand dynamics. *Journal of Molecular Biology*, *199*(1), 195–211. https://doi.org/10.1016/0022-2836(88)90389-0.

Tilton, R. F., Singh, U. C., Weiner, S. J., Connolly, M. L., Kuntz, I. D., Kollman, P. A., et al. (1986). Computational studies of the interaction of myoglobin and xenon. *Journal of Molecular Biology*, *192*(2), 443–456. https://doi.org/10.1016/0022-2836(86)90374-8.

Vallone, B., & Brunori, M. (2004). Roles for holes: Are cavities in proteins mere packing defects? *The Italian Journal of Biochemistry*, *53*(1), 46–52. Retrieved from http://www.ncbi.nlm.nih.gov/pubmed/15356962.

van Zijl, P. C. M., & Yadav, N. N. (2011). Chemical exchange saturation transfer (CEST): What is in a name and what isn't? *Magnetic Resonance in Medicine*, *65*(4), 927–948. https://doi.org/10.1002/mrm.22761.

Vitali, J., Robbins, A. H., Almo, S. C., & Tilton, R. F. (1991). Using xenon as a heavy atom for determining phases in sperm whale metmyoglobin. *Journal of Applied Crystallography*, *24*(5), 931–935. https://doi.org/10.1107/S0021889891005897.

Walker, T. G., & Happer, W. (1997). Spin-exchange optical pumping of noble-gas nuclei. *Reviews of Modern Physics*, *69*(2), 629–642. https://doi.org/10.1103/RevModPhys.69.629.

Wang, Y., Roose, B. W., Palovcak, E. J., Carnevale, V., & Dmochowski, I. J. (2016). A genetically encoded β-lactamase reporter for ultrasensitive 129Xe NMR in mammalian cells. *Angewandte Chemie (International Ed. in English)*, *55*(31), 8984–8987. https://doi.org/10.1002/anie.201604055.

Wang, Y., Roose, B. W., Philbin, J. P., Doman, J. L., & Dmochowski, I. J. (2016). Programming a molecular relay for ultrasensitive biodetection through 129Xe NMR. *Angewandte Chemie (International Ed. in English)*, *55*(5), 1733–1736. https://doi.org/10.1002/anie.201508990.

Weathersby, P. K., & Homer, L. D. (1980). Solubility of inert gases in biological fluids and tissues: A review. *Undersea Biomedical Research*, 7(4), 277–296. Retrieved from http://www.ncbi.nlm.nih.gov/pubmed/6262972.

Weinrich, M., & Worcester, D. L. (2013). Xenon and other volatile anesthetics change domain structure in model lipid raft membranes. *The Journal of Physical Chemistry B*, *117*(50), 16141–16147. https://doi.org/10.1021/jp411261g.

Winkler, D. A., Thornton, A., Farjot, G., & Katz, I. (2016). The diverse biological properties of the chemically inert noble gases. *Pharmacology and Therapeutics*, *160*, 44–64. https://doi.org/10.1016/j.pharmthera.2016.02.002.

Winn, M. D., Ballard, C. C., Cowtan, K. D., Dodson, E. J., Emsley, P., Evans, P. R., et al. (2011). Overview of the CCP4 suite and current developments. *Acta Crystallographica, Section D: Biological Crystallography*, *67*(4), 235–242. https://doi.org/10.1107/S0907444910045749.

Witte, C., Kunth, M., Döpfert, J., Rossella, F., & Schröder, L. (2012). Hyperpolarized xenon for NMR and MRI applications. *Journal of Visualized Experiments*, *67*, 4268. https://doi.org/10.3791/4268.

Wolfenden, R., & Radzicka, A. (1994). On the probability of finding a water molecule in a nonpolar cavity. *Science*, *265*(5174), 936–937. Retrieved from http://www.ncbi.nlm.nih.gov/pubmed/8052849.

Yamamoto, E., Akimoto, T., Shimizu, H., Hirano, Y., Yasui, M., & Yasuoka, K. (2012). Diffusive nature of xenon anesthetic changes properties of a lipid bilayer: Molecular dynamics simulations. *The Journal of Physical Chemistry. B*, *116*(30), 8989–8995. https://doi.org/10.1021/jp303330c.

Zhang, L., Zhang, Y., Cheng, J., Wang, L., Wang, X., Zhang, M., et al. (2017). Inert gas deactivates protein activity by aggregation. *Scientific Reports*, 7(1), 10176. https://doi.org/10.1038/s41598-017-10678-3.

FURTHER READING

Hermans, J., & Shankar, S. (1986). The free energy of xenon binding to myoglobin from molecular dynamics simulation. *Israel Journal of Chemistry*, *27*(2), 225–227. https://doi.org/10.1002/ijch.198600032.

Rubin, S. M., Spence, M. M., Goodson, B. M., Wemmer, D. E., & Pines, A. (2000). Evidence of nonspecific surface interactions between laser-polarized xenon and myoglobin in solution. *Proceedings of the National Academy of Sciences*, *97*(17), 9472–9475. https://doi.org/10.1073/pnas.170278897.

Schoenborn, B. P., Watson, H. C., & Kendrew, J. C. (1965). Binding of xenon to sperm whale myoglobin. *Nature*, *207*(992), 28–30. https://doi.org/10.1038/207028a0.

Schröder, L., Lowery, T. J., Hilty, C., Wemmer, D. E., & Pines, A. (2006). Molecular imaging using a targeted magnetic resonance hyperpolarized biosensor. *Science*, *314*(5798), 446–449. https://doi.org/10.1126/science.1131847.

CHAPTER FOURTEEN

Methods for Defining the Neuroprotective Properties of Xenon

Richie Robel*, Peter Caroccio*, Mervyn Maze†,[1]
*NeuroproteXeon, Orchard Park, NY, United States
†University of California San Francisco, San Francisco, CA, United States
[1]Corresponding author: e-mail address: mervyn.maze@ucsf.edu

Contents

Abstract

Xenon has features that make it an ideal general anesthetic agent; cost and scarcity mitigate xenon's widespread use in the operating room. Discovery of xenon's cytoprotective properties resulted in its application to thwart ongoing acute neurologic injury, an unmet clinical need. The discovery that xenon's neuroprotective effect interacts synergistically with targeted temperature management (TTM) led to its investigation in clinical settings, including in the management of the postcardiac arrest syndrome, in which TTM is indicated. Following successful demonstration of xenon's efficacy in combination with TTM in a preclinical model of porcine cardiac arrest, xenon plus TTM was shown to significantly decrease an imaging biomarker of brain injury for out of hospital cardiac arrest victims that had been successfully resuscitated. With the development of an efficient delivery system the stage is now set to investigate whether xenon improves survival, with good clinical outcome, for successfully resuscitated victims of a cardiac arrest.

Methods in Enzymology, Volume 602
ISSN 0076-6879
https://doi.org/10.1016/bs.mie.2018.01.010

1. INTRODUCTION

Both high price and low availability have limited xenon's penetration as a general anesthetic for routine surgical interventions despite satisfying almost all the other criteria of an ideal anesthetic agent. While xenon's utility as a general anesthetic may be waning, interest in its neuroprotective, cardioprotective, and renoprotective applications is waxing based upon initial preclinical observations (Khajuria, Tay, Shi, Zhao, & Ma, 2014; Maze, 2016; Preckel, Weber, Sanders, Maze, & Schlack, 2006) that are now beginning to be transformed into clinical utility. For this reason, the authors will focus on the methods involved in defining xenon's preclinical and clinical neuroprotective properties and detail a novel device that efficiently delivers xenon, which will soon be used in a Phase 3 pivotal trial.

2. BIOLOGICAL ACTION OF A CHEMICALLY NONREACTIVE MONOATOMIC GAS

Xenon's chemical inertness is due to a full complement of electrons in its outermost shell rendering it incapable of covalent bonding and forming adducts under biological conditions. However, because of xenon's polarizability (Dmochowski, 2009), it has affinity for amino acid residues surrounding preformed hydrophobic cavities thereby changing the functional properties of neighboring proteins by London dispersion forces. In this manner xenon has been shown to have activity on many proteins governed mostly by the size and shape of xenon atoms; examples of these will be provided.

Absent chemical reactivity, xenon is not biotransformed, which results in two important features governing its future clinical use. Most xenobiotic drugs are converted into metabolites that may be toxic; as xenon is not metabolized, the dangers posed by toxic metabolites are obviated. Furthermore, xenon in the inspired and expired gases are identical, permitting recirculation of exhaled xenon and thereby limiting the need for a fresh supply of this scarce resource. A method exploiting this property will be provided later.

3. METHODS TO EXPLORE THE ACTION OF XENON ON THE *N*-METHYL-D-ASPARTATE SUBTYPE OF THE GLUTAMATE RECEPTOR

In a landmark series of studies beginning with electrophysiology (Franks, Dickinson, de Sousa, Hall, & Lieb, 1998) progressing through molecular modeling (Dickinson et al., 2007), and site-directed mutagenesis (Armstrong et al., 2012), and concluding with functional outcome (Harris et al., 2013), Franks group demonstrated that xenon inhibits the *N*-methyl-D-aspartate (NMDA) subtype of the glutamate receptor to produce neuroprotection against acutely injured brain tissue. Because xenon's NMDA action is due to a competitive inhibition of the glycine coactivation site, it does not exhibit the neurotoxic properties of ion-pore blocking NMDA antagonists such as ketamine (Ma, Wilhelm, Maze, & Franks, 2002).

Based upon the fact that the NMDA receptor is involved in the excitotoxic processes that are crucial for the pathophysiology of many forms of neurological injury, the neuroprotective properties of xenon were investigated and confirmed in several injury models in multiple laboratories; these have been recently reviewed (Maze, 2016).

4. METHODS TO DEFINE WHICH NEUROPROTECTIVE INDICATION TO PURSUE

Apart from the NMDA receptor, xenon has activity at several other proteins that are juncture points in injury-reducing actions including two-pore-domain potassium channels (Gruss et al., 2004) and plasmalemmal adenosine triphosphate-sensitive potassium channels (Bantel, Maze, & Trapp, 2010), which were both identified using electrophysiologic techniques.

When considering which of the acute ongoing neurologic indications to pursue it was decided to investigate xenon's interaction with existing neuroprotective strategies. Within the last 15 years, the use of targeted temperature management (TTM), also referred to as therapeutic hypothermia, has been modestly effective in the management of both hypoxic ischemic encephalopathy (Azzopardi et al., 2009) and the postcardiac arrest syndrome (Hypothermia after Cardiac Arrest Study Group, 2002). Remarkably, xenon was shown to have an additive (Sabir et al., 2014) if not synergistic interaction (Ma et al., 2005) with hypothermia in preclinical models of acute neurologic injury in both in vitro and in vivo settings. While several explanations have been mooted, the reason for the hypothermia–xenon interaction has not been resolved. Because of the favorable interaction with TTM, clinical trials were launched to determine the efficacy of xenon for the postcardiac arrest syndrome (Arola et al., 2013, 2017; Laitio et al., 2016) and in hypoxic ischemic encephalopathy (Azzopardi et al., 2016, 2013; Dingley et al., 2014).

5. METHODS TO DEFINE THE NEUROPROTECTIVE PROPERTIES OF XENON IN POSTCARDIAC ARREST SYNDROME

In order to determine the preclinical efficacy of xenon for protection against postcardiac arrest syndrome, it is necessary to understand that apart from the direct injury to the heart and brain there are also systemic perturbations that occur following recovery from cardiac arrest, referred to as the postcardiac arrest syndrome; these reflect the *global* nature of the injury. Indicators of endothelial injury, such as soluble intracellular adhesion molecule I (sICAM-1), are considerably higher in CA patients during reperfusion (Geppert et al., 2003). Heparin-binding protein, a powerful inducer of endothelial damage (Gautam et al., 2001), is released into the circulation in high amounts in postresuscitation cardiac arrest patients (Dankiewicz, Linder, Annborn, Rundgren, & Friberg, 2013). Products of the endothelial glycocalyx are shed at various times during the reperfusion phase with syndecan-1 peaking and glycosaminoglycans peaking at a later time-points following reperfusion (Gabarron, Gomez, Monclus, Rodriguez-Roda, & Comas, 2013). Soluble thrombomodulin, a marker of endothelial injury, increased significantly in resuscitated cardiac arrest patients, peaked at between 6- and 24-h postarrest (Gando, Nanzaki, Morimoto, Kobayashi, & Kemmotsu, 2000). Microparticles isolated from the resuscitated cardiac arrest patients resulted in enhanced endothelial apoptosis ex vivo when compared to the normal controls (Fink et al., 2011).

Indicators of systemic inflammation are also elevated following successful resuscitation. Higher early levels of proinflammatory cytokines (such as IL-6) were seen in patients that had the greatest degree of global ischemia–reperfusion injury (Bro-Jeppesen et al., 2016) and in those that did not survive following initial successful resuscitation (Adrie et al., 2002). Plasma levels of proinflammatory cytokines collected within 12 and 24 h after reperfusion correlated positively with mortality and poor functional outcome. The early increase in IL-6 correlated most strongly with nonsurvival (Peberdy et al., 2016). Rise in procalcitonin levels was noted within 6 h of reperfusion and the highest levels were in the nonsurvivors (Annborn et al., 2013). Postcardiac arrest patients had attenuation of signaling through the pattern recognition receptors on monocytes which was most evident in nonsurvivors (Asmussen et al., 2016). Levels of macrophage migration inhibitory factor, a proinflammatory chemokine, rose significantly during the early reperfusion phase of cardiac arrest victims compared to ICU controls and normal volunteers (Pohl et al., 2016).

Finally, in a systemic context, coagulation is also perturbed in the post-cardiac arrest syndrome. Higher levels of tissue factor (capable of initiating the extrinsic coagulation pathway) were noted in the resuscitated patients, and the patient cohort also had lower levels of tissue factor pathway inhibitor and was associated with worse multiple organ dysfunction syndrome (Gando, Nanzaki, Morimoto, Kobayashi, & Kemmotsu, 1999). Levels of tissue plasminogen activator inhibitor-1 complex (tPAIC), indicative of fibrinolytic shutdown, correlated with development of multiple organ dysfunction syndrome in survivors of cardiac arrest (Wada et al., 2013). Levels of D-dimers (marker of endogenous fibrinolytic activity) were increased only during the reperfusion phase of cardiac arrest victims (Bottiger et al., 1995). Most of the patients who died within 2 days had very high levels of coagulation activity (thrombin–antithrombin complex), reduced anticoagulation (antithrombin, protein C, and protein S), and inhibited antithrombolysis (plasminogen activator inhibitor 1). In those who died there was a secondary downregulation of the thrombomodulin-endothelial protein C receptor pathway (Adrie et al., 2005).

Fries and colleagues implemented a pig cardiac arrest model that simulated a clinical setting of cardiac arrest from a cardiac cause (Fries et al., 2008). Fries and colleagues further refined their model (Fries et al., 2009), and all the important global and local clinical features of postcardiac arrest syndrome were evident in this model that was used as the preclinical underpinnings for the subsequent successful Phase 2 (Laitio et al., 2016) and proposed Phase 3 (NCT03176186) trials.

The study was conducted as follows (Fries et al., 2012). Injury was initiated by a 10-min ventricular fibrillation-induced cardiac and respiratory arrest (termination of mechanical ventilation during general anesthesia with muscle relaxation) after which the animals were resuscitated with compression, with a piston-driven chest compressor, and restoration of mechanical ventilation via a previously inserted endotracheal tube with an FiO_2 of 1.0 and tidal volume of 15 mL/kg. The compression to respiratory rates ratio was set at 30:2. Following 6 min of resuscitation an epinephrine bolus of 30 μg/kg was injected into the right atrium through a previously sited pulmonary artery catheter. After return of spontaneous circulation (ROSC) had occurred, the surviving pigs were randomized into one of three groups ($n = 5$), namely, "Control" in which the pigs received fluid therapy and maintenance of normothermia; mild therapeutic hypothermia (MTH) in which pigs received fluid therapy with surface cooling to achieve a core temperature of 33°C; and mild therapeutic hypothermia with xenon (MTH + Xe) in which pigs also received 1 h of xenon at 70% through a closed system circuit. In the animals that receive hypothermia (i.e., MTH groups), surface cooling lasted for 16 h after which rewarming was controlled to occur at a rate of 1°C/h. Postintervention, the neurologic status was scored with a validated neurologic deficit score in which four items (level of consciousness, respiration, posture, and feeding behavior) were evaluated yielding a score that varies from 0 (no neurologic impairment) to 100 that signifies brain death. Five days after resuscitation the pigs were sacrificed under general anesthesia, the brain was harvested, and histopathology was performed that included assessment of inflammation. Regarding the statistical comparison of the groups, analysis of variance was used for the continuous variables at given time-points, and followed by pair-wise Student's *t* test; Bonferroni corrections were used to adjust for multiple comparisons. Where appropriate, chi-square tests were performed to compare categorical variables.

The groups did not differ in the time to ROSC (11.01 ± 1.41 min for the controls vs 12.39 ± 2.09 min for the MTH group vs 11.37 ± 1.13 min for the MTH + Xe group, $P =$ not significant). During the baseline and the postresuscitation periods, the hemodynamic, respiratory, and acid–base parameters were continuously monitored. The only significant differences between groups occurred for (i) mean arterial pressure (MAP) following rewarming to normothermia in groups rendered hypothermic in which the Xe + MTH group had a lower MAP than control and (ii) at the onset of the hypothermia in which cardiac output was significantly lower in the MTH group (3.12 ± 0.6 L/min) vs Xe + MTH (4.6 ± 0.6 L/min). For those pigs that received hypothermia, a temperature of 33°C was achieved for the 16 h duration. Arterial blood gas measurements were similar in each group at all times. For the histopathological assessments, both treatment groups had significantly less necrotic neurons. Only the xenon-treated pigs exhibited less neuroinflammatory features. Only the pigs that were exposed to xenon showed improved neurological outcome which was durable with the neurological deficit score being statistically significantly different in the final 3 days of evaluation.

What made these data all the more remarkable was that xenon was administered for an extremely short duration (1 h) and only *after* the onset of injury. Furthermore, there were both long-term functional and morphological improvements in the group that received the combination of xenon with MTH.

6. METHODS TO DEFINE THAT XENON PROTECTS AGAINST WHITE MATTER BRAIN INJURY FOLLOWING CARDIAC ARREST (HYPOTHECA-XE NCT00879892)

Based upon successful animal studies in the setting of cardiac arrest by Fries et al. (2012), we undertook a randomized clinical trial (RCT) to compare outcome in patients who had, vs those that did not have, xenon added to therapeutic hypothermia which is also referred to as TTM. We investigated the effect of xenon on the extent of cerebral white matter brain damage as assessed by diffusion tensor magnetic resonance imaging in comatose survivors from out-of-hospital cardiac arrest (OHCA) initially at a single site (University of Turku Hospital) and added a second site in 2013 (University of Helsinki Hospital). In a preliminary report we commented on the feasibility and safety of delivering xenon to OHCA sufferers during therapeutic hypothermia (Arola et al., 2013). A manuscript describing the RCT was recently published (Laitio et al., 2016).

7. METHODS TO DEFINE THAT XENON IMPROVES SURVIVAL WITH GOOD FUNCTIONAL OUTCOME FOLLOWING CARDIAC ARREST (NCT03176186)

Following completion of the above Phase 2 RCT that demonstrated the efficacy of concurrent exposure to xenon during TTM in reducing brain damage in comatose survivors following an OHCA (Laitio et al., 2016), we now propose to undertake a pivotal Phase 3 RCT to test the safety and efficacy of xenon, delivered by a proprietary xenon delivery system (XDS), in comatose survivors following an OHCA. This study is entitled Xenon for Post Out of Hospital Cardiac Arrest Syndrome (XePOHCAS) and is a prospective, randomized, interventional trial in adult subjects with OHCA comparing treatment with standard-of-care postcardiac arrest intensive care TTM to standard-of-care postcardiac arrest intensive care TTM plus xenon by inhalation. The study is to be performed in accordance with Good Clinical Practice, the ethical principles that have their origin in the Declaration of Helsinki, and International Council for Harmonization E6 (Guideline for Good Clinical Practice).

XePOHCAS will enroll 1436 adult subjects with OHCA during which the standard of care for postcardiac arrest intensive care (that includes TTM) will be compared to the same treatment with the addition of 50% xenon.

Inclusion criteria include the following:

1. Age at least 18 years but less than or equal to 80 years.
2. Presumed cardiac cause of arrest.
3. Sustained (>20 min) spontaneous circulation.
4. No response to verbal commands on arrival to emergency department and prior to randomization (Glasgow Coma Scale score of <8).
5. Decision by the attending physician that the patient is eligible for TTM.

Exclusion criteria include the following:

1. A written do not attempt resuscitation is reported to providers before randomization.
2. There is a traumatic etiology of arrest, defined as concomitant blunt, penetrating, or burn-related injury, or uncontrolled bleeding or exsanguination.
3. The subject is suspected or known to be having a stroke or intracranial hemorrhage.
4. The cardiac arrest was unwitnessed.
5. The subject has a no-flow (cardiac arrest to initiation of cardiopulmonary resuscitation/defibrillation) time of greater than 10 min.
6. The duration until sustained return of spontaneous circulation (ROSC) is greater than 30-min postarrest.
7. The interval from arrival in the emergency department to randomization to initiation of intervention is greater than 4 h.
8. The subject is experiencing hypothermia (less than 30°C core temperature).
9. The subject was bedbound prior to cardiac arrest.
10. The subject experienced unconsciousness before cardiac arrest (cerebral trauma, spontaneous cerebral hemorrhage, intoxication, etc.).
11. The subject suffers from coagulopathy.
12. The subject has a systolic arterial pressure less than 80 mmHg or a MAP less than 60 mmHg lasting for more than 30 min after ROSC.
13. The subject is known to be pregnant.
14. The subject has received an investigational drug, device, or biologic product within 30 days.
15. The subject is known to be in the terminal phase of a chronic illness.
16. The subject has experienced hypoxemia ($SaO_2 < 85\%$) for more than 15 min after ROSC.
17. The subject is unable to maintain a SaO_2 of 90% on an FiO_2 of 50%.
18. The subject is known to have any other clinically significant laboratory abnormality, medical condition (such as intrinsic liver disease or severe

chronic obstructive pulmonary disease), or social circumstance that, in the investigator's opinion, makes it inappropriate for the patient to participate in this clinical trial.

19. It is logistically impossible to provide intervention.

20. The subject has any condition that would impact the evaluation of the modified Rankin Scale (mRS).

The primary objective of XePOHCAS is to evaluate at 30- and 90-day postcardiac arrest whether there is a difference in survival and functional outcome in comatose subjects who achieved ROSC within 30 min after cardiac arrest and were treated during TTM with xenon 50% + oxygen vs those receiving similar treatment, but without supplementary xenon.

The secondary objectives are proposed to be:

- evaluation whether there is a difference in percent of subjects surviving to Day 30 and Day 90 following treatment with xenon 50% + oxygen during TTM, compared to treatment with oxygen (without xenon) during TTM;
- comparison of the time to achieve target body temperature in OCHA subjects treated with xenon 50% + oxygen during TTM, compared to treatment with oxygen (without xenon) during TTM;
- testing, estimating, and adjusting for multivariable effects of selected covariates on 30-day and 90-day functional outcome in subjects treated with xenon 50% + oxygen during TTM, compared to treatment with oxygen (without xenon) during TTM;
- conducting sensitivity analyses to investigate the effects of subject dropout on the magnitude of estimates of treatment effects on the primary endpoint;
- evaluating the safety and tolerability of xenon therapy in the treatment of cardiac arrest.

Regarding the study design, the 1st Phase (screening) starts at hospital arrival in the emergency department and continues until randomization ($\leq$4 h). The 2nd Phase (intervention) begins at randomization, and is the start of TTM plus either 50% xenon or standard of care, and lasts for 24 h (or when TTM is stopped). The 3rd Phase (immediate evaluation) is the 72-h period following the completion of TTM. The 4th Phase (intermediate evaluation) lasts from the end of the immediate evaluation phase until hospital discharge. The 5th Phase (long-term evaluation) lasts from discharge until 30 days after cardiac arrest. The 6th and final phase (follow-up) continues from the end of the long-term evaluation phase to 90 days after the OHCA.

The primary efficacy endpoint for both the interim analyses and the final analyses will be the binomial proportion of subjects with good functional outcome on Day 30 and Day 90 post-OHCA. Good functional outcome will be defined as an mRS ≤ 2. Among the secondary endpoints will be the binomial proportion of randomized subjects still alive on study Day 30, the Kaplan–Meier estimate of the proportion of randomized subjects still alive on study Day 30, the subjects' time (from hospital admission) to reach target TTM, the binomial proportion of subjects with mRS ≤ 2 on Day 90, the binomial proportion of subjects alive on Day 90, the subjects' times to death, withdrawal, or other loss to follow-up from Day 1 to study Day 90.

Hypotheses for treatment-related difference in both efficacy endpoints will be tested with two-sided Z-statistics, with alpha levels controlled at the 0.05 level and power at 90%. All efficacy analyses will be performed on intention-to-treat population.

There will be interim and final analyses; the same primary and secondary endpoint analyses will be performed in each. The analyses will use an O'Brien–Fleming sequential design and stopping rules. Stopping for futility will occur if *either* survival *or* good functional are statistically worse and stopping for efficacy will occur if *both* survival and efficacy are improved.

8. METHOD FOR THE EFFICIENT DELIVERY OF XENON DURING THE PHASE 3 XePOHCAS TRIAL

NeuroproteXeon's xenon delivery system (XDS) (Fig. 1) consists of a control unit that mixes two gases (xenon and oxygen) and a disposable circuit that allows for low-flow xenon through a closed system to provide a neuroprotective and nonhypoxic dose of xenon. The device is independent of the ventilator for control. The ventilator provides the motive force for the gas to be moved around the closed circuit, allowing the clinician (respiratory therapist, or nurse, or intensivist) to control the respiratory rate and pressures through the ventilator and the desired concentration of xenon and oxygen through the control unit. The disposable part of the device contains a canister of soda lime (for absorption of CO_2) and a system of valves to ensure that oxygen is added from the ventilator as well as from the control unit.

The electronic control module (ECM) serves as the "brain" of the drug delivery portion of the XDS. The ECM is connected to the xenon supply and controls all dosing of the drug. The ECM will inject gas into the closed-loop breathing circuit and can perform a controlled purging of the circuit. Interaction with this component is necessary to start and stop drug

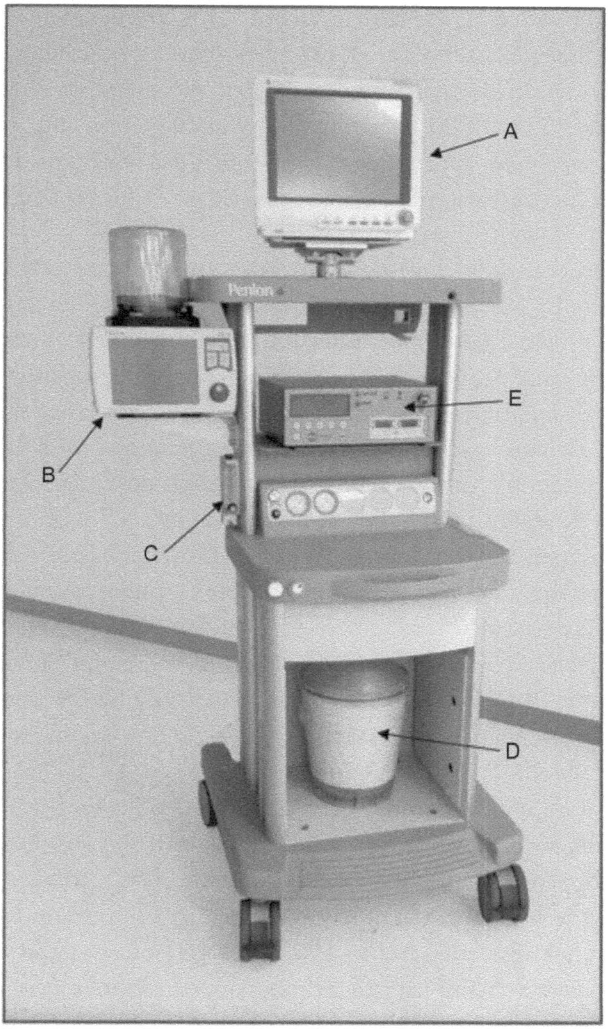

Fig. 1 (A) Optional stand-alone monitor (EKG, NIBP, SPAO2, capnography); (B) ventilator (spirometry calibrated for xenon density); (C) oxygen flush (rotameter up to 10 L/min); (D) 24 h CO_2 absorbent canister; (E) xenon delivery controller.

delivery. The XDS contains xenon and oxygen analyzers as well as other sensors to aid its function in monitoring and maintaining appropriate xenon and oxygen levels within the breathing circuit.

The ECM is activated by pressing the ON/OFF button and turned off by holding down the same button for more than 2 s. The ECM contains an internal battery that can be charged using an external isolated power supply during both powered "OFF" and powered "ON" modes. The circuit can be powered by the external power supply when present; otherwise, the internal battery provides an uninterrupted supply in the event of external power supply failure. The ECM requires a supply of xenon and oxygen at a working pressure of approximately 4 bar (60 psi) through two pressure regulators. The pressure is then limited to 20 psi before reaching the electronically controlled valves.

The microcontroller unit (MCU) on the main board is the processing module in the ECM. The MCU controls the volume of gas passed through each supply pipeline using two independently controlled (normally closed) valves in series that provide dual redundancy protection against unintended singular valve actuation. This arrangement is repeated for the spill function. The MCU actuates each valve. The valve control section is electrically isolated from the rest of the board. All valves are normally closed and will return to a safe state in the event of power failure.

The Emergency-Stop button in the "unlatched" position maintains power to the valve control section. When the "Emergency-Stop" button is "latched," power to this section is removed, and any valve control operations are effectively disabled until it is released. When the Emergency-Stop button is in the "latched" position, a message notifying the operator to the Emergency-Stop state is displayed on the screen, and the multifunction keypad is disabled.

The software within the ECM has been divided into software units. The software units include platform-specific (nonportable) software items and platform-independent (portable) software items. The number of interrupts used within the software system has been kept to a minimum to reduce the risk of unintended system behavior that may arise from asynchronous events. The ECM also contains alarms when concentration levels outside the specified criteria have been detected as an added safety feature to the design.

The XDS has been tested for functionality in four dogs. Anesthesia was induced in dogs using propofol, titrated to effect (6–8 mg/kg) followed by endotracheal intubation, placement in left lateral recumbent position, and connection to the anesthetic rebreathing circuit equipped with a 2-L rebreathing bag. For the first 5–10 min the fresh gas flow was set at 100 mL/kg and the XDS setting was at 50% xenon and 50% oxygen until an end-tidal concentration of 50/50 was reached. The fresh gas flow was then reduced to 60 mL/kg, and the inspired concentration adjusted to maintain the end-tidal setting. This concentration was maintained until 30 min postinduction, at which time the inspired concentration was adjusted to provide an end-tidal level until 45 min postinduction. After 45 min postinduction the end-tidal concentration was maintained for the duration of the test. The XDS was then turned off and the dogs recovered breathing oxygen with the fresh gas flow at 100 mL/kg. The dog's body temperature was maintained at 37–38°C throughout the period by means of external heat as necessary. The endotracheal tube was removed when swallowing reflexes returned, and the dogs were monitored during recovery.

During the period of xenon administration, the XDS setting, fresh gas flow, inspired and end-tidal oxygen, carbon dioxide, and xenon levels were recorded at 5-min intervals. Additionally, the dog's body temperature, respiratory and heart rate oxygen saturation level, (pulse oxygenation monitors on the tongue), and indirect blood pressure (systolic via Doppler monitor) were recorded. The times from turning off the XDS to extubation, and to return of consciousness were recorded. All four subjects recovered with no observed side effects.

The performance of the XDS is depicted in Fig. 2. The XDS was able to maintain the concentration of xenon and oxygen throughout the automode (50% oxygen, 50% xenon). The priming of the circuit was completed within

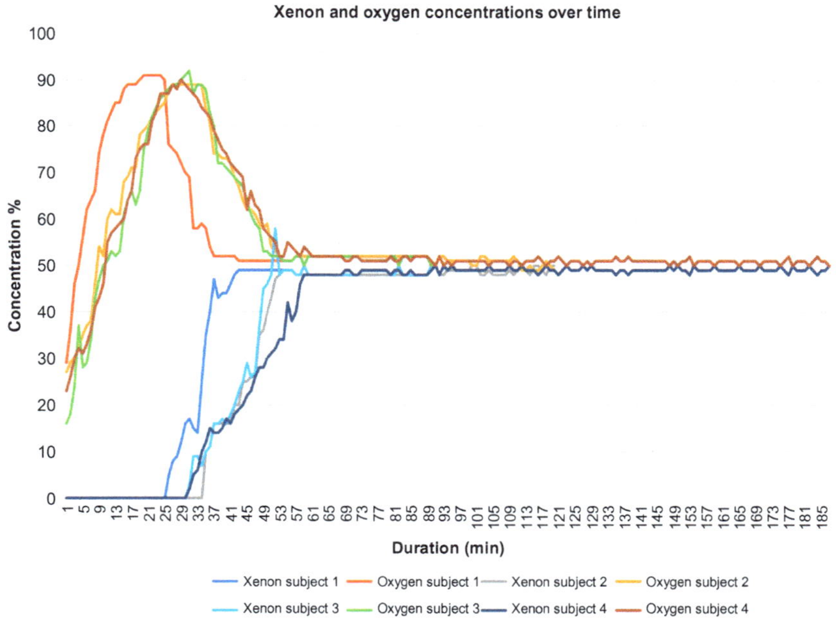

Fig. 2 Xenon and oxygen concentration over time in four propofol-anesthetized dogs using the xenon delivery system.

the desired time frame of 40–45 min. We estimate that with a tight seal by the cuff of the endotracheal tube, leaks from the system will require the addition of between 1 and 2 L/h of "fresh" xenon.

9. CONCLUSION

Using a succession of novel preclinical methods, xenon has been shown to be neuroprotective, synergistic with TTM, and efficacious in a porcine model that closely mimics the clinical setting observed in the post-cardiac arrest syndrome. Building upon this wealth of preclinical studies, a RCT has now reported that less brain is damaged, using an imaging biomarker, when xenon is administered during TTM to resuscitated patients with the postcardiac arrest syndrome. The stage is now set for a Phase 3 RCT to determine whether clinical outcomes are improved using a novel xenon delivery method that should limit the amount of xenon needed for the 24-h administration.

Because of the availability of suitable alternative anesthetics, we do not expect xenon to be used for routine surgical procedures apart from those in which ongoing brain injury is anticipated.

REFERENCES

Adrie, C., Adib-Conquy, M., Laurent, I., Monchi, M., Vinsonneau, C., Fitting, C., et al. (2002). Successful cardiopulmonary resuscitation after cardiac arrest as a "sepsis-like" syndrome. *Circulation*, *106*(5), 562–568.

Adrie, C., Monchi, M., Laurent, I., Um, S., Yan, S. B., Thuong, M., et al. (2005). Coagulopathy after successful cardiopulmonary resuscitation following cardiac arrest: Implication of the protein C anticoagulant pathway. *Journal of the American College of Cardiology*, *46*(1), 21–28.

Annborn, M., Dankiewicz, J., Erlinge, D., Hertel, S., Rundgren, M., Smith, J. G., et al. (2013). Procalcitonin after cardiac arrest—An indicator of severity of illness, ischemia-reperfusion injury and outcome. *Resuscitation*, *84*(6), 782–787.

Armstrong, S. P., Banks, P. J., McKitrick, T. J., Geldart, C. H., Edge, C. J., Babla, R., et al. (2012). Identification of two mutations (F758W and F758Y) in the N-methyl-D-aspartate receptor glycine-binding site that selectively prevent competitive inhibition by xenon without affecting glycine binding. *Anesthesiology*, *117*(1), 38–47.

Arola, O. J., Laitio, R. M., Roine, R. O., Gronlund, J., Saraste, A., Pietila, M., et al. (2013). Feasibility and cardiac safety of inhaled xenon in combination with therapeutic hypothermia following out-of-hospital cardiac arrest. *Critical Care Medicine*, *41*(9), 2116–2124.

Arola, O., Saraste, A., Laitio, R., Airaksinen, J., Hynninen, M., Backlund, M., et al. (2017). Inhaled xenon attenuates myocardial damage in comatose survivors of out-of-hospital cardiac arrest: The Xe-hypotheca trial. *Journal of the American College of Cardiology*, *70*(21), 2652–2660.

Asmussen, A., Fink, K., Busch, H. J., Helbing, T., Bourgeois, N., Bode, C., et al. (2016). Inflammasome and toll-like receptor signaling in human monocytes after successful cardiopulmonary resuscitation. *Critical Care*, *20*(1), 170.

Azzopardi, D., Robertson, N. J., Bainbridge, A., Cady, E., Charles-Edwards, G., Deierl, A., et al. (2016). Moderate hypothermia within 6 h of birth plus inhaled xenon versus moderate hypothermia alone after birth asphyxia (TOBY-Xe): A proof-of-concept, open-label, randomised controlled trial. *Lancet Neurology*, *15*(2), 145–153.

Azzopardi, D., Robertson, N. J., Kapetanakis, A., Griffiths, J., Rennie, J. M., Mathieson, S. R., et al. (2013). Anticonvulsant effect of xenon on neonatal asphyxial seizures. *Archives of Disease in Childhood. Fetal and Neonatal Edition*, *98*(5), F437–9.

Azzopardi, D. V., Strohm, B., Edwards, A. D., Dyet, L., Halliday, H. L., Juszczak, E., et al. (2009). Moderate hypothermia to treat perinatal asphyxial encephalopathy. *The New England Journal of Medicine*, *361*(14), 1349–1358.

Bantel, C., Maze, M., & Trapp, S. (2010). Noble gas xenon is a novel adenosine triphosphate-sensitive potassium channel opener. *Anesthesiology*, *112*(3), 623–630.

Bottiger, B. W., Motsch, J., Bohrer, H., Boker, T., Aulmann, M., Nawroth, P. P., et al. (1995). Activation of blood coagulation after cardiac arrest is not balanced adequately by activation of endogenous fibrinolysis. *Circulation*, *92*(9), 2572–2578.

Bro-Jeppesen, J., Kjaergaard, J., Stammet, P., Wise, M. P., Hovdenes, J., Aneman, A., et al. (2016). Predictive value of interleukin-6 in post-cardiac arrest patients treated with targeted temperature management at 33 degrees C or 36 degrees C. *Resuscitation*, *98*, 1–8.

Dankiewicz, J., Linder, A., Annborn, M., Rundgren, M., & Friberg, H. (2013). Heparin-binding protein: An early indicator of critical illness and predictor of outcome in cardiac arrest. *Resuscitation*, *84*(7), 935–939.

Dickinson, R., Peterson, B. K., Banks, P., Simillis, C., Martin, J. C., Valenzuela, C. A., et al. (2007). Competitive inhibition at the glycine site of the N-methyl-D-aspartate receptor by the anesthetics xenon and isoflurane: Evidence from molecular modeling and electrophysiology. *Anesthesiology*, *107*(5), 756–767.

Dingley, J., Tooley, J., Liu, X., Scull-Brown, E., Elstad, M., Chakkarapani, E., et al. (2014). Xenon ventilation during therapeutic hypothermia in neonatal encephalopathy: A feasibility study. *Pediatrics*, *133*(5), 809–818.

Dmochowski, I. (2009). Xenon out of its shell. *Nature Chemistry*, *1*(3), 250.

Fink, K., Feldbrugge, L., Schwarz, M., Bourgeois, N., Helbing, T., Bode, C., et al. (2011). Circulating annexin V positive microparticles in patients after successful cardiopulmonary resuscitation. *Critical Care*, *15*(5), R251.

Franks, N. P., Dickinson, R., de Sousa, S. L., Hall, A. C., & Lieb, W. R. (1998). How does xenon produce anaesthesia? *Nature*, *396*(6709), 324.

Fries, M., Brucken, A., Cizen, A., Westerkamp, M., Lower, C., Deike-Glindemann, J., et al. (2012). Combining xenon and mild therapeutic hypothermia preserves neurological function after prolonged cardiac arrest in pigs. *Critical Care Medicine*, *40*(4), 1297–1303.

Fries, M., Coburn, M., Nolte, K. W., Timper, A., Kottmann, K., Kuru, T. H., et al. (2009). Early administration of xenon or isoflurane may not improve functional outcome and cerebral alterations in a porcine model of cardiac arrest. *Resuscitation*, *80*(5), 584–590.

Fries, M., Nolte, K., Demir, F., Kottmann, K., Timper, A., Coburn, M., et al. (2008). Neurocognitive performance after cardiopulmonary resuscitation in pigs. *Critical Care Medicine*, *36*(3), 842–847.

Gabarron, S., Gomez, M., Monclus, H., Rodriguez-Roda, I., & Comas, J. (2013). Ragging phenomenon characterisation and impact in a full-scale MBR. *Water Science and Technology*, *67*(4), 810–816.

Gando, S., Nanzaki, S., Morimoto, Y., Kobayashi, S., & Kemmotsu, O. (1999). Tissue factor and tissue factor pathway inhibitor levels during and after cardiopulmonary resuscitation. *Thrombosis Research*, *96*(2), 107–113.

Gando, S., Nanzaki, S., Morimoto, Y., Kobayashi, S., & Kemmotsu, O. (2000). Out-of-hospital cardiac arrest increases soluble vascular endothelial adhesion molecules and neutrophil elastase associated with endothelial injury. *Intensive Care Medicine, 26*(1), 38–44.

Gautam, N., Olofsson, A. M., Herwald, H., Iversen, L. F., Lundgren-Akerlund, E., Hedqvist, P., et al. (2001). Heparin-binding protein (HBP/CAP37): A missing link in neutrophil-evoked alteration of vascular permeability. *Nature Medicine,* 7(10), 1123–1127.

Geppert, A., Zorn, G., Delle-Karth, G., Koreny, M., Siostrzonek, P., Heinz, G., et al. (2003). Plasma concentrations of von Willebrand factor and intracellular adhesion molecule-1 for prediction of outcome after successful cardiopulmonary resuscitation. *Critical Care Medicine, 31*(3), 805–811.

Gruss, M., Bushell, T. J., Bright, D. P., Lieb, W. R., Mathie, A., & Franks, N. P. (2004). Two-pore-domain K+ channels are a novel target for the anesthetic gases xenon, nitrous oxide, and cyclopropane. *Molecular Pharmacology, 65*(2), 443–452.

Harris, K., Armstrong, S. P., Campos-Pires, R., Kiru, L., Franks, N. P., & Dickinson, R. (2013). Neuroprotection against traumatic brain injury by xenon, but not argon, is mediated by inhibition at the N-methyl-D-aspartate receptor glycine site. *Anesthesiology, 119*(5), 1137–1148.

Hypothermia after Cardiac Arrest Study Group. (2002). Mild therapeutic hypothermia to improve the neurologic outcome after cardiac arrest. *The New England Journal of Medicine, 346*(8), 549–556.

Khajuria, A., Tay, C., Shi, J., Zhao, H., & Ma, D. (2014). Anesthetics attenuate ischemia-reperfusion induced renal injury: Effects and mechanisms. *Acta Anaesthesiologica Taiwanica, 52*(4), 176–184.

Laitio, R., Hynninen, M., Arola, O., Virtanen, S., Parkkola, R., Saunavaara, J., et al. (2016). Effect of inhaled xenon on cerebral white matter damage in comatose survivors of out-of-hospital cardiac arrest: A randomized clinical trial. *JAMA, 315*(11), 1120–1128.

Ma, D., Hossain, M., Chow, A., Arshad, M., Battson, R. M., Sanders, R. D., et al. (2005). Xenon and hypothermia combine to provide neuroprotection from neonatal asphyxia. *Annals of Neurology, 58*(2), 182–193.

Ma, D., Wilhelm, S., Maze, M., & Franks, N. P. (2002). Neuroprotective and neurotoxic properties of the 'inert' gas, xenon. *British Journal of Anaesthesia, 89*(5), 739–746.

Maze, M. (2016). Preclinical neuroprotective actions of xenon and possible implications for human therapeutics: A narrative review. *Canadian Journal of Anaesthesia, 63*(2), 212–226.

Peberdy, M. A., Andersen, L. W., Abbate, A., Thacker, L. R., Gaieski, D., Abella, B. S., et al. (2016). Inflammatory markers following resuscitation from out-of-hospital cardiac arrest-A prospective multicenter observational study. *Resuscitation, 103*, 117–124.

Pohl, J., Rammos, C., Totzeck, M., Stock, P., Kelm, M., Rassaf, T., et al. (2016). MIF reflects tissue damage rather than inflammation in post-cardiac arrest syndrome in a real life cohort. *Resuscitation, 100*, 32–37.

Preckel, B., Weber, N. C., Sanders, R. D., Maze, M., & Schlack, W. (2006). Molecular mechanisms transducing the anesthetic, analgesic, and organ-protective actions of xenon. *Anesthesiology, 105*(1), 187–197.

Sabir, H., Walloe, L., Dingley, J., Smit, E., Liu, X., & Thoresen, M. (2014). Combined treatment of xenon and hypothermia in newborn rats—Additive or synergistic effect? *PLoS One, 9*(10), e109845.

Wada, T., Gando, S., Mizugaki, A., Yanagida, Y., Jesmin, S., Yokota, H., et al. (2013). Coagulofibrinolytic changes in patients with disseminated intravascular coagulation associated with post-cardiac arrest syndrome—Fibrinolytic shutdown and insufficient activation of fibrinolysis lead to organ dysfunction. *Thrombosis Research, 132*(1), e64–9.

CHAPTER FIFTEEN

Xenon Anesthesia and CT: Noninvasive Measures of Brain Anesthetic Concentration

Andrew McKinstry-Wu[1], Charles W. Carspecken, Alex Proekt, Max B. Kelz
University of Pennsylvania, Philadelphia, PA, United States
[1]Corresponding author: e-mail address: andrew.mckinstry-wu@uphs.upenn.edu

Contents

Abstract

The existence of a barrier between anesthetic behavioral state transitions has been observed across phyla, but demonstrating that such a barrier exists and is not a pharmacokinetic artifact has not yet been possible in humans. Such an investigation requires temporally precise information regarding the brain concentration of anesthetic in order to demonstrate the specific pharmacokinetic–pharmacodynamic mismatch that is hysteresis. We propose a method to noninvasively determine brain tissue anesthetic concentration using computerized tomography and the radiopaque gaseous anesthetic xenon. Such a technique can be used to investigate pharmacokinetic–pharmacodynamic mismatches in humans.

Methods in Enzymology, Volume 602
ISSN 0076-6879
https://doi.org/10.1016/bs.mie.2018.01.011

1. INTRODUCTION

Any study of pharmacodynamics in vivo requires an exact knowledge of drug concentration in relevant tissue. This is of particular importance in the study of anesthetic effects as (1) not all of the relevant molecular targets are necessarily known and (2) one of the primary effects of the class, hypnosis, appears to be a system-level phenomenon that requires a behavioral endpoint to assess. Detailed models of pharmacokinetics have been critical to the study of anesthetic effects in order to be able to tightly correlate drug concentration with anesthetic action. Validation of these pharmacokinetic models, or study of pharmacodynamic effects that can occur semiindependently of pharmacokinetics (pharmacokinetic–pharmacodynamic mismatches) require temporally precise and accurate measures of drug concentration in relevant tissues. In animal models, this is typically achieved through tissue biopsy or harvest—which is exactly how the first pharmacokinetic models for volatile anesthetics were validated. Alternate approaches, such as microdialysis, are slightly less invasive ways to determine tissue concentration of a drug, though these approaches can be limited in their spatial or temporal resolution, and still require the implantation of a device (Kennedy, 2013). Both direct tissue sampling or a semiinvasive approach such as microdialysis prove inadequate when such approaches are not possible (as in human subject research). These categories of studies require a noninvasive method of determining tissue drug levels with temporal precision.

Theoretically, noninvasive determination of tissue drug levels could be achieved by modifying the drug under investigation. One such modification would be creating a version of the compound with an attached or associated fluorescent marker. This approach will not necessarily work with investigations of small molecules, however, as the tag itself possibly alters the physicochemical properties of the compound—thus making the modified compound an inexact model of the original compound under investigation (Etrych et al., 2016). An alternative approach leaving the chemical composition of the drug unchanged would be incorporation of radioisotopes in the synthesis of the drug. This solution is imperfect, as it would not resolve issues of precision of spatial localization and safety (in the case of human subject research).

We propose a general method of studying the in vivo effects of a general anesthetic, xenon gas, in human volunteers, that allows for temporally and spatially precise noninvasive measurement of drug tissue concentration using computerized tomography (CT). This method allows simultaneous monitoring of behavior, electroencephalography, and brain xenon concentration and could determine whether hysteretic pharmacokinetic–pharmacodynamic mismatch occurs. For example, this approach would allow for investigation of whether the barrier impeding transitions between hypnotic and awake states previously described in multiple animal models, known as neural inertia, also exists in humans (Friedman et al., 2010; Joiner et al., 2013).

2. RELEVANT PROPERTIES AND USES OF XENON

Xenon, a group 18 (noble gas) element, has anesthetic, physicochemical, and radiographic properties making it an ideal agent for the study of anesthetic pharmacokinetic–pharmacodynamic mismatch. Xenon is a naturally occurring, minimally reactive, rare element present in the atmosphere (0.0000087%). Purified xenon is obtained through fractional distillation of liquid air—a process requiring large amounts of energy to produce small quantities of pure gas. Despite being extremely chemically nonreactive, xenon acts as a general anesthetic at concentrations that allow for a nonhypoxic inhaled gas mixture at normal temperature and pressure. The minimum alveolar concentration necessary to produce immobility to a surgical stimulus for xenon in humans has variously been found to be as low as 63% to as high as 71%, while the concentration at which 50% of human subjects regain responsiveness is 33% (Goto et al., 2000; Hecker, Baumert, Horn, & Rossaint, 2004). Xenon is minimally soluble in blood (having a very low blood–gas partition coefficient) and is readily and rapidly diffusing, resulting in extremely fast, predictable kinetics. Unlike other inhaled general anesthetics, it has very little effect on cardiac output as it does not cause vasodilation, nor does it cause significant decrease in renal blood flow compared to other anesthetics (Hecker et al., 2004). Therefore, xenon has little effect on its own kinetics. Xenon is radiodense, having an X-ray mass attenuation coefficient roughly equivalent to iodine (3.633 cm^2/g at 80 keV for xenon vs 3.510 cm^2/g for iodine) (Hubbell & Seltzer, 2004). Thus, increases in xenon concentration will increase radiopacity linearly in a dose-dependent fashion.

Research studies have previously employed xenon's radiographic, physicochemical, or anesthetic properties separately to answer questions unrelated to pharmacokinetic–pharmacodynamic mismatch. Extensive clinical research and overseas use in clinical practice have shown xenon to be an extremely safe and reliable anesthetic (Law, Lo, & Gan, 2016), although studies of xenon as a sole anesthetic are not numerous (due to the high cost of distillation) with other agents frequently used for induction. Many of the anesthetic clinical studies involving xenon administration examined xenon's neuroprotective properties in the face of ischemic injury. A subset of imaging studies employ lower concentration inhalation of a radioactive isotope of xenon, ^{133}Xe, for ventilation/perfusion scans of the lungs, and perfusion imaging of the heart or liver, while both inhaled ^{133}Xe and stable xenon isotopes have been used for studies calculating cerebral blood flow (Coin & Coin, 1980; Drayer et al., 1978; Korhola, Valle, Frick, Wiljasalo, & Riihimäki, 1977; Strieder, Barnes, Aronow, Russell, & Kazemi, 1967; Winkler et al., 1977).

3. XENON DELIVERY

The high cost of xenon gas presents a practical limitation for its use as an inhaled agent. In order to maximally conserve the valuable gas, xenon must be used sparingly and, if possible, scavenged for reuse. Solutions for clinical practice in Europe rely on limiting xenon waste through the use of closed-circuit delivery methods that can also be used for nonclinical applications. Anesthesia workstations with dedicated closed-circuit xenon delivery modes are available commercially from vendors in Europe and are the current best option for xenon delivery (Stephan, Germany; Heinen and Lowenstin, Germany; Air Liquide, France). These workstations have the added benefit of integrated gas analyzers capable of independently verifying delivered xenon concentration. Other xenon delivery systems designed for delivery of xenon during imaging studies or for neuroprotection could also potentially be used for this investigation, but these may not be capable of delivering sufficient concentrations of xenon to produce general anesthesia (often as a safety measure).

Closed-loop delivery systems were designed with endotracheally intubated subjects in mind, but also can be used to deliver the slow, noninvasive, inhaled anesthetic dose–ramp required for studying pharmacokinetic–pharmacodynamic mismatch. Invasive airways, such as endotracheal tubes or laryngeal masks, produce excellent seals that limit gas leakage in an otherwise closed breathing circuit. Unfortunately, these invasive airways are also noxious stimuli that may consequently impede assessments of hypnotic potency. When properly fitted and adjusted, an anesthetic mask with attached head harness can provide an adequate noninvasive seal that allows for xenon administration in a closed circuit. Facial hair can impede the mask's seal, so subjects should shave any facial hair prior to mask fitting. For persistent small leaks, petroleum jelly applied to the portions of the mask in contact with the subject's face can help to minimize gas leaks. The tightly sealed mask also ensures that the subject is inhaling the intended concentration of xenon. The other requirement to control the concentration of inhaled xenon is verifying delivered xenon concentration via gas analysis. While mass spectrometry is the gold standard, it is not a viable solution for obtaining real-time gas measurements due to cost and availability of mass spectrometry devices. To address this, commercial xenon delivery systems use thermal conductivity sensors to analyze xenon concentration. Thermal conductivity allows for measurement of inhaled and end-tidal concentrations of xenon with short latency permitting estimation of equilibrium in blood with stabilization of end-tidal xenon. Thermal conductivity works on the principle that in a mixture of two known gases with different and known thermal conductivities, analyzing the thermal conductivity allows estimation of the relative concentrations of the two gases (King, Bretland, Wilkes, & Dingley, 2005). This functions well for xenon, which is delivered along with a balance of oxygen. When compared to the gold standard of mass spectrometry, thermal conductivity has been shown to be accurate to within 1%–2.5% (Luginbühl, Lauber, Feigenwinter, & Zbinden, 2002). While this level of accuracy limits the value of thermal conductivity as an absolute measure of end-tidal xenon concentration, it is sufficient for the purpose of this study type, as the ultimate measure against which we will compare our various endpoints is tissue xenon concentration as determined by CT.

4. SYNCHRONOUS ELECTROENCEPHALOGRAPHIC (EEG), BEHAVIORAL, AND TISSUE CONCENTRATION MEASURES

The relative speed of obtaining a head CT to determine brain tissue concentration of xenon allows for significant temporal resolution. When coupled with other endpoints that have equal or better temporal resolution, such as behavioral and EEG measures, CT can produce a tight association among brain tissue concentration and those other endpoints. The ideal test would timelock a stimulus with both an EEG and behavioral responses. We recommend an auditory stimulus for ease of administration, nonnoxious nature of the stimulus, and potential for evaluation of response while subject is either conscious or unconscious. Auditory tasks can further be subdivided into active and passive tasks, in order to potentially gain additional information when the subject is unable to provide an active response. For the active task we recommend a go/no go task, where, for example, the presented stimulus could be a recorded "go" or a phase-scrambled version of the same recording (distribution within the frequency domain was identical between the two recordings). This type of task has previously been shown to be sensitive to effects of sedation from both benzodiazepines and ethanol, with increases in latency to response and decreases in response accuracy (Deakin, Aitken, Dowson, Robbins, & Sahakian, 2004; Fillmore & Weafer, 2004). As this task can be mentally demanding after a time, this active task should be limited to a 5-min period where a stimulus is presented once a second. Between these 5 min of active task, it is possible to present a 5-min passive stimulus of a tone at random intervals from 2 to 7 s to provoke an auditory evoked potential on EEG. Response to an auditory stimulus allows time-locking for both the stimulus and response to the subject's EEG, as well as obtain both a discrete (present or absent, correct or incorrect) and continuous (time to response) evaluation of the response for the active task.

5. GENERAL METHOD FOR EVALUATING ANESTHETIC PHARMACOKINETIC–PHARMACODYNAMIC MISMATCH IN HUMANS WITH EEG AND BEHAVIORAL TESTING

Preparation and Baseline Testing

- Subject comes in having refrained from eating for the past 8 h, from most liquids for the past 4 h, and clear liquids for the previous 2 h.
- Given the high emetic potential of xenon, multiple forms of antinausea prophylaxis are recommended. A minimum of two agents with different pharmacologic targets is ideal. In order to minimize agents with psychoactive effects, a 5-HT3 antagonist (such as ondansetron) and NK1 antagonist (such as aprepitant) are recommended.
- After an intravenous line is placed and EEG headcap is affixed, ear buds should be inserted and mask fitted with harness attached, with the subject breathing room air.
- It is important to check the mask for air leaks, followed by connecting subject to basic vital sign monitoring (EKG, pulse oximeter, and blood pressure) prior to beginning continuous monitoring.
- After the subject lies on the CT table, assess the comfort of the subject and add any necessary padding placed to minimize any potential noxious stimuli. If the subject feels subjective increased work of breathing, pressure support ventilation can alleviate this sensation. If pressure support is given, the subject's mask should be rechecked for gas leaks.

- Begin EEG acquisition and record at least 5 min of baseline eyes closed EEG prior to the start of testing.
- Subject should perform baseline passive and active attention tasks—performing multiple rounds of baseline tasks will be useful for later analyses comparing tasks at a given xenon concentration to baseline.
- Acquire baseline CT full head image, choose a plane with minimal to no artifact from EEG cap for all further CT images.

Xenon Administration

- We suggest administering xenon in ascending and descending steps with sufficient time at each step to reach equilibrium prior to testing. It will also be helpful to have increased resolution at concentrations near to predicted loss and reestablishment of consciousness, which generally fall in the range of 30%–45% in healthy subjects. In order to capture the full range of responses, we recommend starting at a dose of 20%, where there is mild impairment with no loss of consciousness and climbing as high as 60% to ensure capturing outliers. Given that we have generally seen equilibration around 10 min, 20 min steps allow for 10 min of testing at each step.
- Xenon concentration is increased or decreased based on the step and subject breathes the new concentration. Serial single slice head CTs should be obtained at regular intervals to establish brain concentration and certify when steady-state concentration is reached. For instance, in a 20-min step, single slice CTs might be taken at minute, 5, 10, 15, and 19 with the expectation that steady state will occur in the vicinity of minute 10.
- The subject should perform the serial tasks, passive and active.

Recovery

- Vital signs continue to be monitored during the recovery period, though subjects return quite rapidly to baseline.
- Once the subject feels fully recovered and vital signs are stable, headcap and intravenous line are removed, and the subject is discharged. Note that due to the possibility of continued fatigue after anesthetic exposure, the subject should not drive themselves.

6. DATA ANALYSIS

The primary focus of the investigation is the detection of possible differences in brain state during periods of identical brain xenon concentration. In order to establish identical concentrations of xenon in the brain, we compare the baseline CT slice to the same slice during the exposure

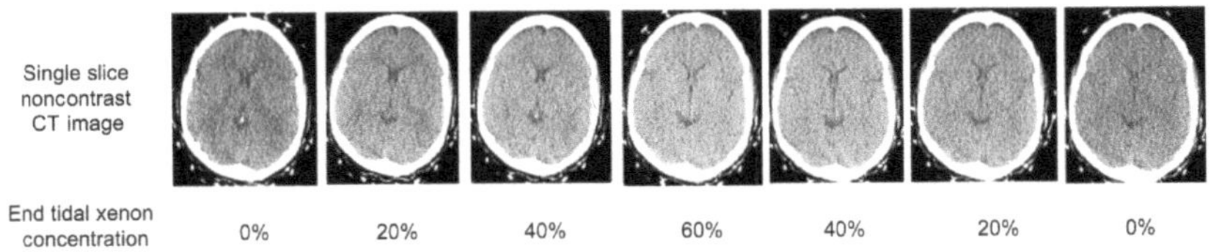

Fig. 1 Increased brain tissue concentration of xenon results in increased radiopacity. These serial images show the same section of brain with increasing and decreasing inhaled xenon concentration.

and calculate the difference in Hounsfield units at an identical region of interest in the two slices (Fig. 1). That difference corresponds linearly to the xenon concentration in tissue and timepoints with an identical difference of Hounsfield units have the same concentration of xenon. We have found that approximately 5 min of inhalation of xenon is sufficient to reach equilibrium in the brain.

Having established identical brain concentrations of xenon, performance at those identical concentrations can be compared, different only in whether the subject was previously awake or anesthetized. Depending on the exact nature of the active task, analysis endpoints can include presence or absence of response, response accuracy, and latency to response. For the suggested passive task, EEG can be used to analyze the auditory evoked potential looking at response amplitude of the evoked peaks and latency to those peaks. Gross and regional spectral differences in the EEG during periods of identical xenon concentration can be compared. Previously described spectral characteristics of xenon hypnosis include frontal–parietal and temporal–occipital increases in relative and total delta power, frontal–parietal and temporal–occipital increases in total theta power, and frontal–parietal and temporal–occipital decrease in relative and total beta power (Laitio et al., 2008). These previous results were recorded with four-channel EEG, however, which lacks resolution for examining any real regional differences in band power. Higher channel count allows for more complex connectivity analyses, including transfer entropy, phase–lag index, and coherence.

7. CHALLENGES AND POTENTIAL PITFALLS

A study of this complexity presents numerous potential challenges. While higher-energy CT scans produce increased image quality and resolution, limiting the overall radiation dose for healthy volunteers is critical.

Such a study can require many CT images, so limiting the image to a single level, and using a lower power CT, at 70 keV, with a larger region of interest to calculate Hounsfield unit change can help to limit radiation dose at the expense of noise and resolution. If the study was focusing on the xenon concentration at a smaller region, a higher energy CT for increased resolution may be necessary. The EEG headcap can additionally cause artifacts in the CT image. In order to minimize those artifacts, a full head CT at baseline can help to choose a level with minimal artifact from the EEG cap for all subsequent scans. Finally, with a subject near the border of consciousness and disinhibited, there is a risk of movement artifact. Acclimatizing the subject to the scanner, mask, and headcap while they are fully awake can minimize provoked anxiety and subsequent movement.

Aside from radiologic issues, the administration of xenon, while extremely safe when compared with most other inhaled anesthetics, does carry some risks. These include not only the disinhibition previously mentioned, but also significant risk of nausea and vomiting. We found that pretreatment with two separate antinausea agents, aprepitant and ondansetron, can prevent nausea and vomiting. Finally, as xenon is more dense than air, there is potential for greater work of breathing at higher inhaled concentrations. Providing mild pressure support during xenon administration proved sufficient to alleviate any subjective increase in work of breathing.

ACKNOWLEDGMENTS

Development and implementation of the described techniques were possible through the generous support of the James S. McDonnell Foundation and the National Institutes of Health, via K08 GM123317. These protocols were developed collaboratively with John Woo at the University of Pennsylvania. Additional thanks to Mathias Bresner for providing neurocognitive testing, W. Andrew Kofke for his expertise in xenon delivery, to George Mashour and Michael Avidan for their support, and for the generous support of the Department of Anesthesiology and Critical Care at the University of Pennsylvania.

REFERENCES

Coin, C. G., & Coin, J. T. (1980). Contact enhancement by xenon gas in computed tomography of the spinal cord and brain: Preliminary observations. *Journal of Computer Assisted Tomography*, *4*(2), 217–221.

Deakin, J. B., Aitken, M. R., Dowson, J. H., Robbins, T. W., & Sahakian, B. J. (2004). Diazepam produces disinhibitory cognitive effects in male volunteers. *Psychopharmacology*, *173*(1–2), 88–97.

Drayer, B. P., Wolfson, S. K., Reinmuth, O. M., Dujovny, M., Boehnke, M., & Cook, E. E. (1978). Xenon enhanced CT for analysis of cerebral integrity, perfusion, and blood flow. *Stroke*, *9*(2), 123–130.

Etrych, T., Lucas, H., Janoušková, O., Chytil, P., Mueller, T., & Mäder, K. (2016). Fluorescence optical imaging in anticancer drug delivery. *Journal of Controlled Release: Official Journal of the Controlled Release Society*, *226*, 168–181.

Fillmore, M. T., & Weafer, J. (2004). Alcohol impairment of behavior in men and women. *Addiction*, *99*(10), 1237–1246.

Friedman, E. B., Sun, Y., Moore, J. T., Hung, H. T., Meng, Q. C., Perera, P., et al. (2010). A conserved behavioral state barrier impedes transitions between anesthetic-induced unconsciousness and wakefulness: Evidence for neural inertia. *PLoS One*, *5*(7), e11903.

Goto, T., Nakata, Y., Ishiguro, Y., Niimi, Y., Suwa, K., & Morita, S. (2000). Minimum alveolar concentration–awake of xenon alone and in combination with isoflurane or sevoflurane. *Anesthesiology*, *93*(5), 1188–1193.

Hecker, K., Baumert, J. H., Horn, N., & Rossaint, R. (2004). Xenon, a modern anaesthesia gas. *Minerva Anestesiologica*, *70*(5), 255–260.

Hubbell, J. H., & Seltzer, S. M. (2004). *Tables of X-ray mass attenuation coefficients and mass energy-absorption coefficients (version 1.4)*. National Institute of Standards and Technology. https://www.nist.gov/pml/x-ray-mass-attenuation-coefficients.

Joiner, W. J., Friedman, E. B., Hung, H. T., Koh, K., Sowcik, M., Sehgal, A., et al. (2013). Genetic and anatomical basis of the barrier separating wakefulness and anesthetic-induced unresponsiveness. *PLoS Genetics*, *9*(9), e1003605.

Kennedy, R. T. (2013). Emerging trends in in vivo neurochemical monitoring by microdialysis. *Current Opinion in Chemical Biology*, *17*, 860–867.

King, R., Bretland, M., Wilkes, A., & Dingley, J. (2005). Xenon measurement in breathing systems: A comparison of ultrasonic and thermal conductivity methods. *Anaesthesia*, *60*(12), 1226–1230.

Korhola, O., Valle, M., Frick, M. H., Wiljasalo, M., & Riihimäki, E. (1977). Regional myocardial perfusion abnormalities on xenon-133 imaging in patients with angina pectoris and normal coronary arteries. *The American Journal of Cardiology*, *39*(3), 355–359.

Laitio, R. M., et al. (2008). Bispectral index, entropy, and quantitative electroencephalogram during single-agent xenon anesthesia. *Anesthesiology*, *108*(1), 63–70.

Law, L., Lo, E., & Gan, T. (2016). Xenon anesthesia: A systematic review and meta-analysis of randomized controlled trials. *Anesthesia and Analgesia*, *122*(3), 678–697.

Luginbühl, M., Lauber, R., Feigenwinter, P., & Zbinden, A. M. (2002). Monitoring xenon in the breathing circuit with a thermal conductivity sensor. Comparison with a mass spectrometer and implications on monitoring other gases. *Journal of Clinical Monitoring and Computing*, *17*(1), 23–30.

Strieder, D. J., Barnes, B. A., Aronow, S., Russell, P. S., & Kazemi, H. (1967). Xenon 133 study of ventilation and perfusion in normal and transplanted dog lungs. *Journal of Applied Physiology*, *23*(3), 359–366.

Winkler, et al. (1977). Xenon inhalation as an adjunct to computerized tomography of the brain: Preliminary study. *Investigative Radiology*, *12*(1), 15–18.

SECTION V

Electrophysiology

CHAPTER SIXTEEN

Native System and Cultured Cell Electrophysiology for Investigating Anesthetic Mechanisms

Paul Hoerbelt, Boris D. Heifets[1]
Stanford University School of Medicine, Stanford, CA, United States
[1]Corresponding author: e-mail address: boris.heifets@gmail.com

Contents

Abstract

Anesthetic agents interact with a variety of ion channels and membrane-bound receptors, often at agent-specific binding sites of a single protein. These molecular-level interactions are ultimately responsible for producing the clinically anesthetized state. Between these two scales of effect, anesthetic agents can be studied in terms of how they impact the physiology of neuronal circuits, individual neurons, and cells expressing individual receptor types. The acutely dissected hippocampal slice is one of the most extensively studied and characterized preparations of intact neural tissue and serves as a highly useful experimental model system to test hypotheses of anesthetic mechanisms. Specific agent–receptor interactions and their effect on excitable membranes can further be defined with molecular precision in cell-based expression systems. We highlight several approaches in these respective systems that we have used and that also have been used by many investigators worldwide. We emphasize economy and quality control, to allow an experimenter to carry out these types of studies in a rigorous and efficient manner.

Methods in Enzymology, Volume 602
ISSN 0076-6879
https://doi.org/10.1016/bs.mie.2018.01.012

1. INTRODUCTION

Electrophysiology is a powerful tool for the study of function and connectivity in the nervous system. For the better part of a century, considerable advances through a series of electrophysiological techniques, from single-channel recording to the monitoring of synchronous activity of intact neuronal ensembles, have prompted an explosion of knowledge in neuroscience. In this chapter, we discuss tried-and-true and state-of-the-art electrophysiological methods suited to revealing the mechanisms of anesthetic drugs.

Classic electrophysiology preparations include the squid giant axon, in which the iconic neuronal action potential was first recorded and defined in the 1930s–1950s by Hodgkin, Huxley, Cole, Curtis, and Katz (Huxley, 2002). The same preparation, which allows for the direct insertion of metallic electrodes into the large axon to measure potential difference across the membrane, was used to discover that local anesthetic drugs act only on the inside of the neuronal membrane to silence action potential firing (Frazier, Narahashi, & Yamada, 1970).

Since the 1970s, consistent advances in the preparation and recording of mammalian nervous tissue, as well as the cloning of mammalian receptors and generation of cDNA libraries, have made it possible to conduct extremely sensitive electrophysiological recordings in vitro and ex vivo. For example, the patch-clamp technique (Sakmann & Neher, 1984), which involves carefully sealing a glass electrode onto a patch of membrane to measure membrane currents under a variety of conditions, remains the gold standard for the study of channels, receptor function, and synaptic biology. Furthermore, equipment to monitor neuronal activity in vivo has improved considerably: probes have become smaller and therefore less invasive, and better amplifier has improved the signal-to-noise ratio.

Electrophysiological investigations have been crucial for developing our current understanding of anesthetics' mechanisms and will continue to drive advances in developing novel anesthetics. Compared to their historical antecedents, such as ether and halothane, modern anesthetics' vastly improved safety profile has undoubtedly fueled the rapid growth in elective surgical procedures worldwide, increasingly performed in sicker patients at the extremes of age. Pushing against these limits highlights areas where anesthetics still fall short, propelling the search for novel, more ideal anesthetic drugs. Current anesthetics are still associated with intraoperative hypotension and hemodynamic instability, postoperative delirium, postoperative cognitive dysfunction, and anesthetic-related neurodevelopmental delay. Efforts to improve anesthetics often begin with a comparison to drugs with known central nervous system receptor pharmacology. However, anesthetics, particularly volatile anesthetics, are notoriously promiscuous in their molecular interactions (Eckenhoff, 2001). Therefore, experimental models that can capture this complexity and recapitulate key features of in vivo neural systems are quite useful, permitting manipulations and analysis at the level of circuit, cell, synapse, and receptor.

In this chapter, we review the major electrophysiological approaches that we and others have successfully used to study anesthetic drugs. We will take the reader through the most critical steps in preparing and executing successful electrophysiology experiments in acutely dissected brain slices and cultured cell systems. Along the way, we will guide the reader toward the minimum essential reagents, equipment, and literary resources needed for the experiments.

2. HIPPOCAMPAL SLICE ELECTROPHYSIOLOGY

We begin with one of the most well-studied intact systems, the acutely dissected rodent hippocampal slice preparation. While anesthetics certainly have targets outside the hippocampus that are responsible for the general anesthetic state, the hippocampus plays a role in anesthetics' amnestic effects (Perouansky & Pearce, 2011) and is capable of numerous forms of synaptic plasticity relevant to learning and memory that may be impacted by anesthetics (Malenka & Bear, 2004). Moreover, this preparation has several highly useful experimental attributes. First, the anatomical organization is easily appreciated by eye or low-power microscope: afferent projections, dendritic fields, and principal cells are organized in well-demarcated zones (Fig. 1A). Second, the "Cornu Ammonis" area 1 (CA1) area of the hippocampus has been extensively described at multiple levels: the anatomical and molecular taxonomy of interneurons and principal cells; the connectivity of afferent inputs to the CA1 principal (pyramidal) cells (PCs) and interneurons (INs); and the connectivity between PCs and INs involved in both "feed-forward" and "feedback" inhibitory circuits. Finally, the various currents within PCs are well understood in terms of the underlying ion channel identity, expressed isoforms, and pharmacological sensitivity (Freund & Buzsáki, 1996; Shepherd, 2005).

Two broad types of electrophysiological recording techniques are often used to study the hippocampus: extracellular field recording and whole-cell recording. Hippocampal field recordings are minimally invasive and measure the aggregate electrical activity of hundreds to thousands of neurons. This set of techniques can be performed with a low power microscope and manually driven micromanipulator. On the other hand, whole-cell recording allows the experimenter to measure and control a single neuron's current or voltage, and requires a high-powered microscope, motorized micromanipulator, and more precise environmental controls. We refer the reader to Molleman (2003), Hille (2001), and Hamill, Marty, Neher, Sakmann, and Sigworth (1981) for other types of recordings, including cell-attached, perforated patch, and inside- or outside-out patch recordings.

2.1 Field Excitatory Postsynaptic Potential and Population Spike Recording in CA1

For the experimenter who is still becoming familiar with electrophysiological techniques in intact systems, we recommend beginning with recording extracellular field potentials in the rodent CA1 hippocampal area, particularly field excitatory postsynaptic potentials (fEPSPs) generated by the Schaffer collateral—CA1 circuit (SchColl-CA1; Fig. 1A–C). Due to the lamellar organization of the hippocampus (Amaral & Witter, 1989), electrical stimulation (via electrode) along the path of the presynaptic SchColl pathway (Fig. 1B) produces a stereotypical, *synchronized* depolarization of the SchColl terminals (fiber volley [FV]), resulting in release of the excitatory neurotransmitter, glutamate. Glutamate release produces a postsynaptic response in CA1 which starts with an *en masse* depolarization of dendrites in *stratum radiatum* (SR) that then travels to the cell bodies in *stratum pyramidale* (SP). If this depolarization (the fEPSP) surpasses the action potential threshold, a synchronous population action potential or "pop spike" is generated in the SP (Fig. 1C–E). The fEPSP is best recorded in SR and quantified by the slope of the waveform's initial rising phase, rather than fEPSP's peak amplitude (Fig. 1C, top). fEPSP amplitude, while potentially useful, is easily contaminated by the pop spike (Fig. 1C, bottom). To understand the significance of the fEPSP slope, it is useful to recall that the aggregate neuronal cell membrane acts like a capacitor in a simple electronic circuit, "buffering" changes in transmembrane voltage against changes in the flow of charge. We can then take the definition of a capacitor, where capacitance (C) is equal to charge (Q) divided by voltage (V), and rearrange it in terms of V, which we are actually measuring with our electrode (1).

$$V = \frac{Q}{C} \tag{1}$$

Expressing (1) as a differential equation (2), we now have a term describing the change in charge over time, which is the definition of current (I).

$$\frac{dV}{dT} = \frac{1}{C} \times \frac{dQ}{dT} \tag{2}$$

Simplifying (2), we see that the change in voltage over time (i.e., the slope of the fEPSP) is equal to the current, I, divided by the constant for capacitance, C.

$$\frac{dV}{dT} = \frac{I}{C} \tag{3}$$

This final expression is known as "Ohm's law" for a capacitor. Thus, fEPSP slope is a measure of aggregate transmembrane current flow.

The pop spike is best recorded in the SP and is measured by its amplitude (Fig. 1D and E). As hippocampal slices deteriorate over time, or when inhibition is pharmacologically blocked (e.g., with picrotoxin [PTx]), a $GABA_A$ receptor ($GABA_AR$) antagonist, the pop spike may get larger, broader, and may be followed by repetitive spiking due to recurrent excitatory connections in CA1 (Fig. 1E). These changes represent hyperexcitability, either through an increasingly depolarized membrane potential or gradual loss of inhibitory INs. Hyperexcitable SP layer neurons may lose their synchrony, broadening the pop spike, and may demonstrate reverberating action potentials in an epileptiform pattern.

fEPSP/pop spike recordings yield useful mechanistic information in their own right, but this simple technique is also a useful quality control measure when we are having problems with a more advanced recording technique, like whole-cell patch clamp. In a healthy slice, fEPSP and pop spikes should have the characteristics shown in Fig. 1C, top and E, top: slope of 200–800 mV/ms range, a small FV, paired pulse facilitation of fEPSPs, and a pop spike in the 10 mV range (see 2.1.3). Common problems arise with excessive tissue manipulation, external solutions (i.e., artificial cerebrospinal fluid [ACSF]), ACSF oxygenation and flow rate, and temperature variability. The experimenter should not have difficulty recording stable fEPSPs or pop spikes for at least an hour.

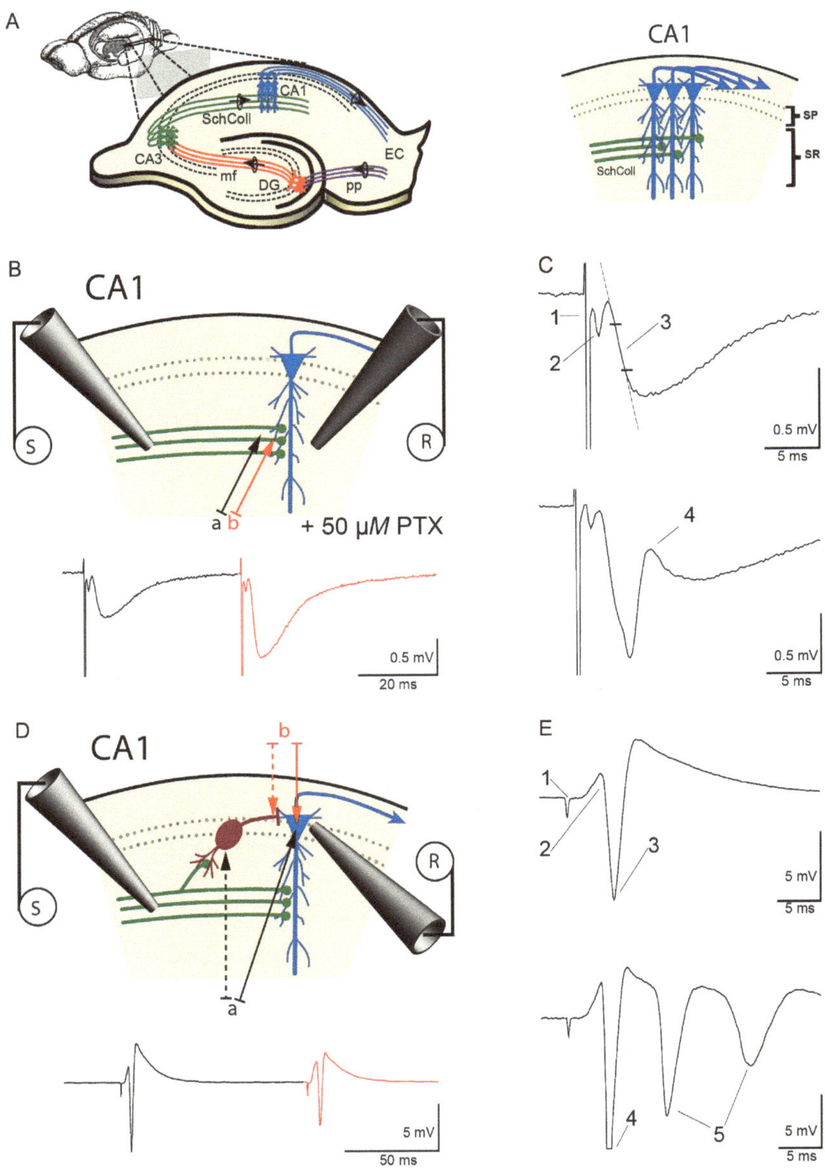

Fig. 1 Hippocampal anatomy and electrode configurations for recording fEPSPs and pop spikes. (A) *Left*, cross section of rodent hippocampus, characteristic of the lamellar organization of this structure. Within the hippocampal slice, major afferent pathways (*lower case*) and subregions with cell layers (*upper case*) are identified. *DG*, dentate gyrus; *EC*, entorhinal cortex; *mf*, mossy fiber; *pp*, performant path. Other areas as listed in main text. *Right*, magnified representation of the CA1 subregion. SchColl afferents from CA3 synapse onto dendrites of CA1 principal cells in the SR. CA1 PC cell bodies are found in

For a more detailed practical consideration of the equipment, solutions, and methods involved in slice electrophysiology, we highly recommend the very accessible book "Patch Clamping: An Introductory Guide to Patch Clamp Electrophysiology" (Molleman, 2003). The protocol below incorporates our preferences, and many substitutions are acceptable.

We prepare our tissue in a manner nearly identical to that described in Mathis, Furman, and Norris (2011). For mice older than 3 months, we typically perform intracardiac perfusion with a modified ACSF (sucrose cutting solution), prior to decapitation and brain dissection. The sucrose solution has been shown to improve viability of INs of the hippocampus (Kuenzi, Fitzjohn, Morton, Collingridge, & Seabrook, 2000) as well as other sensitive brain regions (Aghajanian & Rasmussen, 1989). We generally prefer cutting coronal slices of hemispheric blocks rather than slicing dissected hippocampus. Dissecting hippocampus out and cutting sections in a "jelly roll" fashion (Mathis et al., 2011) has advantages for pathways that must retain their lamellar architecture (e.g., mf–CA3 pathway), at the expense of significantly more tissue manipulation. As an alternative to storing brain slices "free floating" in a slice-holding chamber, one can lay each slice onto a small circle of Whatman Filter paper—this method is appropriate

the SP. (B) Recording configuration and typical responses for fEPSPs. Stimulating (S) and recording (R) electrodes are placed in the SR (*top*). Excitatory synaptic transmission is isolated by continuous presence of 50 μ*M* PTx, a $GABA_AR$ antagonist. The first of the paired stimuli (a, *black*) typically produces a smaller response than the second stimulus (b, *red*). (C) *Top*, magnified fEPSP waveform. *1*, stimulus artifact; *2*, fiber volley; *3*, rising phase of fEPSP. Slope of the fEPSP (mV/ms) is measured on the linear portion as shown, typically taken between 20% and 80% of the fEPSP peak. *Bottom*, fEPSP interrupted by pop spike (*4*). (D) Recording configuration and typical responses for pop spike recordings. Recording electrode is now placed in the SP (*top*). The first stimulus (a, *black*) activates the SchColl pathways and produces a measurable response in CA1 PCs (a, *solid arrow*), while also activating a feed forward IN (*purple*) in the stimulation path (a, *dashed arrow*), whose response is not yet measured. The second stimulus (b, *red*) again activates CA1 PCs through the SchColl pathway (b, *solid arrow*), whose response is now modulated by the inhibitory feed forward IN (b, *dashed arrow*), resulting in paired-pulse depression of the pop spike (*bottom*). (E) *Top*, magnified pop spike waveform. *1*, stimulus artifact; *2*, CA1 dendritic depolarization (fEPSP), in opposite direction with respect to fEPSP shown in **C3**, due to the site of recording; *3*, pop spike. *Bottom*, epileptiform activity in PC layer; *4*, pop spike size increases, amplifier saturation artifact visible; *5*, reverberating pop spikes. *Panel (A): Modified with permission from Amaral, D.G., & Witter, M.P. (1989). The three-dimensional organization of the hippocampal formation: A review of anatomical data.* Neuroscience, 31, *571–591. Panel (E): Pop spike traces provided courtesy of Noëlie Cayla and M. Bruce MacIver.*

for field recordings where transillumination of the slice for microscopy is unnecessary. This method allows the experimenter to transfer the filter paper and slice to the recording chamber without touching the tissue itself. The tissue is adherent enough to the filter paper that additional means of securing the slice (e.g., with a "harp", see below) are unnecessary. Once the hippocampal slices have been prepared and placed into the slice-holding chamber, they should be recovered for 30 min at 35°C and returned to room temperature. We typically begin recording 30 min to 1 h after this time.

For the experimenter who only wishes to perform field recordings, we have assembled a low-cost list of equipment below, schematized in Fig. 2. For those planning to perform whole-cell patch-clamp experiments, additional components are necessary (Fig. 4A–D).

2.1.1 Equipment (Field Recording)

- Antivibration table.
- Dissecting microscope.
- Slice chamber.
- "Harp" to stabilize slice on slice chamber (Fig. 4C) or Whatman Filter Paper Grade 2.
- Carbogen (95% O_2/5% CO_2), H-cylinder with flow regulator.
- Peristaltic perfusion pump capable of 0.5–10 mL/min or inverted separator funnel for gravity perfusion (Fig. 2C).
- Temperature controller (CL-100, Warner Instruments).
- Micromanipulator (MN-153, Narishige).
- Glass capillary pipette puller (PC-10, Narishige) and capillary glass.
- Electrically isolated voltage stimulator (Digitimer DS-2A).
- Bipolar- or monopolar-stimulating electrode: either type can be easily fabricated in the lab.

  Bipolar (Fig. 2E):
  - Pull a glass capillary pipette as if for recording (3–5 MΩ).
  - Using a quarter (Fig. 2F) break the tip until two stainless steel or tungsten-stimulating electrodes (8–15 kΩ impedance, AM Systems) can be passed through the tip (Fig. 2G).
  - Using epoxy or cyanoacrylate glue, join the electrodes.

  Monopolar:
  - Under a dissecting microscope, pass a Kim-wipe or a quarter across the tip of a pipette until it breaks to an approximate 20–50 μm width. Backfill the pipette with ACSF, and place one end of a stainless steel wire (stripped at both ends) into the pipette, the other end connected

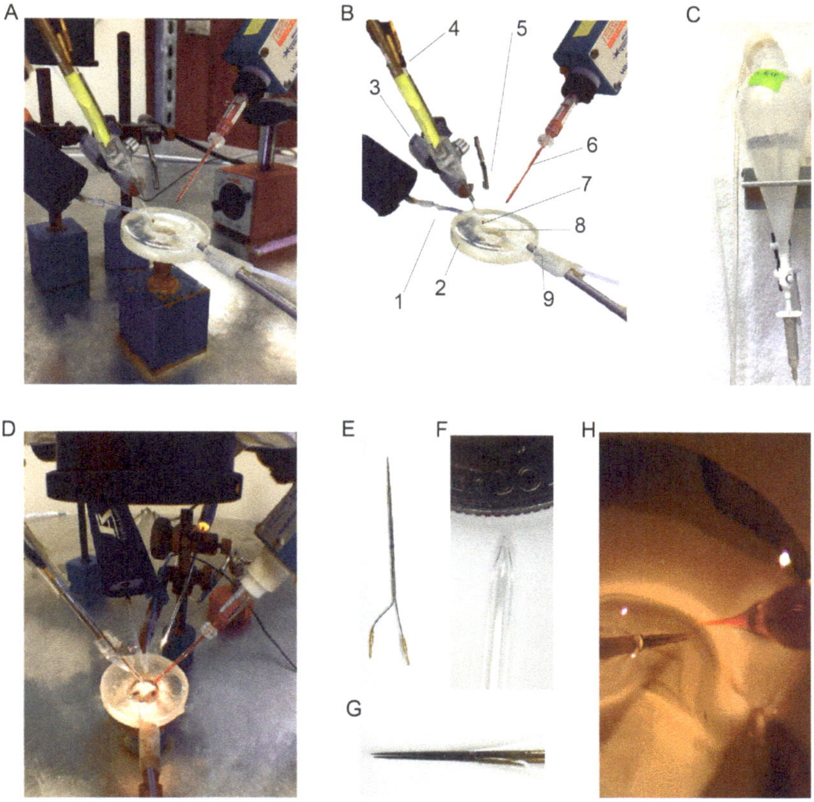

Fig. 2 Experimental setup for field potential recordings. (A and B) Component map. *1*, ACSF inlet through in-line heater; *2*, slice recording platform; *3*, Headstage-grounding pellet holder; *4*, bipolar-stimulating electrode mounted on a micromanipulator; *5*, fiberoptic illuminator; *6*, recording pipette mounted on headstage and filled with ACSF and 0.2% phenol red indicator; *7*, AgCl-grounding pellet; *8*, slice-recording chamber; *9*, ACSF outlet with suction. (C) Separation funnel used for gravity perfusion, tubing inserted at top for bubbling with carbogen. (D) Experimental setup during recording. Dissecting microscope visible at *top*. (E)–(G). Bipolar-stimulating electrode. Two monopolar-stimulating electrodes are passed through a capillary pipette that has been broken against a quarter's edge (F), and glued together. (H) Pop-spike recording setup. Bipolar stimulator positioned in inner SR of CA1, recording electrode in SP.

to the "+" pin of the voltage stimulator. Complete the circuit with another stripped stainless steel wire which should connect the "−" pin and the bath ACSF.

- Fiber-optic light source to illuminate slice.
- Pulse programmer (Master 8, A.M.P.I), although these functions can be recreated through software such as Igor (Wavemetrics) or pCLAMP (Axon).

- Oscilloscope (TDS 220, Tektronix).
- Amplifier with headstage and pipette holder adapter (Axoclamp 700B). AgCl electrodes should be used, stripped where it interfaces with the gold pin of the pipette-holder adapter. An AgCl-grounding pellet should connect the headstage to the ACSF in the perfusion chamber.
- Data acquisition device for analog to digital conversion and analog voltage output (ITC-18, Heka; also feasible with USB-6008, National Instruments).
- Acquisition software (custom written using Igor Pro by Wavemetrics).

2.1.2 Buffered Solutions and Reagents

- ACSF, ideally made the day of the experiment. Our solution contains (in m*M*): 124 NaCl, 26 $NaHCO_3$, 10 glucose, 2.5 KCl, 1 NaH_2PO_4, 1.3 $MgSO_4$, 2.5 $CaCl_2$. Carbogenated, pH is 7.3–7.4, and osmolarity from 280 to 310.
- Sucrose cutting solution may be used up to 1 week. Our solution contains (in m*M*): 215 sucrose, 26 $NaHCO_3$, 20 glucose, 2.5 KCl, 1.6 NaH_2PO_4, 4 $MgCl_2$, 1 $CaCl_2$, 4 $MgSO_4$. Carbogenated, the pH is 7.3–7.4, and osmolarity should be 5%–10% higher than ACSF.
- Any drugs that will be superfused for the duration of recording (e.g., PTx) should ideally be dissolved into ACSF directly on the day of recording.
- Liquid soluble test compounds (e.g., midazolam) should be prepared as stock solutions of 1:1000 or 1:2000. Dissolving compounds in distilled deionized water (ddH_2O) is preferred, although DMSO, Dextran, or Tween-80 may be needed. Concentrated stock solutions limit exposure of the slice to these solvents. Drugs should be pipetted directly into the carbogenated ACSF reservoir (not the slice recording chamber) as needed.
- For volatile liquid anesthetics, e.g., isoflurane, a gas-specific vaporizer is interposed in between the carbogen line and the ACSF perfusion reservoir.
- Alternately, the volatile liquid anesthetic may be added directly into oxygenated ACSF, and then vigorously vortexed and sonicated (Franks & Lieb, 1996).

2.1.3 Procedure for Measuring Pop Spikes and fEPSPs

Paired-pulse pop spike inhibition is very sensitive to changes in feed-forward inhibition (Fig. 1E), expertly reviewed in MacIver (2014).

No blockers of synaptic transmission are added to ACSF, leaving excitatory and inhibitory pathways functionally intact. fEPSPs in the SR layer of CA1 are sensitive to changes in glutamate release from SchColl terminals, and to changes in the complement of AMPA-type glutamate receptors. fEPSPs are typically recorded in the presence of PTx (50 μ*M*) to block synaptic inhibition and assess isolated excitatory synaptic transmission. The procedure for recording these responses differs only in the placement of the recording electrode and in the waveform parameters measured.

1. Start perfusing ACSF through the slice chamber at a rate between 1 and 4 mL/min, ensuring a stable fluid level in the recording chamber.
2. Mount the stimulating electrode on the pipette holder/micromanipulator assembly. The stimulating electrode can be reused many times.
3. Fill a recording pipette (3–5 MΩ) with 3 *M* NaCl (ACSF is also acceptable) and mount it on the headstage/micromanipulator assembly.
4. Using a transfer pipette with the tip cut off, gently aspirate ACSF along with the hippocampal slice from the slice-holding chamber. Alternately, if using a filter paper-adherent tissue preparation, place the filter paper disc into the recording chamber.
5. With perfusion and slice chamber suction temporarily OFF, gently float the slice onto the perfusion chamber surface and secure the slice with a harp, ensuring that the harp's wires are clear of the area to be recorded.
6. Resume slice chamber perfusion and suction.
7. Advance the stimulating electrode diagonally into the slice. The angle should be fixed (15–35 degrees), and identical to the angle of the electrode holder's axis on the micromanipulator, to minimize tissue shearing. Make note of when the electrode tip first deforms the hippocampal surface.
8. If using a sufficiently thin stimulating electrode, advance the electrode tip into the slice, approximately 50–100 μm below the surface. The tip should reside in the inner SR.
9. As before, advance the recording pipette diagonally. When the tip touches tissue, a brief "bounce" should be seen on the oscilloscope. Advance the recording pipette to a similar depth as the stimulating electrode. The stimulating and recording electrodes should be 150–300 μm apart to produce adequate signal without excessive stimulation intensity.
   - For pop spike recording: place the tip in the PC layer.
   - For SchColl—CA1 fEPSP recording: place the tip in the SR.

10. Every 15–20 s, apply a pair of stimuli, separated by 50 ms. Standard pulse widths are 100–200 μs.
11. Calibrate stimulus intensity to produce a third to a half-maximal response.
12. Details of data acquisition depend on the platform used. For these experiments, we typically collect 0.5–1 s of data per "sweep," Bessel filtered at 4 kHz and digitized at 10 kHz. Measurements are taken "online" to monitor the experiment's stability, and include the amplitude of the first and second pop spikes, or slope of the first and second fEPSP.
13. When measuring drug effects on the pop spike or fEPSP, first collect a stable baseline (<10% variation from beginning to end). We typically collect a baseline that is one-third the duration of the "readout" phase, i.e., if we expect a drug effect to stabilize at 30 min postwash-in, a 10 min baseline is appropriate.
14. After the baseline is collected, apply drug at 1:1000 or 1:2000 dilution of the stock concentration.
15. In order to rule out "run-down" or "run-up" of the measured parameter, we attempt to wash out the drug with normal ACSF, anticipating a return of measured parameters to predrug levels.

2.1.4 Notes

- If cost is an issue, these experiments can be performed without a temperature controller, and gravity perfusion may be used in place of a peristaltic pump, provided that both parameters remain stable over the course of the experiment. Some aspects of hippocampal physiology have a notable temperature dependence (Klyachko & Stevens, 2006).
- If bubbles form on the stimulation electrode with each stimulus, the stimulation intensity should be reduced.
- Recording electrodes can be easily made with a short length of silver wire that is kept in household bleach (sodium hypochlorite) for a few hours, which chloridates the silver surface.
- Placing stimulating and recording electrodes too close together, or stimulating at too high an intensity, can produce a direct dendritic depolarization artifact (Fig. 3D), which complicates waveform interpretation.
- Some degree of baseline drift is normal in these recordings, and the potential offset can be electronically nulled during an experiment to stay within the amplifier's dynamic range. Large drift may indicate unstable

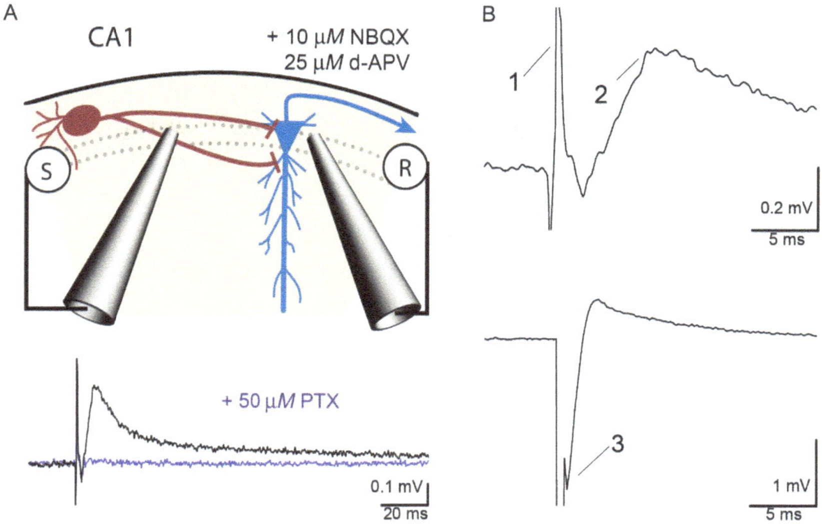

Fig. 3 Recording fIPSPs. (A) Configuration and typical responses for fIPSP. Both stimulating and recording electrodes are in the SP (*top*), aiming to excite perisomatic-targeting basket cell INs and record the coherent hyperpolarizing response of CA1 PCs. Synaptic inhibition is isolated by recording in continuous presence of NBQX and d-APV, antagonists of AMPA and NMDA receptors, respectively. The fIPSP (*B*, *black*) is smaller than fEPSPs and pop spikes, and should be completely blocked by 50 µM PTx (*blue trace*). (B) *Top*, there should be a clear separation and latency between the stimulus artifact (*1*) and the fIPSP (*2*). *Bottom*, high-intensity stimulation can produce direct dendritic depolarization, which has a large downward component (*3*) that runs into a standard 200 µs stimulus, and produces a "peak" at shorter latency than a true fIPSP. This artifact does not disappear in PTx.

fluid levels in the perfusion chamber, or an insufficiently chloridated AgCl-recording electrode.

- Monopolar and bipolar stimulation electrodes are both widely used. In theory, bipolar electrodes allow for more focal stimulation, as current spread should be restricted to the area between anode and cathode. In practice, we can easily elicit similar responses with either. Monopolar electrodes can more easily be advanced into tissue.

2.2 Field Inhibitory Potential Recording in CA1

Many anesthetic drugs are ligands for the $GABA_A$Rs (see Section 3 for more detailed discussion). Drug effects on the fEPSP and pop spike give insight into how they affect excitatory synaptic transmission, feed forward inhibition, and CA1 PC excitability, but these techniques do not directly measure $GABA_A$R-mediated responses. The field IPSP (fIPSP) is a sensitive measure of aggregate $GABA_AR$-mediated potentials arising from the CA1 PC layer and can be assayed in the same (relatively) inexpensive setup used to study fEPSP and pop spikes. As such, it is a useful intermediate technique to study net inhibitory responses in the PC layer without resorting to the more technically demanding whole-cell recording process (Arai, Silberg, & Lynch, 1995; Heifets, Chevaleyre, & Castillo, 2008; Lambert, Borroni, Grover, & Teyler, 1991).

The fIPSP is somewhat more challenging to elicit than fEPSP and pop spikes (Fig. 3). GABAergic INs as a class are famously diverse in their dendritic and axonal arborization, and do not have the same well-defined organization as the excitatory projections within the hippocampus. Fortunately, one class of GABAergic INs has particularly extensive and powerful synaptic connections within the PC layer—these are perisomatic-targeting basket cells. The axons of these basket cells run within the PC layer in a lamellar pattern (much like the excitatory pathways of hippocampus), and their discharge leads to a coordinated hyperpolarization of thousands of CA1 PCs, giving rise to an organized extracellular field.

2.2.1 Equipment

- We recommend use of monopolar electrodes composed of stainless steel or tungsten (commercially available), or the experimenter can fabricate custom-size tips using glass capillary pipettes and stainless steel or tungsten wire (see Section 2.1.1.)
- Other equipment as described in Section 2.1.1.

2.2.2 Buffered Solutions and Reagents

Stimulating the PC layer will also activate recurrent excitatory collaterals to CA1, generating fEPSPs and pop spikes, so one must record the fIPSP while continuously superfusing blockers of synaptic transmission.

- Prepare stock concentration of NBQX disodium (Tocris), a selective AMPA receptor antagonist, at 20 m*M* in ddH_2O.
- Prepare stock concentration of d-APV, a selective NMDA receptor antagonist, at 50 m*M* in ddH_2O.
- Prepare stock concentration of PTx at 100 m*M* in DMSO.
- Add 10 μ*M* NBQX and 25 μ*M* d-APV from stock solution to ACSF reservoir (both diluted 1:2000) at the beginning of the recording session.
- A less expensive (though less selective) alternative is to use kynurenic acid (1 m*M*), which also blocks AMPA and NMDA receptors.

Kynurenic acid can be made in stock solutions of 100 m*M* in ddH_2O for use at 1:100 dilution in ACSF.

2.2.3 Procedure for Measuring Field IPSPs in CA1

Follow procedure outlined in Section 2.1.3, with the following modifications.

1. Begin experiment with NBQX and d-APV in superfused ACSF.
2. Advance monopolar-stimulating electrode about 50–100 μm into tissue with the tip residing in the PC layer.
3. Advance the recording electrode until it 50–100 μm deep into tissue with the tip in the PC layer. The tips of the stimulating and recording electrode should be between 50 and 150 μm apart.
4. Every 15–20 s, stimulate single pulses (Fig. 3B).
5. At the conclusion of the experiment (e.g., peak drug effect or drug wash-out), apply 100 μ*M* PTx (1:2000) to verify that the fIPSP is purely GABAergic (Fig. 3A) and is not contaminated by artifact (Fig. 3B).

2.2.4 Notes

- Because of the proximity of stimulating and recording electrodes, a direct dendritic depolarization artifact is more likely to occur (Fig. 3B, bottom). Note the scale differences in the panels of Fig. 3, as well as the apparent "peak" in Fig. 3B, bottom, which arrives more quickly than the true fIPSP peak in Fig. 3B, top. If this happens, reduce stimulus intensity and/or move one of the electrodes and retry.

2.3 Whole-Cell Patch-Clamp Recording

Whole-cell patch-clamp recordings allow detailed measures of the rapid ion channel-mediated conductance changes across a neuronal membrane. By inserting a recording electrode in series with a neuron, the experimenter can simultaneously measure and control the current flow and intracellular potential of a neuron, as well as alter the intracellular composition for experimental benefit. The experimenter can choose to follow the neuron's potential by injecting constant (or zero) current—this recording mode is known as "current clamp." Conversely, one can specify a neuron's membrane potential and use the amplifier's "voltage clamp" circuitry to inject or withdraw current in order to maintain that membrane potential. For a more detailed yet accessible treatment of the "current clamp" and "voltage clamp" recording modes, we refer the reader to Molleman (2003). For either of these purposes, the composition of the "internal" solution used to fill the patch pipette is critical, as this internal solution dialyses the cell. It is important to note that dialyzing the internal contents of a neuron may alter its physiology in important ways. A well-known example of this phenomenon is that long-term potentiation of the SchColl-CA1 synapse must be induced within a few minutes of gaining intracellular access (Malinow & Tsien, 1990). This limitation can be circumvented to an extent with perforated patch recording, a specialized technique described in Molleman (2003) and beyond the scope of this chapter.

Whole-cell recording requires the experimenter to patch on to the membrane of a single neuron with a patch pipette, remove the small patch with suction, and maintain a seal with the neuron for the duration of recording. This process requires that some pieces of equipment listed earlier be more specialized to allow for cell visualization and pipette stability.

2.3.1 Equipment (Whole-Cell Recording): See Fig. 4 for Components

- Upright microscope with a 4 × and 40 × objective, able to perform DIC imaging for visualizing cell bodies at depth, and a camera attached to a video monitor for real-time video output. An infrared filter is preferred to more clearly image deep into slices.
- Precision motorized micromanipulator (MPC-200, Sutter Instruments).
- Airtight tubing connecting the pipette holder adapter (Fig. 4A) to a mouth suction device, typically a 1-cc syringe chamber.
- Peristaltic perfusion pump and temperature control are strongly recommended, as variability in ACSF flow and temperature can disrupt a pipette's seal with a neuron.
- Other components as described in Section 2.1.1.

2.3.2 Buffered Solutions and Reagents

The recording pipette solution that will fill the recorded cell (internal solution) should be made in batches of 50 mL or more, aliquoted at 0.25–1 mL, and stored at −20°C for up to 1 year. As a general practice, it is helpful to maintain several aliquots of an internal solution that is known to work well, especially when using a new internal solution, in order to troubleshoot problems with whole-cell recording.

A primary consideration in composing an internal solution is whether the experimenter plans on mimicking the neuron's internal milieu and measuring intracellular potentials (e.g., action potentials),

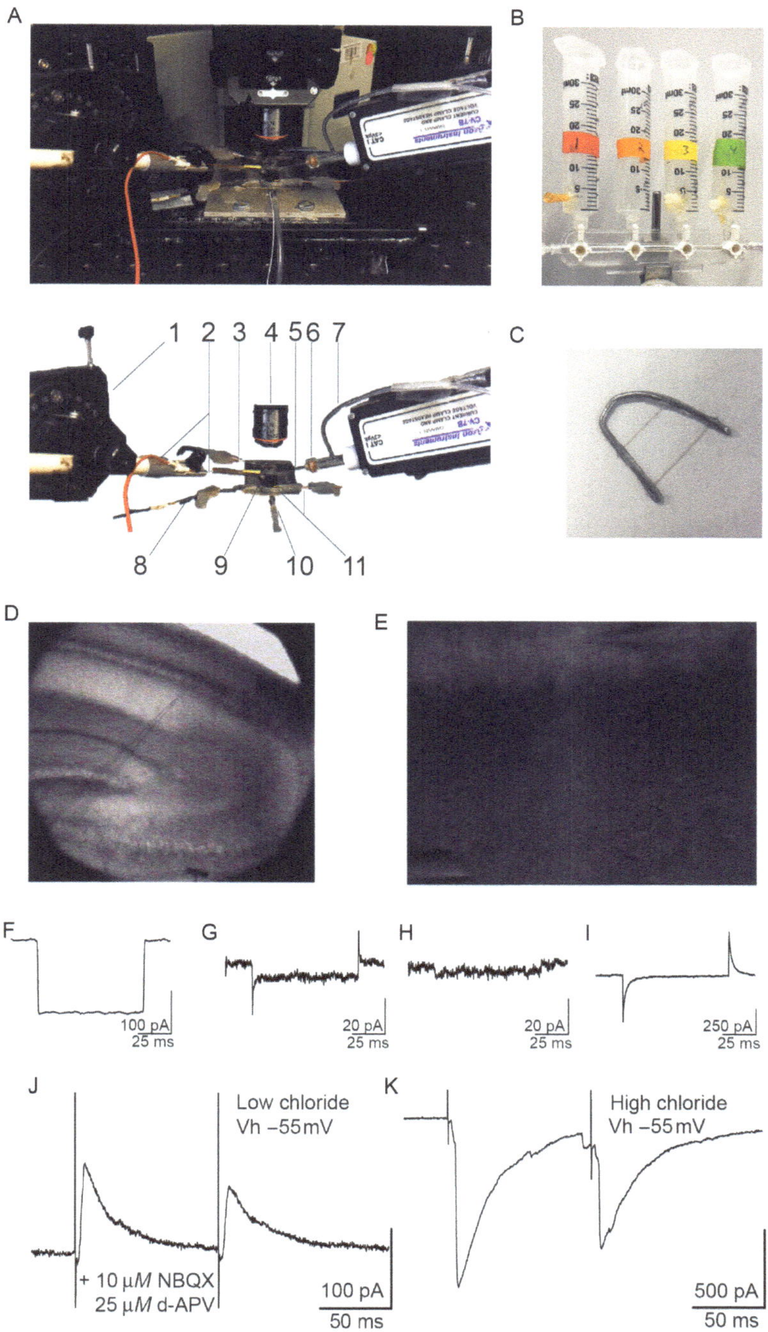

Fig. 4 Experimental setup for whole-cell patch clamping. (A) Component map. *1*, motorized micromanipulator for moving both stimulating and recording electrodes; *2*, monopolar stainless steel-stimulating electrode; *3*, ACSF outlet with suction (bent 20-g syringe needle); *4*, microscope objective lens; *5*, recording electrode; *6*, screw gasket *(Continued)*

typically a current-clamp experiment, vs controlling the neuron's electrical properties in order to measure specific conductances, e.g., $GABA_AR$-mediated inhibitory postsynaptic currents (IPSCs), typically a voltage-clamp experiment. Table 1 lists the reagents and target concentrations of three standard internal solutions we have used. Briefly, cesium-based internal solutions block many potassium channels within the neuron, and thus make the neuron electrically more "compact," improving the ability to voltage-clamp distant parts of the neuron (e.g., dendrites). Using a low chloride internal solution approximates the native intracellular chloride concentration. The resulting gradient of external-to-internal chloride produces an inward, hyperpolarizing flow of chloride ions when $GABA_AR$ channels open. While this solution is more "physiological," this gradient produces

Fig. 4—cont'd tightens to form an airtight connection between recording electrode and pipette holder; *7*, tubing in series with electrode, attached to pipette adapter and sealed with vacuum grease. Used for mouth suction during seal process; *8*, Cathode for monopolar-stimulating electrode. Terminates in stainless steel wire in contact with ACSF at edge of slice-recording chamber; *9*, AgCl-grounding pellet; *10*, ACSF inlet (in-line heater not visible); *11*, thermistor in contact with ACSF for temperature regulation. (B) ACSF reservoir. Bubbled with carbogen via 26-g syringe needle epoxy glued near base of 30-mL syringe. Manifold pipes ACSF to peristaltic pump and in-line heater. (C) Harp for restraining slice onto recording chamber. Nylon fibers (or human hair) are glued to a small, bent, and flattened length of platinum. (D) View of hippocampal slice at 4×. Glass monopolar-stimulating electrode in SR, recording electrode in SP. (E) 40× magnification. Broken glass tip of stimulating electrode visible at bottom (~20 μm). Recording electrode in SP of CA1. Densely packed cells limit visibility of tip, requiring a blind-patch technique. (F) 5 mV test pulse used to detect changes in recording electrode tip resistance (R_{series} or R_s). Prior to entering tissue, $R_s \sim 3\,M\Omega$, which is the pipette tip resistance. (G) A bounce in the test pulse has been detected, positive pressure is removed and sealing has begun. At $R_s \sim 500\,M\Omega$, capacitative transients are evident at either end of test pulse. (H) Transients have been electronically nulled, and $R_s \sim 1\,G\Omega$. (I) Suction is applied by mouth, and intracellular access is obtained. The initial peak produced by the test pulse measures the R_s when patched, typically 3–4 times the starting pipette tip resistance (here, $R_s \sim 12\,M\Omega$). Resistance to synaptic input (R_{in}) is measured when the test pulse is at steady state, before the test pulse offset transient. Quality recordings have R_{in} 10× greater than R_s (here, $R_{in} \sim 200\,M\Omega$). Perforated or leaky cells have much lower R_{in}. (J and K) Whole-cell recording of inhibitory postsynaptic currents (IPSCs). Synaptic inhibition is isolated by NBQX and d-APV, as in Fig. 3. IPSCs are recorded in voltage-clamp mode, so inward (depolarizing) currents are compensated with negative current. The direction of current flow through $GABA_ARs$ depends is determined by the intracellular to extracellular gradient of the chloride ion. Manipulating the internal solution's chloride composition relative to ACSF chloride concentration can therefore change the direction of IPSC even at equivalent holding potential (Vh).

Table 1 Example Composition of Three Typical Recording Pipette Solutions

		Cs-Based Low Chloride Internal	Cs-Based High Chloride Internal	K-Based Internal	
Reagent	**MW**	**Final Concentration**	**Final Concentration**	**Final Concentration**	**Comment**
D-Gluconic acid (50 wt% in H_2O, or 3.15 *M*)	196.16	131	70		Liquid reagent
CsOH (50 wt% in H_2O, or 5.73 *M*)	149.91	131	75		Liquid reagent
CsCl	168.36		50		
K-Gluconate	234.2			100	
NaCl	58.44			10	
$CaCl_2$	147.02	2	0.1		
EGTA	380.4	10	1	0.2	See note for Cs-based solutions
HEPES	238.3	10	10	10	
Glucose	180.2	10	10	10	
Biocytin	372.5	0.2%	0.2%	0.2%	Optional, for posthoc cell reconstruction
QX-314	343.3	5	5		Blocks action potentials

Continued

Table 1 Example Composition of Three Typical Recording Pipette Solutions—cont'd

		Cs-Based Low Chloride Internal	Cs-Based High Chloride Internal	K-Based Internal	
Reagent	**MW**	**Final Concentration**	**Final Concentration**	**Final Concentration**	**Comment**
PhosphoCreatine * diTris	453.38	10	10	10	Components for regenerating ATP and GTP
MgATP	507.18	5	5	5	
Na_3GTP	523.2	0.4	0.4	0.4	
Acidify with		D-Gluconic acid	D-Gluconic acid	D-Gluconic acid	
Basify with		CsOH	CsOH	KOH	
Lower osmolarity with		ddH_2O	ddH_2O	ddH_2O	
Raise osmolarity with		Sucrose	Sucrose	Sucrose	

The first two solutions replace intracellular K with Cs, which reduces the "leakiness" of a cell by blocking many of its K channels from the inside. The improved electrical "compactness" allows more effective voltage clamp of the cell body and dendrites. By the same mechanism, neurons filled with Cs-based solutions will depolarize unless voltage is clamped by an amplifier. Therefore, Cs-based solutions are best suited for voltage-clamp experiments. The third solution, K-based, approximates the normal intracellular concentrations of the major ionic species. This type of solution is ideal for recording action potentials. *Note*: For making pipette solutions with higher EGTA concentrations, we find it useful to store EGTA and CsOH as a stock solution of EGTA 200 m*M*, CsOH 573 m*M*, made by adding 760 mg EGTA to 8 mL of ddH_2O. Once dissolved, an additional 1 mL CsOH (50 wt% in H_2O) and 1 mL ddH_2O is added. Slices containing biocytin should be fixed in 4% paraformaldehyde before antibody staining/imaging.

relatively smaller currents at a normal resting potential (~–65 mV). Thus, a high chloride internal solution is often used for long-term recording of IPSCs near the resting potential: the chloride gradient is larger but reversed, producing an outward, depolarizing chloride ion flow. Cesium-containing internal solutions may be desired when studying smaller currents—therefore QX-314 is also added, to prevent action potentials that would otherwise obscure stimulus-evoked currents. Potassium-based internal solutions are used for measuring spiking activity and are appropriate for current-clamp experiments.

2.3.3 Procedure for Sealing Onto Neurons and Patch Clamping

Similar to learning how to drive a car (or any complex task that operates somewhat on "feel") learning how to form a quality seal and break-in can be considered more art than science. The process of sealing is described in Molleman (2003) but we will detail our own experience below for the reader. The most important resources are calmness and an abundance of patience—the data will almost certainly come, no matter how slowly they may appear to arrive. In electrophysiology, quality is always more important than quantity. Regardless of whether one is performing voltage clamp, current clamp, cell attached, or perforated patch recordings, the process of sealing is generally identical with minor tweaks.

Initial setup of this experiment is similar to that described in Section 2.1.3.

1. At 40 × magnification, visually identify a field of neurons suitable for recording.
2. At 4 × magnification, place the stimulating electrode in SR of CA1.
3. Pull a pipette to the correct tip diameter, usually ~1–5 μm and with a resistance of 2–8 MΩ, depending on the application.
4. Backfill with filtered solution (internal solution for whole-cell recordings or outside-out single-channel recordings, external solution for cell-attached or inside-out recordings).
5. Attach pipette via a screw gasket to the headstage apparatus, to form an airtight seal.
6. Apply gentle positive pressure to the pipette through a mouth pipette or a syringe, with a lockable luer connection.
7. At 4 × magnification, lower the pipette into the solution above the neuron to be recorded using coarse movements.
8. When the pipette is well above the tissue layer, release any air bubbles in the pipette with a large syringe and reapply gentle positive pressure.

9. Lower the pipette to just above the tissue layer and release the pressure.
10. Center the microscope field of view over the recording pipette tip, switch to 40 × magnification, and carefully locate your pipette tip.
11. In voltage clamp mode, run a seal test—typically 5 mV pulses applied at 10–100 Hz.
12. Offset the pipette voltage to account for confounding resistance measurements based on each pipette's unique shape.
13. Now apply medium positive pressure and hold with the tongue held over the mouth or a luer lock on a syringe as the pipette is very slowly lowered into the tissue layer or near the cell using the fine knobs or settings on the micromanipulator.
14. Target the center of the neuron and make adjustments as necessary to account for neuron motion as the pipette is lowered.
15. Watch for a dimple to appear in the surface of the cell, and the seal test resistance to rise about 10% of the value of the pipette resistance.
    a. If "blind patching" in the dense CA1 PC layer, one relies solely on the change in pipette resistance while slowly advancing the recording electrode diagonally (Fig. 4F–H).
16. Right at this moment, gently apply consistent negative pressure (suction) to form a seal (this is the most difficult thing to learn but will become simple with practice). A light suction is better than a strong one so as to not rupture the cell or drag it into the pipette.
17. Allow the resistance to rise to 1 GΩ, which may take 2–3 min (anything longer is not worth recording).
18. During this rise, if the seal is forming well, negative pressure should not be needed, but a combination of applying small amounts of negative or positive pressure may encourage seal formation.
19. Slowly reducing the command voltage from 0 to the neuron's expected resting membrane potential (e.g., −65 mV) will encourage seal formation.
20. Once >1 GΩ is reached, a "gigaseal" has formed, and holding potential approximates membrane potential. The experimenter must null fast and slow capacitive transients observed at the beginning of the seal test pulse.
21. Break into the cell by applying a brief strong suction to the pipette, or by "zapping" with a brief, strong current.

Stimulate every 15–20 s. Once the cell is patched onto and the experiment has begun, measuring current or voltage before and after drug application requires that the recording parameters remain stable, especially the resistance between the pipette tip and the neuron ("series" or "access" resistance, R_s), and the resistance of the neuron to synaptic input ("input" resistance, R_{in}). Thus, in voltage-clamp recordings, each sweep should include an 80 ms, 5 mV test pulse to assess the R_s and R_{in} (Fig. 4I). Responses to paired stimulus pulses are typically acquired with a 40–50 ms interpulse interval (Fig. 4J and K); these allow for the assessment of presynaptic neurotransmitter release properties (Zucker & Regehr, 2002).

3. CULTURED CELL ELECTROPHYSIOLOGY

The benefits of studying the actions of drugs and anesthetics in cultured cells (including either neurons or transfected human embryonic kidney [HEK293] cells) include the ease of the experiment and the highly controlled environment. Indeed, the first evidence that volatile anesthetics could potentiate $GABA_AR$ activity was shown in cultured neurons from dorsal root ganglion (Nakahiro, Yeh, Brunner, & Narahashi, 1989). In the early 1990s, Bekkers and Stevens (1991) found that autapses (synapses made from one neuron onto itself) which form in cultured hippocampal neurons provide excellent homogeneity for highly controlled studies of excitatory or inhibitory synaptic physiology. The preparation easily allows for studying the actions of drugs on synapses in which proteins that are integral to synaptic development and function have been mutated, deleted, or overexpressed. However, this approach is limited in its ability to assay the specific kinetic, biophysical, and pharmacological properties of drugs at defined receptors. For these latter types of studies, transfection of receptor subunits in cultured HEK293 cells is one clear solution for characterizing these properties.

Anesthetics have extensively studied effects on ion channels of various types—and a staggering variety of channels exists. At a minimum, transmembrane channels can be gated by voltage changes, mechanical stress, concentration gradients, G protein-coupled receptor activation, posttranslational modification, e.g., phosphorylation, and/or endogenous (neurotransmitter) or exogenous (drug) ligands. To study the specific properties of these channels (also termed receptor channels or just receptors), which include gating, pharmacological mechanisms, or effects of mutagenesis, cDNA-encoding mammalian or human receptors can be expressed in heterologous systems including *Xenopus laevis* oocytes, Chinese hamster ovary (CHO) cells, or HEK293 cells. In this section, we focus on $GABA_ARs$ expressed heterologously in HEK293 cells, as $GABA_ARs$ are established targets of general anesthetics. Furthermore, HEK293 cells are widely available and sharable; simple to culture, transfect, and patch clamp; and allow for rapid kinetic analysis of receptor function relying on mammalian cellular machinery. In particular, we will focus on whole-cell patch clamp of HEK293 cells here. These general methods can be employed for many applications involving expression of most types of receptor indicated earlier.

As revealed by electrophysiology in HEK293 cells, the volatile anesthetics isoflurane, desflurane, and sevoflurane, as well as the injectable anesthetic drugs propofol, etomidate, barbiturates, and neurosteroids (Davies, Kirkness, & Hales, 1997; Garcia, Kolesky, & Jenkins, 2010) all either enhance or activate specific $GABA_A$Rs. $GABA_A$Rs are transmembrane pentameric ligand-gated chloride channels that generally inhibit neuronal excitability (Sieghart, 1995). Pharmacological binding sites on $GABA_A$Rs are present at the extracellular domain at the subunit interface for hydrophilic molecules like GABA and benzodiazepines, or at transmembrane domain (TMD) locations for lipophilic molecules including general anesthetics. In particular, the volatile anesthetics may potentiate $GABA_A$R function by binding at several intersubunit interfaces (Woll et al., 2018), α Ser270 being identified early on via mutagenesis (Mihic et al., 1997; Nishikawa & Harrison, 2003). Propofol, etomidate, and barbiturates may produce their effects via subtly different occupancy of TMD subunit interfaces (Chiara et al., 2013, 2016; Hamouda et al., 2014), and unlike the volatile drugs, are able to directly gate $GABA_A$Rs in the absence of GABA. Finally, neurosteroids such as alfaxalone binds interfaces with the α subunits at distinct sites; the presence of a δ subunit, observed in the majority of extrasynaptic $GABA_A$Rs in hippocampus and other forebrain structures, potentiates these effects (Eaton et al., 2014; Hosie, Wilkins, da Silva, & Smart, 2006). Due to the recent development of more favorable intravenous vehicles, alfaxalone may return to clinical use as a result of its excellent side effect profile (Monagle, Siu, Worrell, Goodchild, & Serrao, 2015), thus, electrophysiological studies of alfaxalone and related drugs may soon be in high demand.

In this section, we will discuss culture of HEK293 cells followed by transfection of $GABA_A$Rs and whole-cell recordings in either neurons or HEK293 cells. For a detailed description of the procedure to isolate and culture primary rat hippocampal neurons, we refer readers to Mah, Fleck, and Lindsley (2011), based on the protocol by Kaech and Banker (2006). The electrophysiological recording of neurons from primary culture is in many ways similar to that of HEK293 cells (below), except that in primary culture, blockers of voltage-gated K^+ or Na^+ channels may have to be added into the extracellular solution depending on the needs of the experimenter. In our hands, if synaptic activity is blocked pharmacologically during cultured neuronal recordings, it is not necessary to use a bicarbonate-based extracellular recording buffer; a HEPES buffer (similar to that used with HEK293 cells, with the addition of 10 m*M* D-glucose) will suffice for short-term 10–20 min recordings. Excellent references that detail the concentrations to test for anesthetic drugs, as well as other general information outside the scope of this chapter, include Franks and Lieb (1996) and Garcia et al. (2010).

3.1 Obtaining and Preparing Recombinant cDNA

Arising from the success of the genome project, complete sequence data have been obtained for the three most common species used to generate receptor cDNA for heterologous system electrophysiology: mouse (2002), rat (2004), and human (completed in 2006). Recombinant receptor cDNA can often be freely obtained by contacting laboratories that have originally cloned or heavily utilized a particular receptor, and by accepting a material transfer agreement, which is generally free of charge. Alternatively, cDNA can be obtained through several online vendors relatively inexpensively (<$100 per sample). Depending on the requirements of the experimenter, the cDNA can often be used as-is after some optimization, or mutated through a variety of mechanisms to probe structure–function relationship of specific receptor motifs. Particularly important for identification of transfected cells during electrophysiology experiments is the addition of fluorophore cDNA (such as GFP) to the receptor subunit cDNA, often at the C-terminus, to create a chimeric receptor.

3.1.1 Equipment

- Computer with DNA-editing software (e.g., A Plasmid Editor [ApE], SnapGene, DNA Strider).
- Software for viewing receptor protein tertiary structure (e.g., Swiss-PdbViewer).
- Polymerase chain reaction (PCR) machine (i.e., thermal cycler).
- PCR tubes (200 or 50 μL volume).
- Small-volume spectrophotometer (e.g., NanoDrop) and associated software.

- Electrophoresis equipment (including lane dividers/combs, power supply, and gel beds for agarose gels).
- Ultraviolet (UV) light source; both long- and short-wave if possible.
- Gel extraction kit for DNA (e.g., QIAquick or GeneJET).
- DNA purification kits for miniprep and maxiprep (from Qiagen, Thermo Fisher Scientific, or Zymo).
- T4 DNA ligation kit.
- Water or dry bath/heating block (for temperatures of 42°C or 55°C).
- Bacterial shaker.
- 14-mL culture tubes.
- 100- to 250-mL Erlenmeyer flasks.
- Snap-top and screw-top ultracentrifuge tubes (1.5 mL).
- −80°C freezer.
- Autoclave.
- Refrigerated floor centrifuge (Sorvall or Beckmann) with large tubes and rotors for preparation of bacterial pellets and DNA pellets.
- DNA SpeedVac.
- Gel imager and computer.

3.1.2 Buffers and Reagents

- TAE buffer (40 m*M* Tris, 20 m*M* acetic acid, 1 m*M* EDTA; pH 8.3; made from 50 × stock).
- Agarose gel, made fresh (0.8%–1.2% w/v agarose/TAE buffer, depending on length of DNA).
- Appropriate restriction enzymes (e.g., from New England Biolabs or Thermo Fisher Scientific).
- Enzyme buffers (usually provided with enzyme purchase).
- dNTP mix (deoxynucleotide solution mix).
- DNA ladder (usually from 6 × stock).
- DNA loading buffer.
- ddH_2O, sterile.
- Subcloning efficiency *E. coli* (e.g., DH5α); or higher efficiency chemically competent cells.
- High-fidelity polymerase.
- DNA gel stain (SYBR Safe or ethidium bromide).
- Luria broth (LB), autoclaved.
- Bacterial plates (100 μg/mL ampicillin or 50 μg/mL kanamycin in LB agar).
- Mammalian expression cDNA vectors (e.g., pRK5 or pRK7).
- Primers (15–40 bp).

3.1.3 Procedure

1. If cDNA is obtained dry, place filter paper containing cDNA sample (at least 10 ng) in a 1.5-mL tube and extract cDNA from paper by soaking overnight in sterile ddH_2O (no less than a concentration of 0.1 μg/mL) at 4°C.
2. If cDNA is obtained in bacteria, such as a bacterial stab, follow instructions in the product insert.
3. Transformation of bacteria with cDNA.
   a. Place agar plates on the benchtop upside down to warm.
   b. Remove DH5α competent *E. coli* from −80°C freezer and thaw on ice.
   c. Label 14-mL culture tubes, chill on ice.
   d. Prepare 0.01 μg/μL cDNA solutions.
   e. To 14-mL tubes, add 100 μL bacteria and 100 ng DNA.
   f. Incubate on ice 30 min.
   g. Heat shock at 42°C for 45 s.
   h. Incubate on ice for 2 min.
   i. Add 900 μL of LB broth.
   j. Shake at 37°C for 30 min (225 rpm).
   k. Spread 100 μL on plates and allow the agar to absorb the solution.
   l. Incubate overnight upside down at 37°C (no longer than 16 h).
   m. With sterile pipette tips, pick 1–3 colonies from different clusters on each plate and separately inoculate 2 mL LB + ampicillin culture.
   n. After colony picking, store plates at 4°C upside down and wrapped in parafilm.
4. Purify plasmid DNA using kit instructions (obtain 100 ng–10 μg).
5. Use PCR-driven overlapping extension (Heckman & Pease, 2007) to mutate cDNA or to create chimeras.
   a. Obtain external and internal primers to produce desired mutation (Thermo Fisher Scientific, Elim, or other vendor).
   b. Perform PCR.
   c. Extract desired size cDNA bands from gel.
   d. Extract cDNA.
   e. Ligate cDNA fragments together into full plasmid.
   f. Transform bacteria (see above) and miniprep cDNA.
6. Confirm mutation and proper open reading frame orientation with restriction enzyme digest, and then sequencing (e.g., through Genewiz or Elim) following miniprep.
7. After sequence is confirmed, perform maxiprep to prepare for HEK293 cell transfection with purified plasmid cDNA.

3.1.4 Notes

- cDNA plasmids obtained from vendors or laboratories should be sequenced prior to use to ensure quality and sequence fidelity.
- For long-term storage (years), solubilized cDNA can be frozen at −20°C in TE (Tris–EDTA) buffer at pH 8.0.
- For short-term storage (1 or 2 months), cDNA in our hands is stable in water or TE buffer at 4°C. The optimal concentration for storage is 1 mg/mL, which can be obtained if cDNA is first dissolved in water; TE buffer should not be concentrated.
- cDNA should never be vigorously vortexed, as this may cause shearing.
- Prior to PCR or transformation, it is preferable to have dilute DNA in water instead of TE buffer as EDTA may interfere with these reactions.
- Exposure to cDNA gel stains may be hazardous and appropriate precautions should be taken.
- UV glasses and UV face shield are essential for working with UV light.
- Long-wave UV light is preferable for extracting cDNA from gels, as it is less likely to cause damage to cDNA.
- PCR protocols may require substantial optimization before ideal products are obtained.

3.2 HEK293 Cell Culture and Transfection

HEK293 cells represent ideal template systems for the expression and study of membrane proteins, particularly ion channel receptors. These cells are relatively simple to grow, transfect, and patch clamp. As mammalian cells (in contrast to *Xenopus* oocytes), HEK293 contain ideal molecular machinery to export synthesized receptors to the cell membrane, specifically including optimized codon requirements for transcription and translation. We have studied the kinetic and pharmacological properties of $GABA_A$, glycine, nicotinic acetylcholine, and glutamate receptors in HEK293 cells, whereas others have successfully used these cells to reveal the actions of local anesthetics on voltage-gated sodium channels. HEK293 cells must be cultured in monolayers and consistently have their population thinned out (i.e., passaged) approximately once to twice per week to avoid confluency, less than 10 times total from a starter culture aliquot. That is, a new starter aliquot must be obtained from frozen stock every month or two. Newly plated cells can be transfected 12–24 h after plating, once they have adhered to the culture dish, and then recorded from generally 2–4 days after plating (1–3 days following transfection).

For transfection of receptors including $GABA_A$ and others that require multiple subunit assembly, it is essential to determine optimal ratios for transfection. This can be done by performing a dose–response curve of transfection ratios and then assessing the addition of all subunits through functional analysis (patch clamp) or through immunocytochemistry. For example, canonical $GABA_A$Rs are composed of α, β, and γ subunits, although α and β subunits readily form functional receptors. Therefore, a typical ratio to ensure insertion of γ subunits into most receptor complexes is 1:1:2 α:β:γ. Benzodiazepines employed during patch clamp, or a γ-specific antibody used in immunocytochemistry, can ensure insertion of the third subunit. However, other "third" subunits, such as δ, ε, or θ, may require the use of other drugs such as THIP, or GABA dose–response curves in combination with saturating concentrations of a barbiturate, to ensure subunit insertion. In our hands, the simplest approach is often to do a functional patch clamp assay, because the immunocytochemical expression of a particular subunit in a cell may not necessarily mean that it is located on the cell surface as part of a functional receptor unit. Alternatively, codon optimization may be required to express certain less-studied subunits as part of functional receptors. Importantly, it is essential that at least one of the subunits (likely the α or β, which are required to form functional $GABA_A$Rs) possesses a fluorophore like GFP to be able to identify transfected HEK293 cells. Generally, only a fraction of cells (40%–80%) will express the receptor of interest, so patching onto cells blindly would be inefficient.

3.2.1 Equipment

- Biosafety cabinet for sterile work involving recombinant DNA (rDNA).
- Cell culture incubator (5% CO_2) at 37°C.
- 10 cm diameter culture dishes.
- Nunc 35-mm cell culture dishes (Nunclon Δ [Delta] treated for cell adhesion) for transfection and recording.
- Long glass pipettes with flame-polished tips.
- Vacuum line in biosafety cabinet.
- Cell counting equipment, including hemocytometer.
- Cell culture microscope.
- Liquid nitrogen cold storage unit (vapor phase).
- Refrigerated tabletop centrifuge (4°C) for 15- and 50-mL tubes.

3.2.2 Buffers and Reagents

- HEK293 cells from ATCC.
- DMEM low glucose with pyruvate and glutamax, generally from Gibco.

- Fetal calf serum (FCS) 10% in DMEM (50 mL added to 450 mL; save the removed 50 mL for transfection).
- Trypsin–EDTA (0.05%).
- 70% Ethanol.
- Cationic lipid transfection reagent, e.g., Lipofectamine-2000.

3.2.3 Procedure

1. Obtain and store HEK293 cells, following product insert.
2. Spray all surfaces and pipetters with 70% ethanol to sterilize, use ethanol-sprayed gloves.
3. Thaw aliquot of cells and plate onto 10-cm dishes at $0.1–1 \times 10^6$/mL seeding density, in 10 mL total 37°C warmed DMEM with 10% FCS that has been "oxygenated" in incubator for 1 h prior to adding.
4. Passage at least once before splitting into 35-mm dishes for transfection.
   a. In cabinet, carefully aspirate media from 10-cm dish with vacuum attached to glass pipette.
   b. Add 3 mL warmed/oxygenated Trypsin–EDTA 0.05%.
   c. Place in incubator for 2–4 min or until very gentle tapping causes cells to break free from surface.
   d. Add 7 mL warmed/oxygenated DMEM/FCS, gently pipette up and down at least 10 times, ensuring homogenous mixture and the break-up of clumps.
   e. Place in sterile 15- or 50-mL tube and centrifuge at 4°C at 1500 × *g* for 1 min.
   f. Ensure that there is a robust pellet and aspirate off solution.
   g. Gently rinse pellet 2 × with fresh DMEM/FCS.
   h. Gently mix pellet into 1 mL DMEM/FCS by pipetting up and down at least 10 times, ensuring homogenous mixture and the break-up of clumps.
   i. Count cells.
   j. Plate at $0.1–1 \times 10^6$/mL seeding density in DMEM/FCS.
5. Split into 35-mm dishes for transfection
   a. Plate at the same seeding density as earlier.
   b. After 12–24 h, examine adherence and if cells are adhered and healthy, proceed to transfection.
6. Transfect cells
   a. Prepare MEM media (no serum), warm to 37°C and oxygenate in incubator.

b. Follow the product insert for cationic lipid transfection, e.g., Lipofectamine 2000.
c. Mix cDNA with MEM, and separately mix Lipofectamine with MEM.
d. Incubate for 10 min.
e. Can transfect cDNA from multiple subunits; 1 μg total cDNA per dish.
f. Mix together cDNA/MEM and Lipofectamine/MEM solutions.
g. Incubate mixture for 1 h before adding to cells.
h. Add 50–200 μL mixture dropwise gently onto surface of solution in 35 mm dishes to give final concentration of 1 μg cDNA/dish.

3.2.4 Notes

- Many others use penicillin/streptomycin in cell culture media to inhibit bacterial growth, but we find this is unnecessary if sterile procedure is closely followed and incubator trips are minimized.
- Mold growth is an issue in incubators and around biosafety cabinets and can be minimized by adding <1 g/L of copper sulfate to the water bath in the incubator. The solution should be a very light blue and changed every 2 weeks. Care must be taken to avoid corrosion of the water bath.
- For cotransfection of multiple receptor subunits, a small volume of the co-cDNA to be cotransfected should be gently mixed prior to adding MEM. We have achieved more consistent transfection of all cells in a dish with this approach, as cDNA may clump inappropriately in media if not mixed initially.
- cDNA for transfection should ideally be in water, as TE buffer inhibits transfection. Alternatively, cDNA should be diluted in water if possible during transfection.
- For $GABA_A$Rs, we find that subunit ratios may have to be optimized to produce the expected receptors.
- Some subunits are very difficult to assemble into functional receptors following HEK293 transfection; thus, other cell lines, transfection methods, or electrophysiological approaches may need to be considered.

3.3 Perfusion and Recording of Cultured Cells

Healthy adherent HEK293 cells should express transfected cDNA within 24–72 h after transfection; some cDNA may take slightly longer to express. Alternatively, stably transfected cell lines can be used which do not require routine transfection, but depending on the transfected construct, cell health may suffer over the long-term due to the persistent effects of the DNA products. Cells can be gently moved the day of recordings to an incubator near the electrophysiology rig if needed. Notably, HEK293 cells are simpler to seal and patch than cultured neurons and neurons in slice. They have stronger membranes and are larger than neurons, so it is advisable to use lower-resistance pipettes of 1.5–4 MΩ. As with cultured neurons, we have found that the use of a microforge to fire-polish and smooth the tips of electrodes into an hourglass shape after pulling (and immediately before recording) is helpful to seal and break into cells.

Unlike neurons, HEK293 cells do not have processes that extend for several 100 μm from the soma, and do not form synapses. Thus, due to the shape of the cells, all receptors expressed on HEK293 cells are somatic (located on the cell body), such that one can rapidly switch solution flow (with millisecond-order precision) to assess fast kinetic and pharmacological properties of a large proportion of expressed receptors simultaneously (Fig. 5). This would be impossible to assess in neurons given that synaptic receptors are not openly targetable by rapid solution exchange due to their

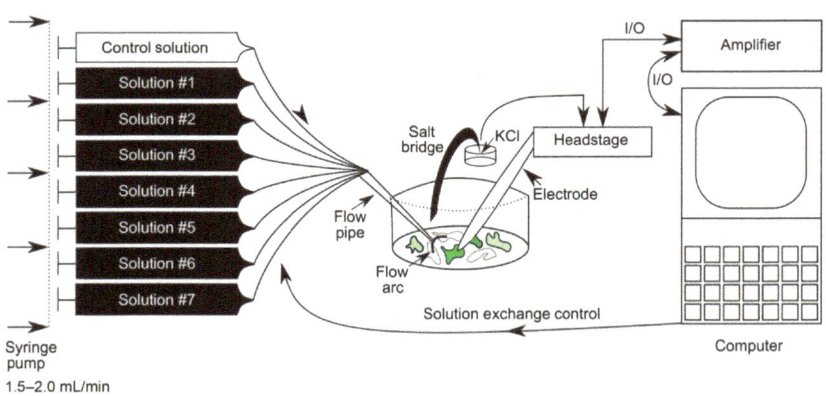

Fig. 5 Electrophysiological fast-perfusion setup. Solution superfusion and whole-cell patch clamp is schematized. Solutions can be simultaneously driven by a syringe pump and controlled for individual exit from a shared flow pipe. Solution delivery is rapid, local, and restricted to a ~0.2 mm flow arc for cell superfusion. A complete circuit is formed across the membrane of a recorded cell (*green*). Circuit components consist of headstage amplifier, Ag–AgCl electrode, and salt bridge-KCl connection back to the headstage. The headstage signal is boosted by an external patch amplifier, which is analyzed by computer. Signals are converted (analog to digital) and filtered (eight-pole Bessel at 4 kHz). *Modified with permission from Hoerbelt, P. (2015). Novel interactions of biogenic amines with GABAA receptors (doctoral thesis). Albany Medical College of Union University, ProQuest Dissertations Publishing, 3663393.*

obscured, protected locations. Furthermore, synaptic compartmentalization of specific receptor subtypes would ultimately bias the location of these subtypes, e.g., synaptic $GABA_ARs$ often contain the $\gamma 2$ subunit, whereas extrasynaptic receptors may not, instead possessing a δ subunit. Theoretically, then, rapid drug application onto an entire neuron may not provide useful information about the actions of a drug on a specific receptor subtype. By the same token, the specific receptor proportions expressed by a neuron in an intact biological system, whether at synaptic, perisynaptic, or extrasynaptic sites, may not be achievable in the more "artificial" environment of a HEK293 cell. Thus, the actions of a drug on a specific receptor subtype often do not provide adequate information regarding the relationship of that drug to changes in neuronal activity in an intact system.

3.3.1 Equipment (See Fig. 5 for a Schematic of the Setup)

- Inverted fluorescence microscope, at least 10 × and 40 × objectives, with a custom-designed depression to fit a 35-mm Nunc cell culture dish.
- Cultured HEK293 cells plated in Nunc dishes, or cultured neurons on glass coverslips (Hoerbelt, Lindsley, & Fleck, 2015).
- Nunc dishes.
- Fine tweezers.
- Fire-polishing apparatus, such as a Narishige MF-900 microforge.
- Rapid perfusion system, consisting of either a piezotranslator such as the Warner VC-77SP for ~100 μs solution exchange, sufficient for studying all ligand-gated channels including glutamate receptors; or an in-house made solenoid valve system for 5–20 ms exchange (Fleck, 2002), sufficient for $GABA_ARs$.
- Recording amplifier and headstage.
- Micromanipulator.

3.3.2 Buffers and Reagents

- HEPES-based external solution for short-term recordings, pH 7.3 (290–305 mOsm), with 0.1 mg/mL phenol red indicator and (in m*M*): 145 NaCl, 3 KCl, 5 HEPES, 1.8 $CaCl_2$, 1 $MgCl_2$, and, for neurons, 10 glucose.
- High-chloride internal solution (for the study of $GABA_ARs$). For neurons (in m*M*): 145 CsCl, 5 EGTA, 10 HEPES, 0.5 $CaCl_2$, 3 ATP-Mg^{2+}, and 10 phosphocreatine, pH 7.3 (280–290 mOsm). For HEK293 cells (in m*M*): 135 CsCl, 10 CsF, 5 EGTA, 10 HEPES, 0.5 $CaCl_2$, and 1 $MgCl_2$, pH 7.3 (295 mOsm).

- Drug solutions in appropriate external solution, made fresh from frozen concentrated stocks or powder.
- Vacuum grease.

3.3.3 Procedure

1. For whole-cell patch clamp, the procedure is very similar to Section 2.3.3 with the following minor changes.
2. Pull pipettes; fire-polish with microforge just before recording.
3. For HEK293 cells, gently exchange culture media in Nunc dishes with HEPES external solution just before placing in microscope.
4. For neurons, quickly stick a glass coverslip neuron-side-up into the center of a Nunc dish with vacuum grease and immediately but gently add external solution with a pipette.
5. Transfer cells to microscope stage, find appropriate cell, and approach the cell as in Section 2.3.3.
6. Obtain a gigaseal and break-in.
7. Record $GABA_AR$ currents at −60 to −80 mV in high chloride internal solution; if using a low chloride solution, which better approximates a neuron's intracellular:extracellular chloride gradient and chloride reversal potential, record at 0 mV.
8. For neurons, add 0.5 μ*M* tetrodotoxin in external solution to block synaptic events if desired.
9. Rapidly apply drug compounds for 5–20 ms (if studying glutamate receptors) or several seconds (for $GABA_ARs$) with washout in between for at least 10 s.
10. Preapply antagonists and continue application with anesthetic drug.
11. For long-term (>20 min) recording of neurons or any-length recording of action potentials in neurons, consistent superfusion of a bicarbonate-based external solution at 2 mL/min is necessary; otherwise, HEPES buffer may be used.

3.3.4 Notes

- Different drugs have different solubilities and tendencies for breakdown in solution; this must be taken into account.
- As with slice recordings, series resistance and access resistance must be monitored throughout recordings (Molleman, 2003).
- When recording from certain $GABA_ARs$ or other ligand-gated ion channels, particularly including the $GABA_AR$ β3 subunit, trace Zn^{2+} in buffer solutions will block these receptors, significantly affecting

recorded currents (Dunne et al., 2002; Hoerbelt et al., 2015, 2016; Hosie, Dunne, Harvey, & Smart, 2003; Wooltorton, McDonald, Moss, & Smart, 1997). Using mutated β3 cDNA may alleviate this issue (Dunne et al., 2002).

4. SUMMARY AND CONCLUSION

In this chapter, we have identified several methods to study the mechanisms by which anesthetic agents alter signaling in native neuronal systems and in defined receptors that are expressed in those systems. Specifically, we have guided the reader toward accomplishing successful field recordings and whole-cell patch clamp recordings in hippocampus, a common and versatile preparation. We have also outlined how we prepare and transfect cDNA into HEK293 cells, a useful starting point for identifying drug actions at $GABA_AR$s, one of the most notable classes of general anesthetic targets. Furthermore, we have discussed patch clamp recording in transfected, cultured cells as well as dissociated primary cultures of neurons. Along the way, we have highlighted many experimental tweaks and methodological constraints that we have found enhance our experiments. In closing, we hope that the approaches we have described will provide a useful set of protocols for researchers hoping to interrogate the short- or long-term functions of novel (or even commonly known) anesthetic drugs in native systems.

REFERENCES

Aghajanian, G. K., & Rasmussen, K. (1989). Intracellular studies in the facial nucleus illustrating a simple new method for obtaining viable motoneurons in adult rat brain slices. *Synapse (New York, NY)*, *3*, 331–338.

Amaral, D. G., & Witter, M. P. (1989). The three-dimensional organization of the hippocampal formation: A review of anatomical data. *Neuroscience*, *31*, 571–591.

Arai, A., Silberg, J., & Lynch, G. (1995). Differences in the refractory properties of two distinct inhibitory circuitries in field CA1 of the hippocampus. *Brain Research*, *704*, 298–306.

Bekkers, J. M., & Stevens, C. F. (1991). Excitatory and inhibitory autaptic currents in isolated hippocampal neurons maintained in cell culture. *Proceedings of the National Academy of Sciences of the United States of America*, *88*, 7834–7838.

Chiara, D. C., Jayakar, S. S., Zhou, X., Zhang, X., Savechenkov, P. Y., Bruzik, K. S., et al. (2013). Specificity of intersubunit general anesthetic-binding sites in the transmembrane domain of the human α1β3γ2 γ-aminobutyric acid type A (GABAA) receptor. *The Journal of Biological Chemistry*, *288*(27), 19343–19357.

Chiara, D. C., Jounaidi, Y., Zhou, X., Savechenkov, P. Y., Bruzik, K. S., Miller, K. W., et al. (2016). General anesthetic binding sites in human α4β3δ γ-aminobutyric acid type A receptors (GABAARs). *The Journal of Biological Chemistry*, *291*(51), 26529–26539.

Davies, P. A., Kirkness, E. F., & Hales, T. G. (1997). Modulation by general anaesthetics of rat $GABA_A$ receptors comprised of alpha 1 beta 3 and beta 3 subunits expressed in human embryonic kidney 293 cells. *British Journal of Pharmacology, 120*, 899–909.

Dunne, E. L., Hosie, A. M., Wooltorton, J. R., Duguid, I. C., Harvey, K., Moss, S. J., et al. (2002). An N-terminal histidine regulates Zn(2+) inhibition on the murine GABA(A) receptor beta3 subunit. *British Journal of Pharmacology, 137*, 29–38.

Eaton, J. B., Lucero, L. M., Stratton, H., Chang, Y., Cooper, J. F., Lindstrom, J. M., et al. (2014). The unique alpha4 +/−alpha4 agonist binding site in (alpha4)3(beta2)2 subtype nicotinic acetylcholine receptors permits differential agonist desensitization pharmacology versus the (alpha4)2(beta2)3 subtype. *The Journal of Pharmacology and Experimental Therapeutics, 348*, 46–58.

Eckenhoff, R. G. (2001). Promiscuous ligands and attractive cavities: How do the inhaled anesthetics work? *Molecular Interventions, 1*(5), 258–268.

Fleck, M. W. (2002). Molecular actions of (S)-desmethylzopiclone (SEP-174559), an anxiolytic metabolite of zopiclone. *The Journal of Pharmacology and Experimental Therapeutics, 302*, 612–618.

Franks, N. P., & Lieb, W. R. (1996). An anesthetic-sensitive superfamily of neurotransmitter-gated ion channels. *Journal of Clinical Anesthesia, 8*, 3s–7s.

Frazier, D. T., Narahashi, T., & Yamada, M. (1970). The site of action and active form of local anesthetics. II. Experiments with quaternary compounds. *The Journal of Pharmacology and Experimental Therapeutics, 171*, 45–51.

Freund, T. F., & Buzsáki, G. (1996). Interneurons of the hippocampus. *Hippocampus, 6*, 347–470.

Garcia, P. S., Kolesky, S. E., & Jenkins, A. (2010). General anesthetic actions on $GABA_A$ receptors. *Current Neuropharmacology, 8*, 2–9.

Hamill, O. P., Marty, A., Neher, E., Sakmann, B., & Sigworth, F. J. (1981). Improved patch-clamp techniques for high-resolution current recording from cells and cell-free membrane patches. *Pflugers Archiv: European Journal of Physiology, 391*, 85–100.

Hamouda, A. K., Stewart, D. S., Chiara, D. C., Savechenkov, P. Y., Bruzik, K. S., & Cohen, J. B. (2014). Identifying barbiturate binding sites in a nicotinic acetylcholine receptor with [3H]allyl m-trifluoromethyldiazirine mephobarbital, a photoreactive barbiturate. *Molecular Pharmacology, 85*(5), 735–746.

Heckman, K. L., & Pease, L. R. (2007). Gene splicing and mutagenesis by PCR-driven overlap extension. *Nature Protocols, 2*, 924–932.

Heifets, B. D., Chevaleyre, V., & Castillo, P. E. (2008). Interneuron activity controls endocannabinoid-mediated presynaptic plasticity through calcineurin. *Proceedings of the National Academy of Sciences of the United States of America, 105*, 10250–10255.

Hille, B. (2001). *Ion channels of excitable membranes* (3rd ed.). MA, USA: Sinauer Associates, Inc.

Hoerbelt, P., Lindsley, T. A., & Fleck, M. W. (2015). Dopamine directly modulates $GABA_A$ receptors. *The Journal of Neuroscience: The Official Journal of the Society for Neuroscience, 35*, 3525–3536.

Hoerbelt, P., Ramerstorfer, J., Ernst, M., Sieghart, W., Thomson, J. L., Hough, L. B., et al. (2016). Mutagenesis and computational docking studies support the existence of a histamine binding site at the extracellular beta3 + beta3- interface of homooligomeric beta3 $GABA_A$ receptors. *Neuropharmacology, 108*, 252–263.

Hosie, A. M., Dunne, E. L., Harvey, R. J., & Smart, T. G. (2003). Zinc-mediated inhibition of GABA(A) receptors: Discrete binding sites underlie subtype specificity. *Nature Neuroscience, 6*, 362–369.

Hosie, A. M., Wilkins, M. E., da Silva, H. M., & Smart, T. G. (2006). Endogenous neurosteroids regulate $GABA_A$ receptors through two discrete transmembrane sites. *Nature, 444*, 486–489.

Huxley, A. F. (2002). Hodgkin and the action potential 1935-1952. *The Journal of Physiology, 538*, 2.
Kaech, S., & Banker, G. (2006). Culturing hippocampal neurons. *Nature Protocols, 1*, 2406–2415.
Klyachko, V. A., & Stevens, C. F. (2006). Temperature-dependent shift of balance among the components of short-term plasticity in hippocampal synapses. *The Journal of Neuroscience: The Official Journal of the Society for Neuroscience, 26*, 6945–6957.
Kuenzi, F. M., Fitzjohn, S. M., Morton, R. A., Collingridge, G. L., & Seabrook, G. R. (2000). Reduced long-term potentiation in hippocampal slices prepared using sucrose-based artificial cerebrospinal fluid. *Journal of Neuroscience Methods, 100*, 117–122.
Lambert, N. A., Borroni, A. M., Grover, L. M., & Teyler, T. J. (1991). Hyperpolarizing and depolarizing $GABA_A$ receptor-mediated dendritic inhibition in area CA1 of the rat hippocampus. *Journal of Neurophysiology, 66*, 1538–1548.
MacIver, M. B. (2014). Anesthetic agent-specific effects on synaptic inhibition. *Anesthesia and Analgesia, 119*, 558–569.
Mah, S. J., Fleck, M. W., & Lindsley, T. A. (2011). Ethanol alters calcium signaling in axonal growth cones. *Neuroscience, 189*, 384–396.
Malenka, R. C., & Bear, M. F. (2004). LTP and LTD: An embarrassment of riches. *Neuron, 44*, 5–21.
Malinow, R., & Tsien, R. W. (1990). Presynaptic enhancement shown by whole-cell recordings of long-term potentiation in hippocampal slices. *Nature, 346*, 177–180.
Mathis, D. M., Furman, J. L., & Norris, C. M. (2011). Preparation of acute hippocampal slices from rats and transgenic mice for the study of synaptic alterations during aging and amyloid pathology. *Journal of Visualized Experiments, 49*, pii: 2330. https://doi.org/10.3791/2330.
Mihic, S. J., Ye, Q., Wick, M. J., Koltchine, V. V., Krasowski, M. D., Finn, S. E., et al. (1997). Sites of alcohol and volatile anaesthetic action on GABA(A) and glycine receptors. *Nature, 389*, 385–389.
Molleman, A. (2003). The practice of patch clamping. In A. Molleman (Ed.), *Patch clamping* (pp. 95–114). John Wiley & Sons, Ltd.
Monagle, J., Siu, L., Worrell, J., Goodchild, C. S., & Serrao, J. M. (2015). A phase 1c trial comparing the efficacy and safety of a new aqueous formulation of alphaxalone with propofol. *Anesthesia and Analgesia, 121*, 914–924.
Nakahiro, M., Yeh, J. Z., Brunner, E., & Narahashi, T. (1989). General anesthetics modulate GABA receptor channel complex in rat dorsal root ganglion neurons. *FASEB Journal: Official Publication of the Federation of American Societies for Experimental Biology, 3*, 1850–1854.
Nishikawa, K., & Harrison, N. L. (2003). The actions of sevoflurane and desflurane on the gamma-aminobutyric acid receptor type A: Effects of TM2 mutations in the alpha and beta subunits. *Anesthesiology, 99*, 678–684.
Perouansky, M., & Pearce, R. A. (2011). How we recall (or don't): The hippocampal memory machine and anesthetic amnesia. *Canadian Journal of Anaesthesia = Journal canadien d'anesthesie, 58*, 157–166.
Sakmann, B., & Neher, E. (1984). Patch clamp techniques for studying ionic channels in excitable membranes. *Annual Review of Physiology, 46*, 455–472.
Shepherd, G. (2005). The synaptic organization of the brain. Fifth Edition. *The Quarterly Review of Biology, 80*, 502.
Sieghart, W. (1995). Structure and pharmacology of gamma-aminobutyric acid(A) receptor subtypes. *Pharmacological Reviews, 47*, 181–234.
Woll, K.A., Zhou, X., Bhanu, N.V., Garcia, B.A., Covarrubias, M., Miller, K.W., et al. (2018) Identification of binding sites contributing to volatile anesthetic effects on GABA

type A receptors. *FASEB Journal: Official Publication of the Federation of American Societies for Experimental Biology*, fj201701347R. https://doi.org/10.1096/fj.201701347R.
Wooltorton, J. R., McDonald, B. J., Moss, S. J., & Smart, T. G. (1997). Identification of a Zn2+ binding site on the murine $GABA_A$ receptor complex: Dependence on the second transmembrane domain of beta subunits. *Journal of Physiology*, *505*, 633–640.
Zucker, R. S., & Regehr, W. G. (2002). Short-term synaptic plasticity. *Annual Review of Physiology*, *64*, 355–405.

FURTHER READING

Belelli, D., Lambert, J. J., Peters, J. A., Wafford, K. A., & Whiting, P. J. (1997). The interaction of the general anesthetic etomidate with the gamma-aminobutyric acid type A receptor is influenced by a single amino acid. *Proceedings of the National Academy of Sciences of the United States of America*, *94*, 11031–11036.
Cestari, I. N., Uchida, I., Li, L., Burt, D., & Yang, J. (1996). The agonistic action of pentobarbital on GABAA beta-subunit homomeric receptors. *Neuroreport*, 7, 943–947.
Jayakar, S. S., Zhou, X., Chiara, D. C., Dostalova, Z., Savechenkov, P. Y., Bruzik, K. S., et al. (2014). Multiple propofol-binding sites in a γ-aminobutyric acid type A receptor (GABAAR) identified using a photoreactive propofol analog. *The Journal of Biological Chemistry*, *289*(40), 27456–27468.
Krasowski, M. D., Nishikawa, K., Nikolaeva, N., Lin, A., & Harrison, N. L. (2001). Methionine 286 in transmembrane domain 3 of the $GABA_A$ receptor beta subunit controls a binding cavity for propofol and other alkylphenol general anesthetics. *Neuropharmacology*, *41*, 952–964.
Li, G. D., Chiara, D. C., Sawyer, G. W., Husain, S. S., Olsen, R. W., & Cohen, J. B. (2006). Identification of a $GABA_A$ receptor anesthetic binding site at subunit interfaces by photolabeling with an etomidate analog. *The Journal of Neuroscience: The Official Journal of the Society for Neuroscience*, *26*, 11599–11605.
Yip, G. M., Chen, Z. W., Edge, C. J., Smith, E. H., Dickinson, R., Hohenester, E., et al. (2013). A propofol binding site on mammalian $GABA_A$ receptors identified by photolabeling. *Nature Chemical Biology*, *9*, 715–720.

CHAPTER SEVENTEEN

Electrophysiological Analysis of Voltage-Gated Ion Channel Modulation by General Anesthetics

Elaine Yang[1], Lianteng Zhi, Qiansheng Liang, Manuel Covarrubias
Vickie and Jack Farber Institute for Neuroscience, Sidney Kimmel Medical College, Jefferson College of Biomedical Sciences, Thomas Jefferson University, Philadelphia, PA, United States
[1]Corresponding author: e-mail address: Elaine.Yang@jefferson.edu

Contents

Abstract

Voltage-gated ion channels (VGICs) of excitable tissues are emerging as targets likely involved in both the therapeutic and toxic effects of inhaled and intravenous general anesthetics. Whereas sevoflurane and propofol inhibit voltage-gated Na^+ channels

Methods in Enzymology, Volume 602
ISSN 0076-6879
https://doi.org/10.1016/bs.mie.2018.01.013

(Navs), sevoflurane potentiates certain voltage-gated K^+ channels (Kvs). The combination of these effects would dampen neural excitability and, therefore, might contribute to the clinical endpoints of general anesthesia. As the body of work regarding the interaction of general anesthetics with VGICs continues to grow, a multidisciplinary approach involving functional, biochemical, structural, and computational techniques, many of which are detailed in other chapters, has increasingly become necessary to solve the molecular mechanism of general anesthetic action on VGICs. Here, we focus on electrophysiological and modeling approaches and methodologies to describe how our work has elucidated the biophysical basis of the inhibition Navs by propofol and the potentiation of Kvs by sevoflurane.

1. INTRODUCTION

General anesthesia results from promiscuous interactions of a diverse group of nonspecific drugs with multiple targets in the nervous system (Covarrubias, Barber, Carnevale, Treptow, & Eckenhoff, 2015; Eckenhoff, Zheng, & Kelz, 2008; Franks, 2008). Ligand-gated and voltage-gated ion channels responsible for electrical signaling are major targets of general anesthetics (Covarrubias et al., 2015; Franks, 2006, 2008; Hemmings et al., 2005; Herold & Hemmings, 2012). However, a complete understanding of the molecular mechanisms of general anesthetic action on these ion channels has remained elusive. To tackle this problem, anesthesia researchers have used a reductionist approach by studying individual ion channel subtypes in isolation. The general approach relies on complementary methodologies, including heterologous expression, electrophysiology, mutational analysis, computer modeling, protein biochemistry, and structural biology. Such a concerted effort might ultimately render a sharper picture of how anesthetic binding to ion channels alters their function (Covarrubias et al., 2015; Woll, Peng, Liang, et al., 2017; Woll, Skinner, Gianti, et al., 2017; Woll, Weiser, Liang, et al., 2015). We have investigated individual voltage-gated Na^+ (Nav) and K^+ (Kv) channels using heterologous expression in mammalian cells and *Xenopus* oocytes and electrophysiological methods (patch clamping and two-electrode voltage clamping) in concert with mutational analysis and kinetic modeling (Barber, Carnevale, Klein, Eckenhoff, & Covarrubias, 2014; Barber, Liang, Amaral, Treptow, & Covarrubias, 2011; Barber, Liang, & Covarrubias, 2012; Covarrubias et al., 2015; Liang, Anderson, Jones, et al., 2015). An investigation of the molecular mechanism of general anesthetic action on Navs or Kvs usually begins with these approaches to address several fundamental questions: (1) Does the anesthetic at clinically relevant concentrations modulate ion channel function? (2) Does the anesthetic modulate ion channel gating and/or permeation? (3) Does the anesthetic block the pore? (4) Does the anesthetic selectively promote or stabilize a particular functional state of the ion channel? Answers to these questions help generate hypotheses about possible structural determinants of anesthetic action, which can be investigated further by mutational analysis of functional domains associated with gating and/or permeation. In an iterative process, the electrophysiological, mutational, biochemical, and structural approaches help refine the models of the molecular mechanisms underlying general anesthetic action of Navs and Kvs (Fig. 1). In this chapter, we describe the aforementioned electrophysiological approaches to investigate the interactions of

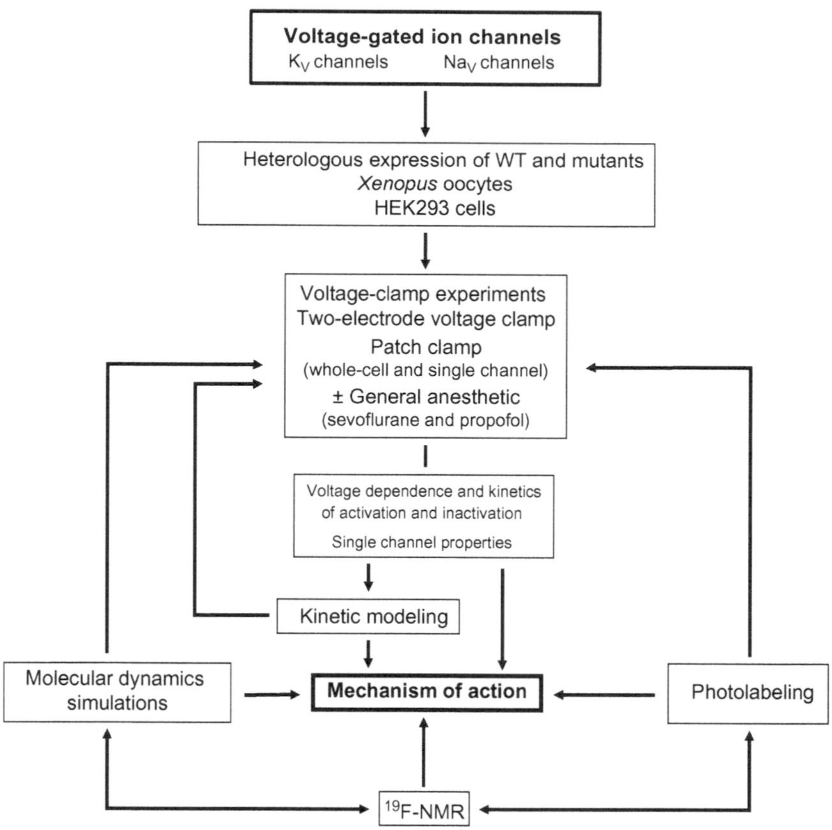

Fig. 1 Workflow of a general approach to investigate the molecular mechanisms of general anesthetic action on Nav and Kv channels. While this chapter focuses on heterologous expression and electrophysiological approaches, these methodologies do not occur in isolation. They are part of a multidisciplinary plan that includes biochemical, biophysical, and computational approaches, which complement one another to ultimately solve the physical basis of general anesthesia at the molecular level.

propofol with two bacterial Navs (NaChBac and NavMs) and the interactions of sevoflurane with *Drosophila* and mammalian Kv channels (K-Shaw and Kv1.2, respectively).

2. INVESTIGATING THE MODULATION OF NAV CHANNELS BY PROPOFOL

In contrast to their eukaryotic counterparts, which are comprised of four homologous and interconnected domains, bacterial Navs (BacNavs) are homotetramers with four identical subunits, each of which is equivalent and homologous to one eukaryotic domain. BacNavs lack certain structural elements, such as cytoplasmic domains involved in eukaryotic Nav fast inactivation (Ahern, Payandeh, Bosmans, & Chanda, 2016; Payandeh & Minor, 2015; Yu & Catterall, 2004). However, key structural elements are conserved between bacterial and eukaryotic Navs, including shared voltage-gated ion channel (VGIC) superfamily architecture and functionally critical residues, which has permitted extensive structure–function investigations of ion channel selectivity and conductance, gating, and pharmacology (Ahern et al., 2016; Charalambous & Wallace, 2011; Naylor, Bagneris, DeCaen, et al., 2016; Payandeh & Minor, 2015; Yu & Catterall, 2004). Importantly, there is extensive evidence that BacNavs closely recapitulate the pharmacological responses seen in eukaryotic Navs. The comparatively simple architecture of bacterial Navs has additionally allowed the determination of several X-ray crystallographic structures (McCusker, Bagnéris, Naylor, et al., 2012; Payandeh, Gamal El-Din, Scheuer, Zheng, & Catterall, 2012; Payandeh, Scheuer, Zheng, & Catterall, 2011; Sula et al., 2017; Tsai, Tani, Irie, et al., 2013; Zhang, Ren, DeCaen, et al., 2012), which have formed the basis for computational studies.

Experimentally, BacNavs like NaChBac, from *Bacillus halodurans* (Ren et al., 2001), and NavMs, from *Magnetococcus marinus* (McCusker et al., 2012), offer many advantages over their eukaryotic counterparts. Because eukaryotic Nav domains are homologous but different, mutagenesis can be significantly more complicated compared to BacNavs, to which point mutations are applied in all four subunits simultaneously. Moreover, BacNavs can reliably be expressed in a variety of heterologous expression systems, which will be discussed later, and high levels of expression are routinely achieved in several *E. coli* strains for high-resolution structural experiments.

With recent advances in structural biology leading to the cryo-electron microscopy structures of NavPaS from the American cockroach *Periplaneta americana* (Shen et al., 2017) and EeNav1.4-β1 from electric eel *Electrophorus electricus* (Yan, Zhou, Wang, et al., 2017), determination of the full-length structure of a mammalian Nav is expected in the near future. Nonetheless, given the wealth of knowledge they have yielded and the clear advantages they offer experimentally, it is undeniable that bacterial Navs are invaluable tools in structure–function studies to define key principles underlying Nav function. Here, we describe techniques and experimental approaches to investigate the biophysical basis of general anesthetic action on Navs, focusing on the mechanisms of propofol action on NaChBac and NavMs.

2.1 Selection of Expression System for Transient Expression

NaChBac is a versatile tool for examining Nav function because it can be robustly and reliably expressed in its functional form in a number of expression systems, including *Xenopus laevis* oocytes (Lee, Goodchild, & Ahern, 2012), African green monkey kidney cells (COS) (Kuzmenkin, Bezanilla, & Correa, 2004), Chinese hamster ovary (CHO) cells (Ren et al., 2001), and human embryonic kidney (HEK) cells (Barber et al., 2014; Chahine, Pilote, Pouliot, Takami, & Sato, 2004; Ouyang, Jih, Zhang, Correa, & Hemmings, 2007; Sand, Gingrich, Macharadze, Herold, & Hemmings, 2017). NavMs can similarly be expressed in HEK cells for functional studies (Bagneris, DeCaen, Naylor, et al., 2014; McCusker et al., 2012). HEK cells, HEK293 for NaChBac and HEK293T for NavMs, have been our primary expression system for a number of reasons. First, in our experience, the expression of NaChBac in oocytes yields ionic currents that are often too large for accurate two-electrode voltage clamp studies. Second, mammalian expression systems permit high-resolution patch clamp experiments for accurate investigation of gating kinetics. While this may not be a problem in NaChBac, which has kinetics roughly one-tenth the speed of eukaryotic channels (Ren et al., 2001), it is an important consideration in faster channels like NavMs. And lastly, though NaChBac and NavMs are both BacNavs, mammalian expression systems provide cellular environments that will facilitate transition to future investigations of mammalian Navs.

2.2 Transient Transfection Methods

For our voltage clamp experiments, NaChBac and NavMs cDNA are subcloned into the pTracer-CMV2 mammalian expression vector (Invitrogen), which are efficiently amplified in *E. coli* and purified for transient transfection using commercially available plasmid purification kits (e.g., QIAGEN or ThermoFisher Scientific). Cells are transiently transfected with plasmid DNA using either the calcium phosphate method or transfection reagents, which are both detailed here. In the appropriate expression system, both approaches are quite reproducible and effective, and while both are relatively simple, transfection reagents such as Lipofectamine 2000 from Invitrogen require significantly less effort and are thus our transfection method of choice.

2.2.1 Calcium Phosphate Transfection Protocol

1. Cells should be plated on 6-cm culture dishes in 5 mL of growth medium such that they are 40%–50% confluent at the time of transfection. Using this method, transfection efficiency is highly dependent on the total cell surface area exposed to the precipitate.
2. Add 18 μL of 2 *M* $CaCl_2$ and 10 μg of DNA to a sterile microcentrifuge tube. Add sterile, nuclease-free water to the tube to a final volume of 150 μL. In a second sterile tube, add 150 μL of 2XHEPES buffered saline (HBS)[a].
3. Add the DNA–$CaCl_2$ solution *dropwise* to the HBS. The size and density of the calcium phosphate precipitate is key to a successful transfection, as clumped DNA will not adhere to or enter cells as efficiently. Bubbling air through the HBS with a Pasteur pipette while adding the DNA–$CaCl_2$ solution may be used to mix the solution in a way that promotes formation of an appropriately fine precipitate. Alternatively, intermittent vortexing may also be used.
4. Incubate at room temperature for 30 min to allow precipitate formation. At this time, exchange the growth medium on the cells with fresh medium. Incubating the cells in fresh growth medium for at least 30 min prior to transfection will improve transfection efficiency.
5. Add the precipitate *dropwise* to the cells and incubate for 4–6 h at 37°C, 5% CO_2.
6. Remove the transfection medium and wash with PBS. Dissociate the cells with the appropriate dissociation reagent (e.g., trypsin/EDTA) and plate the cells on coverslips for patch clamp experiments.

[a] These are amounts for the transfection of a single 6-cm culture plate and should be scaled appropriately for the transfection of additional, smaller, or larger plates. For example, volumes should be doubled for a 10-cm dish containing 10 mL of culture medium.

2.2.2 Lipofectamine 2000 Transfection Protocol

1. Cells should be plated on 3.5-cm culture dishes in 2 mL of growth medium with a confluency of 70%–90% at the time of transfection[b].
2. Replace growth medium on the cells with reduced-serum (Opti-MEM, ThermoFisher Scintific) and/or antibiotic-free medium. This step, while not required, may improve transfection efficiency.
3. Add 250 μL of Opti-MEM to two separate sterile tubes. To one tube, add 5–12.5 μL of Lipofectamine reagent. To the other, add 2.5–5 μg of DNA[c]. Gently mix the contents of each tube and incubate at room temperature for 5 min.
4. Combine the DNA and Lipofectamine mixtures. Incubate at room temperature for 20 min.
5. Add the DNA–Lipofectamine mixture to the cells and incubate for 4–6 h at 37°C, 5% CO_2.
6. Remove the transfection medium and wash with PBS. Dissociate the cells with the appropriate dissociation reagent (e.g., trypsin/EDTA) and plate the cells on coverslips for patch clamp experiments.

In general, higher amounts of DNA lead to higher Nav expression the following day, but each transfection reaction and expression system is unique and wide variations are often observed. Nav expression will continue to increase in the days after transfection to a point where currents may saturate the amplifier in the whole-cell patch clamp configuration. In this case, it may be necessary to perform these experiments within 24 h after transfection.

2.3 Propofol Preparation and Delivery

General anesthetic agents are highly lipophilic. Thus, when preparing anesthetic solutions, care must be taken to dilute the anesthetic to appropriate concentrations in recording solutions to ensure that the anesthetic dissolves rather than emulsifying. Additionally, glass vials and bottles should be used whenever possible to minimize adsorption of the anesthetic agent to storage containers.

With a clogP of 3.79 (Hall, Xi, Lor, et al., 2010; Woll et al., 2015), propofol (2,6-diisopropylphenol) has a maximum solubility of approximately 600 μ*M* in aqueous solution and can be stored at 4°C for approximately 1 week at 400 μ*M* (K. Woll, personal communication, October 18, 2016).

[b] Higher confluency is advantageous when using Lipofectamine, as some cell death is expected.

[c] The optimal Lipofectamine:DNA ratio is highly dependent on cell type and will require initial optimization.

Nonetheless, we prefer to prepare fresh stock and working propofol solutions on each day of use. Propofol stock solution is prepared in extracellular bath solution (see Section 2.4 for recipe) by alternating sonication and vortexing, in 30-s increments for a total of 4 min. Sonication should always precede vortexing to avoid adsorption of the anesthetic to the top of the container. Stock solution is then diluted in extracellular recording solution to working concentrations and shaken vigorously before use. While other groups working with propofol have shown that effects are reversible upon washout (Ouyang, Wang, & Hemmings, 2003), we prefer to expose each cell to only one concentration and avoid using washout data to eliminate any potential inaccuracies caused by membrane lipid retention of propofol molecules (e.g., cumulative effects).

2.4 Patch Clamp Electrophysiology of Nav Channels

Patch clamp electrophysiology is the preferred method that we use to study the effects of general anesthetics on Nav function. The following is a brief description of the specific methods we have employed to study the effects of propofol on NaChBac and NavMs.

Microelectrodes (patch pipettes) are pulled from borosilicate capillary glass (LA16, Dagan Corporation) with a micropipette puller (e.g., HEKA PIP6, HEKA). Prior to recording, patch pipettes are fire polished to a final resistance of 1.5–2.3 MΩ. Patch clamp recordings of NaChBac and NavMs are performed in the whole-cell patch clamp configuration using an Axopatch 200B amplifier (Molecular Devices) and Digidata 1440A analog-to-digital converter (Molecular Devices). Passive leak current and capacitive transients are subtracted online by standard P/4 protocol (NaChBac) or offline using the passive leak subtraction feature in pCLAMP (NavMs). All recordings are low-pass Bessel-filtered at 2 kHz and digitized at 15.4 kHz. Clampex 10 (pCLAMP 10, Molecular Devices) is used to control voltage protocols and for data acquisition.

For NaChBac recordings, the extracellular bath solution contains (in m*M*): 140 NaCl, 4 KCl, 1.5 $CaCl_2$, 1.5 $MgCl_2$, 10 HEPES, and 5 D-glucose; pH 7.3 adjusted with NaOH. The intracellular pipette solution contains (in m*M*): 15 NaCl, 80 CsF, 40 CsCl, 10 EGTA, and 10 HEPES; pH 7.3 adjusted with CsOH. For NavMs recordings, the extracellular bath solution contains (in m*M*): 150 NaCl, 1.8 $CaCl_2$, 1 $MgCl_2$, and 10 HEPES; pH 7.4 adjusted with NaOH. The intracellular pipette solution contains (in m*M*): 30 NaCl, 110 CsF, 1 KCl, 10 HEPES, and 5 EGTA; pH 7.4 adjusted with CsOH.

During patch clamp recordings, cells are continuously perfused with bath solution at room temperature (22–25°C). In all experiments, all control recordings are collected first, prior to any anesthetic exposure. Following control recordings, propofol is perfused for ~3 min before collecting paired anesthetic recordings and continuously thereafter.

2.5 Investigating General Anesthetic Modulation of Macroscopic Nav Gating

Gating of VGICs can generally be described by a linear kinetic scheme (Fig. 2A). In response to membrane depolarization, typical Navs transition from the resting closed state to intermediate activated states ($C_0 \rightarrow \cdots \rightarrow C_n$), open ($C_n \rightarrow O$), then inactivate ($O \rightarrow I$). Transitions between states are functions of both time and voltage. Upon hyperpolarization of the membrane voltage, channels return to the closed resting state. To investigate the biophysical basis of general anesthetics in VGICs, the voltage dependence and kinetics of each of these processes are examined by patch clamp electrophysiology in the absence and then the presence of anesthetic. Investigation of both macroscopic ionic and gating currents helps to establish a detailed mechanistic picture of channel gating.

The voltage dependence of activation is assessed by determining the conductance–voltage (*G–V*) relationship. The membrane potential is first clamped at a strongly hyperpolarized holding voltage (V_h), where most Navs are in the resting closed state. A series of depolarizing test pulses is applied in 5–10 mV increments to activate and open channels, returning the membrane to V_h between pulses, which generates a family of inward Na^+ currents (Fig. 2B and C, top). Peak current amplitudes are converted to peak chord conductances and plotted vs test voltage to generate the *G–V* relationship (Barber et al., 2014, 2012; Liang et al., 2015) (Fig. 2B and C, bottom). The best-fit Boltzmann function is then used to estimate macroscopic gating parameters, such as maximum conductance (G_{max}), effective gating charge of activation (z), and midpoint voltage ($V_{1/2}$) of activation, which are compared in the absence and presence of anesthetic.

To examine macroscopic activation kinetics ($C_n \leftrightarrow O$), we measure the time constants of activation and deactivation (Fig. 3A and B). Inward Na^+ currents are elicited as described in the previous paragraph. Deactivation tail currents are elicited by hyperpolarizing/repolarizing steps (in 5 or 10 mV increments) immediately after a short depolarizing pulse that fully activates the channels. Time constants of activation and deactivation are derived from the rising phase of the inward Na^+ currents (Fig. 3A) and from tail current

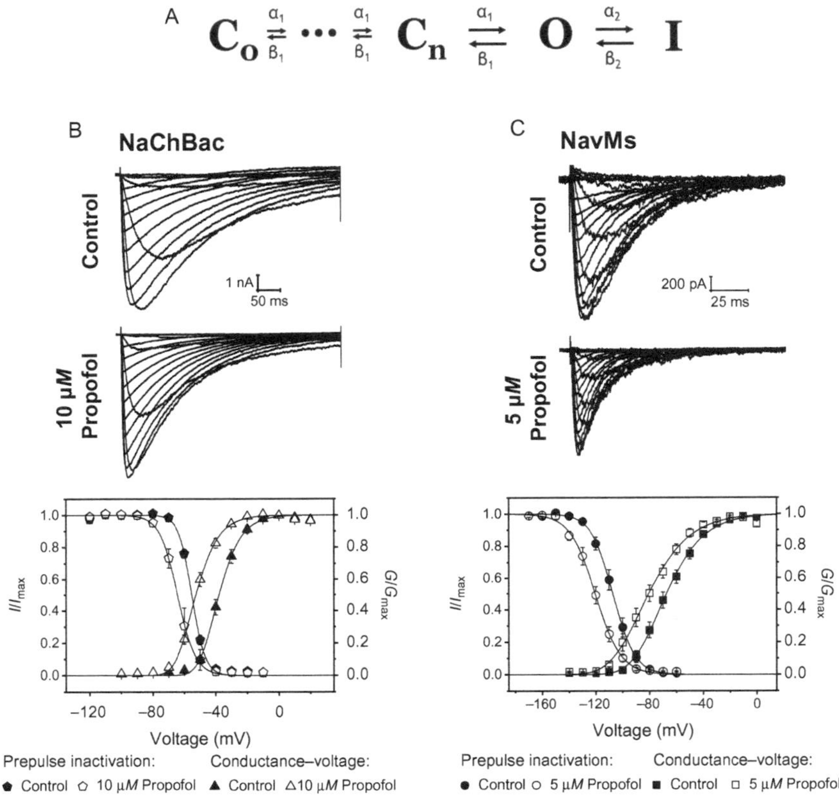

Fig. 2 (A) Sequential state kinetic scheme of voltage-gated ion channels. (B and C, *top*) Representative paired current families from (B) NaChBac and (C) NavMs, in the absence and presence of propofol at the indicated concentrations. NaChBac currents are evoked by 500-ms depolarizing steps (−100 to +60 mV, $\Delta V = 10$ mV) from a holding potential of −120 mV. NavMs currents are evoked by 200-ms depolarizing steps (−140 to +40 mV, $\Delta V = 10$ mV) from a holding potential of −180 mV. (B and C, *bottom*) *G–V* and prepulse inactivation relationships in the absence and presence of propofol at the indicated concentrations. Prepulse inactivation in NaChBac is assessed with a two-pulse protocol: (1) a 2-s conditioning pulse (−120 to −10 mV, $\Delta V = 10$ mV), followed immediately by (2) a 50-ms test pulse to +10 mV; the holding potential is −120 mV. In NavMs, the two-pulse protocol is (1) a 1-s conditioning pulse (−170 to −60 mV, $\Delta V = 10$ mV), followed immediately by (2) a 50-ms test pulse to −30 mV; the holding potential is −180 mV.

relaxations (Fig. 3B), respectively, using the best-fit single exponential or sum of exponentials. This analysis does not take into consideration the lag that typically precedes the rising phase of current activation, which gives these currents their characteristic sigmoidal shape. The lag is due to the

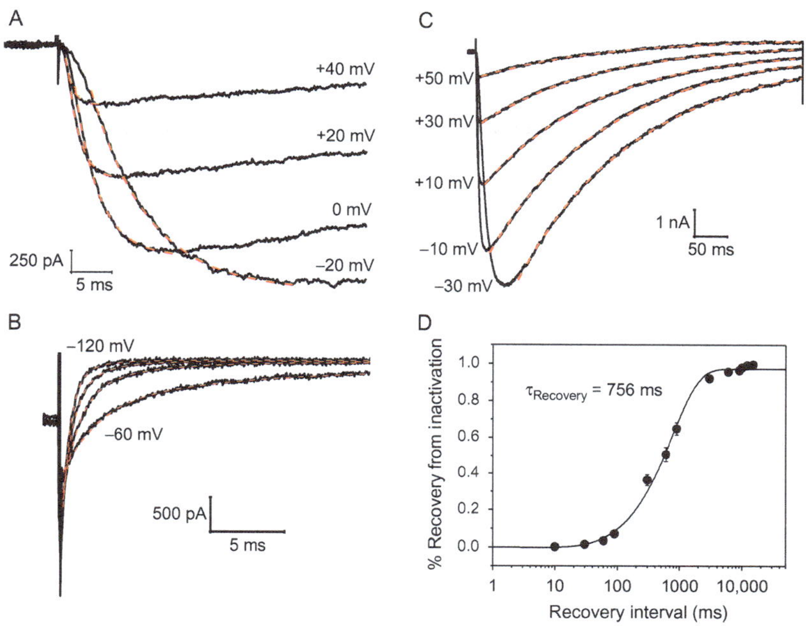

Fig. 3 Examination of macroscopic NaChBac gating kinetics under control conditions. Time constants of (A) activation, (B) deactivation, and (C) inactivation are derived from the rising phase of inward Na^+ currents, from tail current relaxations, and from the decaying component of inward Na^+ currents, respectively. In NaChBac, inward Na^+ currents are evoked as described in Fig. 2B, and tail current relaxations are elicited by hyperpolarizing/repolarizing steps (−120 to −60 mV, $\Delta V = 10$ mV) immediately after a short depolarizing pulse to −10 mV. All are well described by single or sum of two exponential functions; best-fit single exponential or sum of exponentials are shown for select voltages as *red dashed lines*. (D) The time constant of recovery from inactivation is derived from the time course of recovery. In NaChBac, the time course of recovery is generated using a two-pulse protocol: (1) a 2-s conditioning pulse at −10 mV, followed by (2) a 50-ms test pulse to −10 mV. The pulses are separated by a variable recovery interval ($\Delta t = 10$ ms to 12 s) at −120 mV. The time course of recovery is well described by a single exponential function.

electrically silent transitions between intermediate activated but closed states that occur before the channel pore actually opens ($C_0 \leftrightarrow \cdots \leftrightarrow C_n$).

Gating currents are capacitive currents that result from charge displacement, namely, movement of the voltage sensors, across the membrane electric field. Activation and deactivation are associated with the ON- and OFF-gating currents, respectively. To observe gating currents, the ionic currents, typically orders of magnitude larger than gating currents, must be fully eliminated by substituting all permeant ions (e.g., Na^+) with nonpermeant ions (e.g., *N*-methyl-glucamine) in the recording solutions[d]. Gating currents are elicited with a series of depolarizing voltage steps similar to that described earlier (Kuzmenkin et al., 2004). The time integrals of the ON- and OFF-gating currents are then used to generate the gating charge–voltage (*Q*–*V*) relationship, which is then characterized by the best-fit Boltzmann function to extract the total gating charge (Q_{max}), the effective valence (z_q), and the $V_{1/2,\ q}$ of charge displacement. Because the gating charge is conserved, genuine ON and OFF *Q*–*V* relationships are expected to be identical[e].

The voltage dependence of Nav inactivation is assessed with a two-pulse protocol consisting of: (1) a conditioning pulse lasting several seconds to inactivate channels, followed immediately by (2) a short test pulse to a depolarized voltage to elicit the current generated by available (not inactivated) Navs. Normalized peak current amplitudes of the test pulse (I/I_{max}) are plotted against the conditioning voltages to generate the prepulse inactivation curve (Fig. 2B and C, bottom)[f]. The best-fit Boltzmann function describing the curve is used to derive macroscopic inactivation gating parameters, including the z and $V_{1/2}$ of inactivation.

To examine the kinetics of the transition between the open and inactivated states (O ↔ I), we measure the time constants of inactivation and recovery from inactivation (Fig. 3C and D). Time constants of inactivation are derived from the decaying component of inward Na^+ currents using the best-fit single exponential or sum of exponentials (Fig. 3C). The time constant of recovery from inactivation is derived from the time course of recovery, which is generated with a two-pulse protocol: (1) an activating pulse of sufficient duration to inactivate all channels, followed by (2) a second activating test pulse to assess recovered current. The pulses are separated by a variable recovery interval at a hyperpolarized voltage where recovery from inactivation dominates. The ratio of the current amplitudes of pulse 2 to pulse 1 (fractional recovery) is plotted against the recovery interval to generate the time course of recovery, and the time constant of recovery is derived from the best-fit single exponential or sum of exponentials (Fig. 3D).

[d] Ion conduction can alternatively be blocked with a high-affinity pore blockers or by creating mutations that render the ion channel non-conducting. It is important to note that these approaches may introduce complications because gating and permeation are not entirely independent processes.

[e] A difference suggests charge immobilization possibly associated with inactivation.

[f] If the conditioning pulses are long enough for Nav kinetics to reach steady state, this protocol then generates the steady-state inactivation curve.

2.6 Modulation of Macroscopic Nav Gating by Propofol

In both NaChBac and NavMs, propofol induces hyperpolarizing shifts of the prepulse inactivation and *G–V* relationships (Fig. 2). It also reduces time constants of inactivation, accelerates time constants of activation, and modestly increases the time constant of deactivation, but does not have significant effects on recovery from inactivation. Upon binding, propofol may promote entry into the open and inactivated–open states, which will be the subject of further investigations with iterative electrophysiology and mutational analysis, as well as computational and structural methods.

2.7 Computer Simulations to Generate Working Models and Interpret Experimental Results

Following a comprehensive electrophysiological characterization of ion channel function in the absence and presence of anesthetic, a working model of anesthetic modulation is developed with in silico simulations. IonChannelLab[g] is a robust software suite that uses Markov chain models to simulate dynamic behavior of ligand- and voltage-gated ion channels (De Santiago-Castillo, Covarrubias, Sánchez-Rodríguez, Perez-Cornejo, & Arreola, 2010). While modeling and simulation of ion channels under different experimental conditions typically require building complex kinetic schemes and solving complicated differential equations, IonChannelLab boasts an extensive simulation library that is able to model channels with Ohmic or Goldman–Hodgkin–Katz behavior. Moreover, it can simulate time courses of both ionic and gating currents and single channel behavior, all within a user-friendly and intuitive Graphical User Interface (Fig. 4). In this way, IonChannelLab removes the need for advanced programming experience and allows researchers to focus on model development rather than software development and implementation.

To develop a kinetic model of anesthetic action, we first build a model simulating control conditions, then modify it to reflect functional changes caused by the anesthetic. We begin with one of the preloaded examples or with a previously published kinetic scheme in the kinetic model editor (Fig. 4A). The latter is built by adding new states, defining the permeant ion and their status (open or closed), adding transitions between them, and defining rate constants in the function editor. We then simulate voltage clamp experiments in the simulation editor (Fig. 4B), replicating experimental

[g] http://www.jadesantiago.com/Electrophysiology/IonChannelLab/.

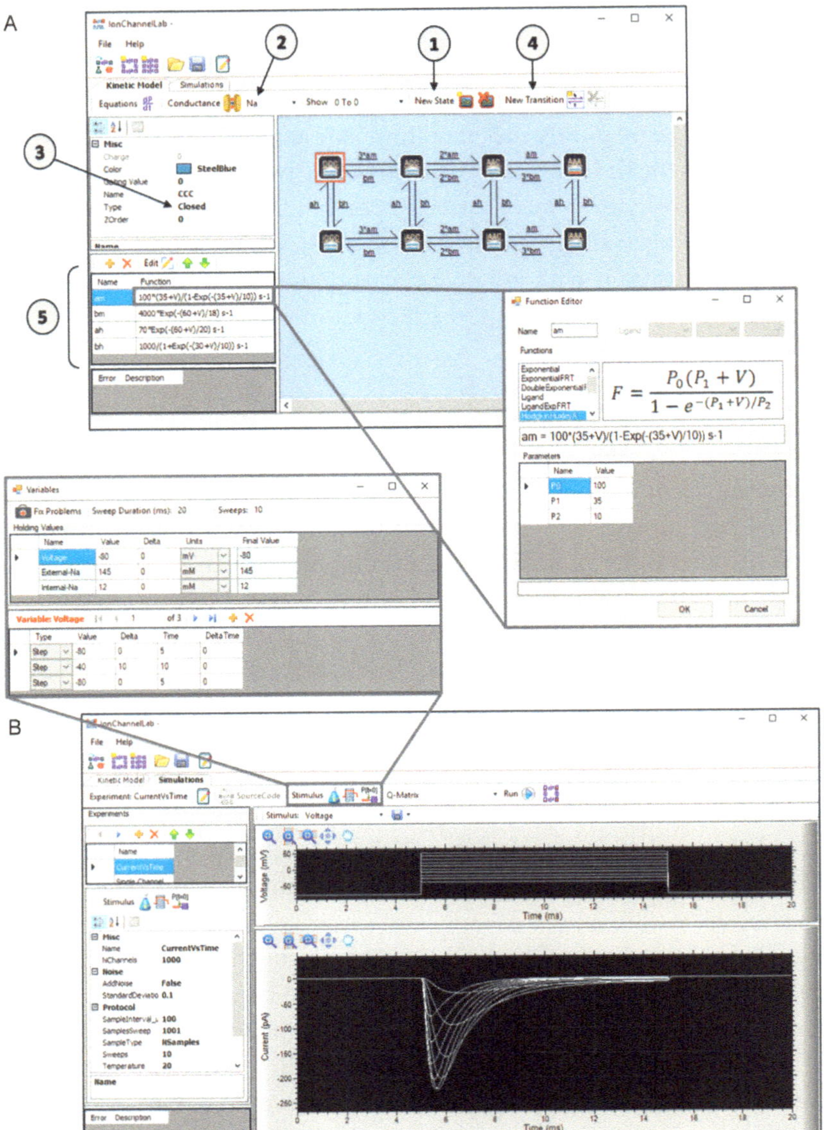

Fig. 4 In silico simulations in IonChannelLab. The intuitive IonChannelLab GUI allows users to build kinetic models in (A) the kinetic model editor and simulate voltage-clamp experiments in the (B) simulation editor. In the kinetic model editor, users can (1) add new states; define the (2) permeant ion and (3) conduction status of each state; and (4) add transitions between states. Double clicking the transition definition on the left-hand side of the GUI opens (5) the function editor, where users can edit rate constants.

protocols and conditions, and analyze the simulated currents in the same manner that experimental currents are analyzed. The kinetic model is updated and simulated currents are analyzed in an iterative process until simulated parameters reflect experimental values. From the final kinetic model under control conditions, the model is then modified (e.g., changing rate constants and/or the state status) to reflect experimentally observed effects induced by the anesthetic.

2.8 Limitations and Enhanced Approaches to Model Ion Channel Gating

As described earlier, the analysis of macroscopic currents obtained from carefully designed voltage protocols interrogating different aspects of gating over a wide range of voltages provides robust qualitative and semiquantitative insights into the mechanisms of gating and the modes of anesthetic action. It cannot, however, produce a well-constrained, global quantitative assessment of gating parameters and rate constants. The ideal data set permitting such an assessment would examine activation, deactivation, inactivation, and recovery from inactivation in each cell, in the absence and presence of the anesthetic. Unfortunately, significant cell-to-cell variability, experimental drift, and insufficient patch stability during extended recordings limit the acquisition of such an ideal data set. Nonetheless, even a partial data set can overcome certain constraints (e.g., activation and deactivation currents) to carry out global kinetic modeling. IChMASCOT[h] is a free, user-friendly software suite created in our laboratory to conduct quantitative global kinetic modeling of ion channel gating (Dougherty, De Santiago-Castillo, & Covarrubias, 2008; Kaulin, De Santiago-Castillo, Rocha, & Covarrubias, 2008; Ritter, Zemel, Lepore, & Covarrubias, 2015).

Kinetic models can be further tested and constrained by conducting single channel analysis, which allows direct estimation of rate constants from the fluctuations and dwell times in recordings of unitary currents. QuB[i] is a particularly powerful free software package designed to conduct direct fitting of kinetic models to single channel recordings. Recent literature emphasizes the value of single channel analysis to understand ion channel dynamics from the historical and scientific perspectives (Mukhtasimova, daCosta, & Sine, 2016; Sivilotti & Colquhoun, 2016).

[h] http://www.jadesantiago.com/Electrophysiology/IChMASCOT/.
[i] https://qub.mandelics.com/.

3. INVESTIGATING THE MODULATION OF KV CHANNELS BY SEVOFLURANE

To investigate the mechanisms of general anesthetic action on Kv channels, we have focused on the interactions of sevoflurane with *Drosophila* K-Shaw and rat Kv1.2 expressed in *X. laevis* oocytes. Earlier work in rats demonstrated that Kv1-type channels expressed in the central medial nucleus of the thalamus may play a role in the endpoints of general anesthesia induced by sevoflurane, a widely used inhaled anesthetic (Alkire, Asher, Franciscus, & Hahn, 2009; Lioudyno, Birch, Tanaka, et al., 2013). Sevoflurane potentiates the *Drosophila* K-Shaw and several Kv1-type channels by promoting the open state in two ways (Fig. 5). It induces a hyperpolarizing shift of the conductance–voltage relationship, suggesting an apparent stabilization of the open state, and also increases the maximum conductance. The latter suggests that sevoflurane also destabilizes a nonconducting state outside of the activation pathway of Kv channels.

Exploiting the experimental advantages of *Xenopus* oocytes as an expression system has permitted comprehensive structure–function investigations of anesthetic modulation of Kv channels. We and others have characterized multiple Kv channel mutants (Barber et al., 2012; Bu, Liang, Zhi, et al., 2018; Liang et al., 2015; Woll, Peng, et al., 2017) and have generated high-resolution recordings of single Kv channel currents (Fig. 6). From these experiments, photoaffinity labeling, and computational approaches, a mechanistic model of general anesthetic action on Kv channels has emerged.

3.1 Heterologous Expression of Kv Channels in *X. laevis* Oocytes

For more than three decades, *X. laevis* oocytes have been a prominent system for studies of ion transport and physiology because they have the capacity to effectively translate exogenous mRNAs. They are also readily harvested with a short surgical procedure and can be maintained in vitro for extended periods of time. Mature *Xenopus* oocytes are large (1 mm in diameter) and physically robust, allowing for manual microinjection of the mRNA. *Xenopus* oocytes are particularly advantageous because they only express small (100–200 pA) endogenous outward currents in the physiological range of membrane potentials. Detailed procedures for the use of *Xenopus* oocytes to study ion channel function have been previously described (Stühmer, 1992, 1998). In this section, we focus on specific procedures we have applied to study general anesthetic action on Kv channels, focusing on K-Shaw and Kv1.2.

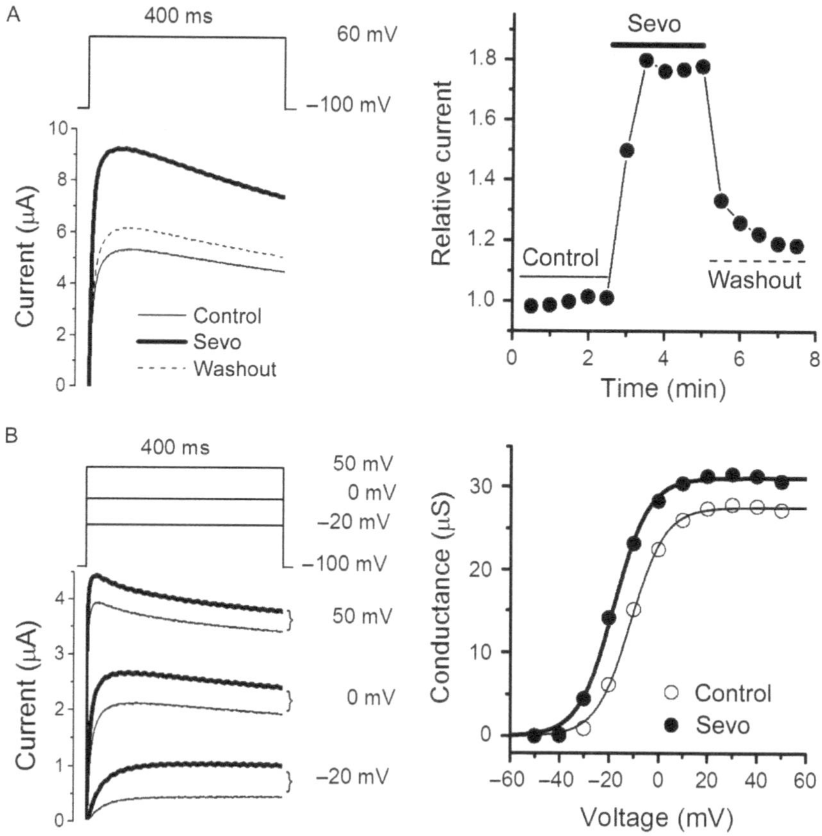

Fig. 5 Two-electrode voltage clamp of K-Shaw and Kv.1.2. (A) K-Shaw currents are potentiated by sevoflurane in a reversible manner. Currents are evoked by repeated depolarizations to +60 mV from a holding potential of −100 mV, in the absence and presence of 1 m*M* sevoflurane. Current amplitude rapidly increases by 80% with application of 1 m*M* sevoflurane, and the effect is reversible, as the current returns to near control values upon washout. (B) Kv.1.2 currents are also potentiated by sevoflurane. Current families are evoked with step depolarizations in 10-mV increments from a holding potential of −100 mV. Select currents evoked at −20, 0, and +50 mV are shown in the absence (*thin line*) and presence (*thick line*) of 1 m*M* sevoflurane (*left*). From these currents, *G–V* relationships are determined in the absence and presence of 1 m*M* sevoflurane; representative paired *G–V* curves are shown (*right*).

3.2 Preparation of cDNA, mRNA, and Oocytes for Electrophysiology Experiments

Kv channel cDNAs are maintained in appropriate expression vectors suitable for in vitro transcription as previously reported (Barber et al., 2012; Liang et al., 2015; Woll, Peng, et al., 2017). These constructs commonly use the T3, T7, or SP6 RNA polymerase promoters. For best results, these constructs also typically include the 5′- and 3′-untranslated regions of *Xenopus* β-globin for efficient translation and a long polyA sequence to enhance mRNA stability. For mutational analysis, we use the QuickChange site-directed mutagenesis protocol (Agilent Technologies) and PCR-based procedures for cassette mutagenesis. Once the plasmid of interest is amplified and purified with a commercially available kit (e.g., QIAGEN or ThermoFisher Scientific), the following steps are taken to prepare the mRNA and *Xenopus* oocytes for electrophysiology experiments:

3.2.1 Preparation of mRNA for Injection

Linearize the plasmid at the 3′-end of the insert with an appropriate restriction endonuclease, and purify the linearized plasmid with a PCR purification kit (e.g., QIAquick by QIAGEN). Using the linearized DNA as a template, synthesize mRNA with an in vitro RNA transcription kit (e.g., mMESSAGE mMACHINE by Ambion, ThermoFisher Scientific). Purify the mRNA using an RNA isolation kit (e.g., RNeasy by QIAGEN) and dilute the resulting mRNA to a final concentration of 10–50 ng/μL with RNase-free water. Precautions must be taken at all times to prevent degradation of mRNA (Goldin & Sumikawa, 1992). To this end, we routinely prepare single-use, small volume (2 μL) aliquots of the mRNA for oocyte injections.

3.2.2 Preparation of Xenopus Oocytes for Injection

To harvest mature (Stage V or VI) oocytes, we use standard collagenase dispersion techniques (Goldin, 1992). First, gently separate oocytes into clusters of 10–20 with forceps. The goal is to separate the oocytes for sufficient collagenase exposure while minimizing overall oocyte lysis. Digest dissociated oocytes with collagenase A in calcium-free[j] ND96 (see recipe in Section 3.4) for 1.5 h and repeat the digestion with fresh collagenase A for an additional 1–1.5 h. Maintain the oocytes at 19°C in maintenance medium: 500 mL Leibovitz's L-15 medium (Gibco) in 220 mL H_2O, supplemented with 10 m*M* HEPES and 0.1 mg/mL gentamicin; pH 7.4 (titrated with NaOH). Individual oocytes lacking the follicular layer of cells are suitable for mRNA injection. If pieces of follicular layer remain, carefully remove them with fine-tip forceps.

[j] Presence of calcium during collagenase digestion is particularly damaging to oocytes, as calcium activates proteases.

3.2.3 Injection of Xenopus *Oocytes for Electrophysiology Experiments*

Pull glass micropipettes for injection with 3-000-203-G/X glass capillaries (Drummond Scientific Company) and, under the microscope, cut the tip with Dumont forceps such that the tip is 10–30 μm in diameter. Backfill the micropipette with mineral oil using a 30-G needle and syringe. Following manufacturer's instructions, use a nanoliter microinjector (e.g., Nanoject by Drummond Scientific Company) to deliver the mRNA to individual oocytes. Pipette 1–2 μL of mRNA onto a clean piece of Parafilm M (Bemis) and draw the mRNA into the micropipette with the microinjector, taking care to avoid air bubbles. Transfer the oocytes to maintenance medium in a 3.5-cm Petri dish with plastic mesh mounted on the bottom to hold the oocytes in place during injection. Adjust the injection volume and inject the mRNA into the oocytes through the dark hemisphere. We typically inject 46–92 nL of mRNA per oocyte. The mRNA concentration and injection volume should be adjusted based upon the expression level desired, which may require optimization.

3.2.4 Maintenance of Injected Xenopus *Oocytes*

Incubate the injected oocytes at 19°C for 12–48 h in maintenance medium before use. Typically, injected oocytes can be kept for 3–5 days. Transfer oocytes to a new Petri dish containing fresh medium each day, taking care to remove dead oocytes.

3.3 Preparation and Delivery of Sevoflurane

Sevoflurane, like most general anesthetics, is lipophilic. Thus, glass vials and syringes as well as polytetrafluoroethylene (PTFE) parts and tubing should be used whenever possible. Moreover, since sevoflurane is a volatile anesthetic, gastight systems should be used and care must be taken to work quickly and limit exposure to open air. While some loss is inevitable, these precautions will help to minimize concentration errors.

Prepare fresh sevoflurane stock and working solutions daily in glass septum vials with open-top screw caps and Tuf-Bond PTFE/silicone discs (Fisher Scientific) that reseal after punctured. First, add 40 μL of sevoflurane (Ultane) to 40 mL of ND96 (see Section 3.4) and immediately seal the vial to make a 7.5 m*M* sevoflurane stock solution. Sonicate and vortex in 30-s intervals, repeating 3–4 times until the sevoflurane is fully dissolved in the ND96. To prepare working solutions, dilute sevoflurane stock solution in ND96. Use gastight syringes (e.g., from Hamilton Company) to transfer the appropriate volume of stock solution to another sealed vial containing the corresponding volume of ND96 through the PTFE/silicone lid. Vials containing anesthetic working solution should always be shaken prior to use.

As previously mentioned, gastight and chemically inert equipment should be used for the delivery of volatile anesthetics such as sevoflurane to minimize anesthetic loss. Under manual control with a 50-mL gastight syringe and PTFE tubing or with a gastight perfusion system, deliver sevoflurane to the recording chamber at a constant speed of approximately 3 mL/min. Sevoflurane concentrations of samples taken directly from the recording chamber should be verified periodically by HPLC. In our experience, losses are typically <15%.

3.4 Two-Electrode Voltage Clamping of Kv Channels

For two-electrode voltage-clamping (TEVC) experiments, place a single oocyte in the recording chamber (RC-3Z; Warner Instruments) containing ND96 (standard frog Ringer's), which is comprised of (in m*M*): 96 NaCl, 2 KCl, 1.8 $CaCl_2$, 1 $MgCl_2$, 5 HEPES, and 2.5 sodium pyruvate; pH 7.4 (titrated with NaOH). Key steps for TEVC experiments are described here.

Voltage clamping and data acquisition are performed with TEVC instrumentation (OC-725C; Warner Instruments) and Digidata 1440A (Molecular Devices), respectively. Voltage clamp protocols and data acquisition are controlled by Clampex 10.3 (pCLAMP 10.3; Molecular Devices).

Pull microelectrodes from borosilicate FLG12 glass capillaries (Dagan Corporation) to a resistance of 0.2–0.5 MΩ when filled with 3 *M* KCl and immersed in ND96. Continuously perfuse the oocyte in the recording chamber with ND96 using a gravity-driven perfusion system at approximately 3 mL/min. Manually control the valve system (VC-6 valve controller; Warner Instruments) to expose a single oocyte to ND96 or anesthetic. Bath drainage is also done by gravity flow.

As previously described (Stühmer, 1992), impale the oocyte with the two microelectrodes and measure the resting membrane potential, which typically ranges between −40 and −50 mV in healthy oocytes. Both microelectrodes should register similar values. All measurements are conducted at room temperature (20–23°C). In voltage clamp mode, determine the level of Kv channel expression by measuring outward currents evoked with repeated 400-ms depolarizations to +60 mV from a holding potential of −100 mV in 20-s intervals. With continuous perfusion of the recording chamber, a stable response is expected, with acceptable levels of expression ranging from 2 to 10 μA. The best results are obtained from impaled oocytes displaying minimal leak current that remains stable for the duration of the recordings (<100 nA at −100 mV).

In a typical experiment, voltage protocols are first run under control conditions (ND96, anesthetic-free) then repeated with anesthetic. All experiments are conducted under constant perfusion. First, assess current magnitude and kinetics under control conditions with 400-ms depolarizations to +60 mV from −100 mV in 20-s intervals. Then, evaluate the peak conductance–voltage (*G–V*) relationship with outward K^+ currents evoked by 400-ms step depolarizations (−60 to +60 mV, $\Delta V = 10$ mV) from a holding potential of −100 mV, with 3-s start-to-start intervals. The membrane leak and capacitive transients resulting from the voltage steps are subtracted online using standard P/4 protocols delivered before the experimental protocol (Steffan, Hennesthal, & Heinemann, 1998). After all control recordings have been measured, expose the oocyte to the anesthetic (e.g., sevoflurane) diluted in ND96 and repeat the voltage protocols to assess effects on current magnitude, current kinetics, and the *G–V* relationship. Protocols may be repeated after washout with ND96 to assess reversibility.

Analysis of the *G–V* relationship is often based on the estimation of the chord conductance ($G_{chord} = I/[V_c - V_r]$), where *I* is the ionic current, V_c is the command voltage, and V_r is the reversal voltage of the current. This estimate is based on the assumption or knowledge of a linear instantaneous current–voltage (*I–V*) relationship, or the open channel *I–V* relationship. The *G–V* relationship can also be obtained with a variation of the voltage protocol described earlier. Rather than returning the membrane to the holding potential after each depolarizing step, repolarize the membrane to a voltage that closes the channels and that also provides enough driving force to generate a tail current (e.g., −60 mV). Measure the amplitude of the instantaneous current resulting from the repolarization, which determines the fraction of ion channels that had been opened by the variable depolarizing steps. This measurement is not affected by the shape of the instantaneous *I–V* relationship because all data points are obtained at a constant voltage. In either case, the best-fit Boltzmann function is used to extract the midpoint voltage ($V_{1/2}$) of activation, the effective gating charge of activation (z), and the maximum conductance (G_{max}). All data are analyzed using Clampfit 10.3 (pCLAMP, Molecular Devices), Excel (Microsoft), and OriginPro 9 (OriginLab).

This analysis helps to assess the relative stability of the Kv channel's open state. Negative and positive shifts in the $V_{1/2}$ (parallel shifts of the *G–V* relationship to the left and right, respectively) correspond to a relative stabilization and destabilization of the open state, respectively. Additional voltage protocols may be implemented depending on the specific experiment and biophysical property to be examined. Examples of these voltage protocols are described in Section 2 and in recent publications by our group (Barber et al., 2014, 2011, 2012; Liang et al., 2015).

3.4.1 Investigating the Macroscopic Effects of Sevoflurane on Kv Channels with TEVC

The sample recordings and corresponding analyses shown here demonstrate typical data from K-Shaw and Kv1.2 and provide just a snapshot of how sevoflurane modulates Kv channels (Liang et al., 2015). After establishing a stable baseline, exposure of the oocyte to 1 m*M* sevoflurane rapidly induces a 1.8-fold potentiation of the whole-oocyte K-Shaw outward current (Fig. 5A). Upon washout with ND96, the current approaches the baseline level (Fig. 5A). As reported previously, the sevoflurane-induced potentiation is due to a hyperpolarizing shift of the G–V relationship and an increase in the G_{max} (Barber et al., 2012; Liang et al., 2015; Woll, Peng, et al., 2017). Like K-Shaw, Kv1.2 is also potentiated by sevoflurane under identical experimental conditions. At different voltages (−20, 0, and +50 mV shown), the Kv1.2 current is consistently potentiated as a result of the hyperpolarized G–V relationship ($\Delta V_{1/2} = -7$ mV) and the 13% increase in G_{max} (Barber et al., 2012; Liang et al., 2015; Woll, Peng, et al., 2017) (Fig. 5B).

3.4.2 Limitations of TEVC and Alternatives

TEVC is intrinsically limited because the oocyte membrane represents a very large surface area. As a result, capacitive currents evoked by voltage steps are large and slow, which limits voltage clamping speed and resolution. This system is also susceptible to series resistance and space clamp errors (Baumgartner, Islas, & Sigworth, 1999). Nonetheless, by using a small volume laminar flow chamber (e.g., RC-3Z, Warner Instruments) with a properly placed virtual ground along the main axis of the chamber and by only accepting currents with amplitudes <20 μA, one can obtain satisfactory and reliable recordings. To monitor errors, the membrane voltage should always be simultaneously recorded via the voltage microelectrode during the delivery of voltage steps. Alternative methods, such as the cut-open oocyte and the macropatch techniques (Siefani & Bezanilla, 1998; Stühmer, 1992, 1998), have been developed for improved voltage clamping. In addition to TEVC, our laboratory also uses patch clamping techniques to record macropatch and single channel currents in *X. laevis* oocytes and mammalian heterologous expression systems.

3.5 Patch Clamp Electrophysiology of Kv Channels

To carry out patch clamping experiments on *X. laevis* oocytes, it is necessary to also remove the vitelline envelope adjacent to the plasma membrane prior to patch clamping. This is typically achieved by carefully peeling the vitelline envelope with fine-tip Dumont microdissection forceps (Roboz). Bathing the oocyte in a hypertonic solution will shrink the oocyte and cause the plasma membrane to separate from the vitelline envelope, facilitating its removal. However, with sufficient practice, this is generally not necessary. After peeling, the oocyte is transferred with a Pasteur pipet to the recording chamber containing bath solution and used immediately. To record macropatch and single channel currents in *X. laevis* oocytes, we use a standard gigaseal patch clamp technique (cell-attached and cell-free patches) as detailed previously by others (Stühmer, 1992, 1998). Here, we summarize the basics of the procedure.

The Axopatch 200B amplifier connected to the Digidata 1440A (Molecular Devices) is used for data acquisition under the control of Clampex 10.3. Typically, currents are low-pass filtered at 2 kHz (8-pole Bessel, −3 dB cut-off frequency) and digitized at 50 μs/point (20 kHz). All experiments are conducted at room temperature (20–23°C).

Fabricate microelectrodes (patch pipettes) from borosilicate glass (G75150T-4, Warner Instruments) using a suitable puller (e.g., P-97 Flaming-Brown puller, Sutter Instruments; PIP-6 vertical puller, HEKA). Coat patch pipettes with Sylgard 184 silicone elastomer (Dow Corning), reaching as close to the tip as possible without blocking the opening. Sylgard coating minimizes the capacitive transient and background noise by reducing stray capacitance from the microelectrode. Fire polish the tip to a resistance of 0.5–2 MΩ immediately prior to use. This range of resistances is adequate for recordings of macroscopic currents. For unitary currents, microelectrodes with a high tip resistance (10–20 MΩ) improve the chance of isolating a single channel.

For the cell-attached patch configuration, the patch pipette contains ND96, and the solution in the recording chamber contains (in m*M*): 140 K-aspartate, 10 KCl, 1.8 $CaCl_2$, 10 HEPES; pH 7.2 (titrated with KOH). This high-potassium extracellular solution clamps the oocyte membrane potential near 0 mV to allow a direct reading of the membrane potential across the patch. For inside-out patches, the patch pipette contains ND96, and the bath solution contains (in m*M*): 98 KCl, 1.5 $MgCl_2$, 1 EGTA, and 10 HEPES; pH 7.2 (titrated with KOH) (Beck & Covarrubias, 2001).

For high-quality macroscopic recordings in the cell-attached patch configuration, the seal resistance should be at least 2 GΩ and remain relatively stable during the experiment. For single channel recordings, lower background noise is necessary to resolve unitary events and transitions, which requires a stable seal resistance of at least 20 GΩ.

3.6 Investigating the Microscopic Effects of Sevoflurane on Kv Channels with Patch Clamp Electrophysiology

Obtained using these methods, the examples provided here illustrate the effects of sevoflurane on K-Shaw at the single channel level in the inside-out configuration. In response to a step depolarization, a single K-Shaw channel opens and transitions stochastically between conducting and nonconducting states for the duration of the step (Fig. 6A, left). To determine the mean unitary amplitude and open probability (P_O), the best-fit sum of two Gaussian functions is used to describe the all-point histogram (APH; Fig. 6A, right). From the current sweep shown, the larger APH peak corresponds to the baseline fluctuations with a mean amplitude of 0 pA (closed state), and the well-resolved smaller APH peak corresponds to fluctuations in the open state with a mean amplitude of 4 pA. The area under the smaller peak divided by the total area under both peaks yields an estimate of the P_O (0.34).

Delivering 1 m*M* sevoflurane to the bath, which exposes the intracellular face of the inside-out patch to anesthetic, induces an apparent increase in P_O (Fig. 6B). A more detailed statistical analysis of dwell times is necessary to determine the changes underlying the increase in P_O, which is outside the scope of this chapter. However, the observed increase in activity is qualitatively consistent with the potentiation of macroscopic currents observed with TEVC (Fig. 5A).

We can also ask whether the sevoflurane-induced potentiation is also due to an increase in unitary conductance. To this end, a ramping voltage protocol (Fig. 6C, left inset) is used to determine the unitary current–voltage relationship, as the current flowing through a single open channel is affected by voltage changing at constant rate. The slope of this relationship yields the unitary slope conductance. Alternatively, the unitary mean amplitude can be measured from AHPs generated at different voltages, as described previously.

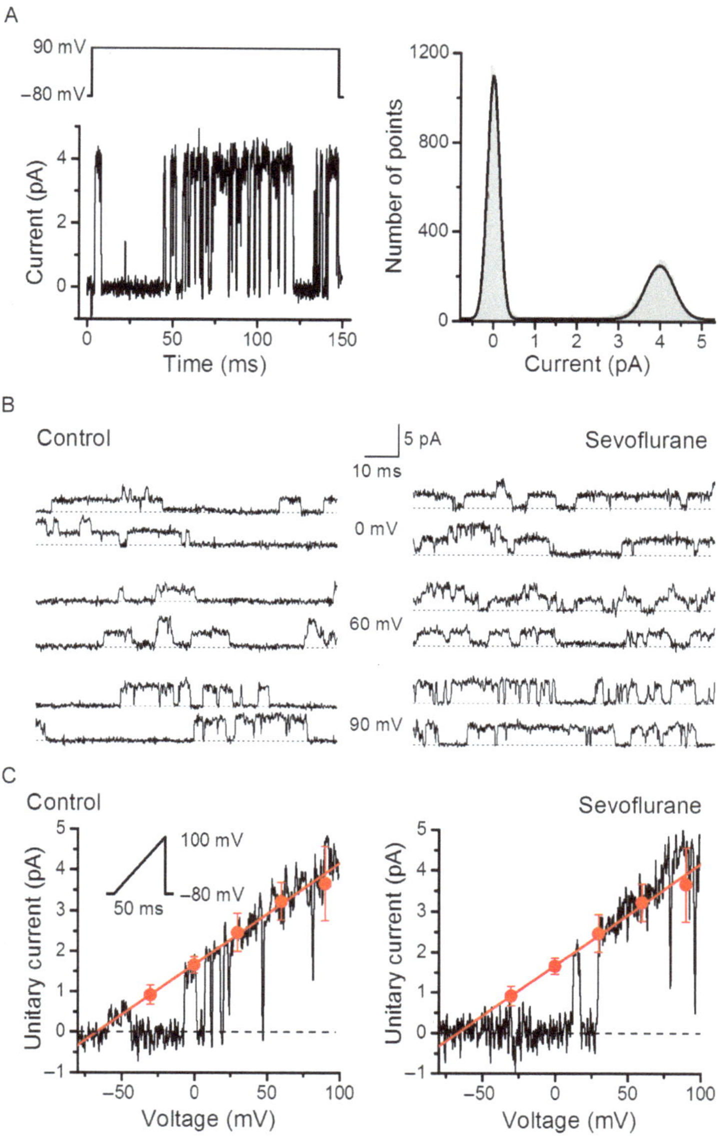

Fig. 6 Single channel analysis of K-Shaw. (A) In response to a step depolarization to +90 mV, a single K-Shaw channel opens and transitions stochastically between conducting (4 pA) and nonconducting (0 pA) states (*left*). The all-point histogram (APH, *right*) is used to determine the mean unitary amplitude and open probability (P_O). (B) Analysis of 100-ms steps to different voltages in the absence (*left*) and presence (*right*) of 1 m*M* sevoflurane using this method suggests that sevoflurane increases P_O. (C) The unitary conductance in the absence (*left*) and presence (*right*) of 1 m*M* sevoflurane is determined using a ramping voltage protocol (*black*) or using individual APHs (*red*). With 1 m*M* sevoflurane, there is no change in the slope of the unitary current–voltage relationship, demonstrating that sevoflurane does not affect unitary conductance. The data included in this figure was generated in part by Shelly T. Jones.

Using this alternative approach, the slope of the relationship between mean unitary amplitudes and voltage also yields the unitary slope conductance. For K-Shaw, both methods produced identical results in the absence and presence of the anesthetic (approximately 25 pS; Fig. 6C), demonstrating that sevoflurane does not affect K^+ conduction through the pore.

4. CONCLUSION AND PERSPECTIVE

Structure–function studies based on electrophysiological, biophysical, and mutational analysis of ion channels have been instrumental in the investigation of mechanisms of anesthetic action. We have elucidated how general anesthetics inhibit bacterial Navs and potentiate certain Kvs, effects that are consistent with a net modulatory action that would dampen neural excitability (Covarrubias et al., 2015). In particular, our findings demonstrate that general anesthetics (e.g., propofol and sevoflurane) might have a special affinity for binding pockets that involve moving parts of the gating machinery (Barber et al., 2014, 2011, 2012; Liang et al., 2015). In concert with powerful photoaffinity labeling techniques and the ever-improving techniques in structural and computational biology (Carnevale & Klein, 2017; Kinde, Bondarenko, Granata, et al., 2016; Raju, Barber, LeBard, Klein, & Carnevale, 2013; Stock, Hosoume, & Treptow, 2017; Tronin, Nordgren, Strzalka, et al., 2014; Woll, Peng, et al., 2017; Woll, Skinner, et al., 2017), iterative functional and mutational analysis will continue to provide critical insight and validation to solve the molecular mechanisms governing the action of general anesthetics on VGICs. Thus, researchers pursuing serious molecular-level mechanistic studies in general anesthesia are bound to apply these methodologies.

Future applications and developments must take into account significant outstanding challenges, however. For instance, candidate target ion channels in their native mammalian environment exist as supramolecular complexes with multiple partners that regulate their function (Ahern et al., 2016; Bouza & Isom, 2017; Dai, Hall, & Hell, 2009; Pongs, Schwarz, Abbott, et al., 2010), such as gating and permeation. These partners (e.g., auxiliary subunits and signaling proteins) might also be targets of anesthetics. Thus, the progress that future structure–function studies bring will depend on the understanding of native ion channel complexes and the implementation of reliable surrogate expression systems to reconstitute them. Fortunately, much progress has been made in this area (Long, 2005; Oiki, 2015; Wu, Yan, Li, et al., 2016; Yan et al., 2017) and should be exploited in the field of general anesthesia while consistently striving to use a multidisciplinary approach to constrain plausible models of general anesthetic action. Such an effort will pave the way to better understand physiological interactions of general anesthetics with ion channel complexes and ultimately help elucidate the physical basis of action under native conditions.

REFERENCES

Ahern, C. A., Payandeh, J., Bosmans, F., & Chanda, B. (2016). The hitchhiker's guide to the voltage-gated sodium channel galaxy. *The Journal of General Physiology, 147*(1), 1–24. https://doi.org/10.1085/jgp.201511492.

Alkire, M. T., Asher, C. D., Franciscus, A. M., & Hahn, E. L. (2009). Thalamic microinfusion of antibody to a voltage-gated potassium channel restores consciousness during anesthesia. *Anesthesiology, 110*(4), 766–773. http://www.ncbi.nlm.nih.gov/pubmed/19322942. Accessed 11 January 2017.

Bagneris, C., DeCaen, P. G., Naylor, C. E., et al. (2014). Prokaryotic NavMs channel as a structural and functional model for eukaryotic sodium channel antagonism. *Proceedings of the National Academy of Sciences of the United States of America, 111*(23), 8428–8433. https://doi.org/10.1073/pnas.1406855111.

Barber, A. F., Carnevale, V., Klein, M. L., Eckenhoff, R. G., & Covarrubias, M. (2014). Modulation of a voltage-gated Na+ channel by sevoflurane involves multiple sites and distinct mechanisms. *Proceedings of the National Academy of Sciences of the United States of America, 111*(18), 6726–6731. https://doi.org/10.1073/pnas.1405768111.

Barber, A. F., Liang, Q., Amaral, C., Treptow, W., & Covarrubias, M. (2011). Molecular mapping of general anesthetic sites in a voltage-gated ion channel. *Biophysical Journal, 101*(7), 1613–1622. https://doi.org/10.1016/j.bpj.2011.08.026.

Barber, A. F., Liang, Q., & Covarrubias, M. (2012). Novel activation of voltage-gated K(+) channels by sevoflurane. *The Journal of Biological Chemistry, 287*(48), 40425–40432. https://doi.org/10.1074/jbc.M112.405787.

Baumgartner, W., Islas, L., & Sigworth, F. J. (1999). Two-microelectrode voltage clamp of Xenopus oocytes: Voltage errors and compensation for local current flow. *Biophysical Journal,* 77(4), 1980–1991. https://doi.org/10.1016/S0006-3495(99)77039-6.

Beck, E. J., & Covarrubias, M. (2001). Kv4 channels exhibit modulation of closed-state inactivation in inside-out patches. *Biophysical Journal, 81*(2), 867–883. https://doi.org/10.1016/S0006-3495(01)75747-5.

Bouza, A. A., & Isom, L. L. (2017). Voltage-gated sodium channel β subunits and their related diseases. In *The handbook of experimental pharmacology.* Berlin, Heidelberg: Springerhttps://doi.org/10.1007/164_2017_48.

Bu, W., Liang, Q., & Zhi, L. (2018). Sites and functional consequence of alkylphenol anesthetic binding to Kv1.2 channels. *Molecular Neurobiology, 55*(2), 1692–1702. https://doi.org/10.1007/s12035-017-0437-2.

Carnevale, V., & Klein, M. L. (2017). Small molecule modulation of voltage gated sodium channels introduction: Physiological role of voltage gated sodium channels (VGSCs). *Current Opinion in Structural Biology, 43,* 156–162. https://doi.org/10.1016/j.sbi.2017.02.002.

Chahine, M., Pilote, S., Pouliot, V., Takami, H., & Sato, C. (2004). Role of arginine residues on the S4 segment of the *Bacillus halodurans* Na+ channel in voltage-sensing. *The Journal of Membrane Biology, 201*(1), 9–24. https://doi.org/10.1007/s00232-004-0701-z.

Charalambous, K., & Wallace, B. A. (2011). NaChBac: The long lost sodium channel ancestor. *Biochemistry, 50*(32), 6742–6752. https://doi.org/10.1021/bi200942y.

Covarrubias, M., Barber, A. F., Carnevale, V., Treptow, W., & Eckenhoff, R. G. (2015). Mechanistic insights into the modulation of voltage-gated ion channels by inhalational anesthetics. *Biophysical Journal, 109*(10), 2003–2011. https://doi.org/10.1016/j.bpj.2015.09.032.

Dai, S., Hall, D. D., & Hell, J. W. (2009). Supramolecular assemblies and localized regulation of voltage-gated ion channels. *Physiological Reviews, 89*(2), 411–452. https://doi.org/10.1152/physrev.00029.2007.

De Santiago-Castillo, J. A., Covarrubias, M., Sánchez-Rodríguez, J. E., Perez-Cornejo, P., & Arreola, J. (2010). Simulating complex ion channel kinetics with IonChannelLab. *Channels (Austin, Tex.), 4*(5), 422–428. https://doi.org/10.4161/chan.4.5.13404.

Dougherty, K., De Santiago-Castillo, J. A., & Covarrubias, M. (2008). Gating charge immobilization in Kv4.2 channels: The basis of closed-state inactivation. *The Journal of General Physiology, 131*(3), 257–273. https://doi.org/10.1085/jgp.200709938.

Eckenhoff, R., Zheng, W., & Kelz, M. (2008). From anesthetic mechanisms research to drug discovery. *Clinical Pharmacology and Therapeutics, 84*(1), 144–148. https://doi.org/10.1038/clpt.2008.77.

Franks, N. P. (2006). Molecular targets underlying general anaesthesia. *British Journal of Pharmacology, 147*(Suppl. 1), S72–S81. https://doi.org/10.1038/sj.bjp.0706441.

Franks, N. P. (2008). General anaesthesia: From molecular targets to neuronal pathways of sleep and arousal. *Nature Reviews. Neuroscience, 9*(5), 370–386. https://doi.org/10.1038/nrn2372.

Goldin, A. L. (1992). Maintenance of *Xenopus laevis* and oocyte injection. *Methods in Enzymology, 207*(C), 266–279. https://doi.org/10.1016/0076-6879(92)07017-I.

Goldin, A. L., & Sumikawa, K. (1992). Preparation of RNA for injection into Xenopus oocytes. *Methods in Enzymology, 207*(C), 279–297. https://doi.org/10.1016/0076-6879(92)07018-J.

Hall, M. A., Xi, J., Lor, C., et al. (2010). m-Azipropofol (AziPm) a photoactive analogue of the intravenous general anesthetic propofol. *Journal of Medicinal Chemistry, 53*(15), 5667–5675. https://doi.org/10.1021/jm1004072.

Hemmings, H. C., Akabas, M. H., Goldstein, P. A., Trudell, J. R., Orser, B. A., & Harrison, N. L. (2005). Emerging molecular mechanisms of general anesthetic action. *Trends in Pharmacological Sciences, 26*(10), 503–510. https://doi.org/10.1016/j.tips.2005.08.006.

Herold, K. F., & Hemmings, H. C., Jr. (2012). Sodium channels as targets for volatile anesthetics. *Frontiers in Pharmacology, 3*, 50. https://doi.org/10.3389/fphar.2012.00050.

Kaulin, Y. A., De Santiago-Castillo, J. A., Rocha, C. A., & Covarrubias, M. (2008). Mechanism of the modulation of Kv4:KChIP-1 channels by external K+. *Biophysical Journal, 94*(4), 1241–1251. https://doi.org/10.1529/biophysj.107.117796.

Kinde, M. N., Bondarenko, V., Granata, D., et al. (2016). Fluorine-19 NMR and computational quantification of isoflurane binding to the voltage-gated sodium channel NaChBac. *Proceedings of the National Academy of Sciences of the United States of America, 113*(48), 13762–13767. https://doi.org/10.1073/pnas.1609939113.

Kuzmenkin, A., Bezanilla, F., & Correa, A. M. (2004). Gating of the bacterial sodium channel, NaChBac: Voltage-dependent charge movement and gating currents. *The Journal of General Physiology, 124*(4), 349–356. https://doi.org/10.1085/jgp.200409139.

Lee, S., Goodchild, S. J., & Ahern, C. A. (2012). Local anesthetic inhibition of a bacterial sodium channel. *The Journal of General Physiology, 139*(6), 507–516. https://doi.org/10.1085/jgp.201210779.

Liang, Q., Anderson, W. D., Jones, S. T., et al. (2015). Positive allosteric modulation of Kv channels by sevoflurane: Insights into the structural basis of inhaled anesthetic action. Rasmusson RL, ed. *PLoS One, 10*(11), e0143363. https://doi.org/10.1371/journal.pone.0143363.

Lioudyno, M. I., Birch, A. M., Tanaka, B. S., et al. (2013). Shaker-related potassium channels in the central medial nucleus of the thalamus are important molecular targets for arousal suppression by volatile general anesthetics. *The Journal of Neuroscience*, *33*(41), 16310–16322. https://doi.org/10.1523/JNEUROSCI.0344-13.2013.

Long, S. B. (2005). Crystal structure of a mammalian voltage-dependent shaker family K+ channel. *Science*, *309*(5736), 897–903. https://doi.org/10.1126/science.1116269.

McCusker, E. C., Bagnéris, C., Naylor, C. E., et al. (2012). Structure of a bacterial voltage-gated sodium channel pore reveals mechanisms of opening and closing. *Nature Communications*, *3*, 1102. https://doi.org/10.1038/ncomms2077.

Mukhtasimova, N., daCosta, C. J. B., & Sine, S. M. (2016). Improved resolution of single channel dwell times reveals mechanisms of binding, priming, and gating in muscle AChR. *The Journal of General Physiology*, *148*(1), 43–63. https://doi.org/10.1085/jgp.201611584.

Naylor, C. E., Bagneris, C., DeCaen, P. G., et al. (2016). Molecular basis of ion permeability in a voltage-gated sodium channel. *The EMBO Journal*, *35*(8), 820–830. https://doi.org/10.15252/embj.201593285.

Oiki, S. (2015). Channel function reconstitution and re-animation: A single-channel strategy in the postcrystal age. *The Journal of Physiology*, *593*(12), 2553–2573. https://doi.org/10.1113/JP270025.

Ouyang, W., Jih, T.-Y., Zhang, T.-T., Correa, A. M., & Hemmings, H. C. (2007). Isoflurane inhibits NaChBac, a prokaryotic voltage-gated sodium channel. *The Journal of Pharmacology and Experimental Therapeutics*, *322*(3), 1076–1083. https://doi.org/10.1124/jpet.107.122929.

Ouyang, W., Wang, G., & Hemmings, H. C. (2003). Isoflurane and propofol inhibit voltage-gated sodium channels in isolated rat neurohypophysial nerve terminals. *Molecular Pharmacology*, *64*(2), 373–381. https://doi.org/10.1124/mol.64.2.373.

Payandeh, J., Gamal El-Din, T. M., Scheuer, T., Zheng, N., & Catterall, W. A. (2012). Crystal structure of a voltage-gated sodium channel in two potentially inactivated states. *Nature*, *486*(7401), 135–139. https://doi.org/10.1038/nature11077.

Payandeh, J., & Minor, D. L. (2015). Bacterial voltage-gated sodium channels (BacNa(V)s) from the soil, sea, and salt lakes enlighten molecular mechanisms of electrical signaling and pharmacology in the brain and heart. *Journal of Molecular Biology*, *427*(1), 3–30. https://doi.org/10.1016/j.jmb.2014.08.010.

Payandeh, J., Scheuer, T., Zheng, N., & Catterall, W. A. (2011). The crystal structure of a voltage-gated sodium channel. *Nature*, *475*(7356), 353–358. https://doi.org/10.1038/nature10238.

Pongs, O., Schwarz, J. R., Abbott, G., et al. (2010). Ancillary subunits associated with voltage-dependent K+ channels. *Physiological Reviews*, *90*(2), 755–796. https://doi.org/10.1152/physrev.00020.2009.

Raju, S. G., Barber, A. F., LeBard, D. N., Klein, M. L., & Carnevale, V. (2013). Exploring volatile general anesthetic binding to a closed membrane-bound bacterial voltage-gated sodium channel via computation. *PLoS Computational Biology*, *9*(6), e1003090. https://doi.org/10.1371/journal.pcbi.1003090.

Ren, D., Navarro, B., Xu, H., Yue, L., Shi, Q., & Clapham, D. E. (2001). A prokaryotic voltage-gated sodium channel. *Science*, *294*(5550), 2372–2375. https://doi.org/10.1126/science.1065635.

Ritter, D. M., Zemel, B. M., Lepore, A. C., & Covarrubias, M. (2015). Kv3.4 channel function and dysfunction in nociceptors. *Channels*, *9*(4), 209–217. https://doi.org/10.1080/19336950.2015.1056949.

Sand, R. M., Gingrich, K. J., Macharadze, T., Herold, K. F., & Hemmings, H. C. (2017). Isoflurane modulates activation and inactivation gating of the prokaryotic Na(+) channel NaChBac. *The Journal of General Physiology*, *149*(6), 623–638. https://doi.org/10.1085/jgp.201611600.

Shen, H., Zhou, Q., Pan, X., Li, Z., Wu, J., & Yan, N. (2017). Structure of a eukaryotic voltage-gated sodium channel at near-atomic resolution. *Science*, *355*(6328), eaal4326. 80. https://doi.org/10.1126/science.aal4326.

Siefani, E., & Bezanilla, F. (1998). Cut-open oocyte voltage-clamp technique. *Methods in Enzymology*, *293*, 300–318. https://doi.org/10.1016/S0076-6879(98)93020-8.

Sivilotti, L., & Colquhoun, D. (2016). In praise of single channel kinetics. *The Journal of General Physiology*, *148*(2), 79–88. https://doi.org/10.1085/jgp.201611649.

Steffan, R., Hennesthal, C., & Heinemann, S. H. (1998). Voltage-dependent ion channels: Analysis of nonideal macroscopic current data. *Methods in Enzymology*, *293*, 391–419. https://doi.org/10.1016/S0076-6879(98)93026-9.

Stock, L., Hosoume, J., & Treptow, W. (2017). Concentration-dependent binding of small ligands to multiple saturable sites in membrane proteins. *Scientific Reports*, 7(1), 5734. https://doi.org/10.1038/s41598-017-05896-8.

Stühmer, W. (1992). Electrophysiological recording from Xenopus oocytes. *Methods in Enzymology*, *207*(C), 319–339. https://doi.org/10.1016/0076-6879(92)07021-F.

Stühmer, W. (1998). Electrophysiologic recordings from Xenopus oocytes. *Methods in Enzymology*, *293*, 280–300. https://doi.org/10.1016/S0076-6879(98)93019-1.

Sula, A., Booker, J., Ng, L. C. T., Naylor, C. E., DeCaen, P. G., & Wallace, B. A. (2017). The complete structure of an activated open sodium channel. *Nature Communications*, *8*, 14205. https://doi.org/10.1038/ncomms14205.

Tronin, A. Y., Nordgren, C. E., Strzalka, J. W., et al. (2014). Direct evidence of conformational changes associated with voltage gating in a voltage sensor protein by time-resolved X-ray/neutron interferometry. *Langmuir*, *30*(16), 4784–4796. https://doi.org/10.1021/la500560w.

Tsai, C. J., Tani, K., Irie, K., et al. (2013). Two alternative conformations of a voltage-gated sodium channel. *Journal of Molecular Biology*, *425*(22), 4074–4088. https://doi.org/10.1016/j.jmb.2013.06.036.

Woll, K. A., Peng, W., Liang, Q., et al. (2017). Photoaffinity ligand for the inhalational anesthetic sevoflurane allows mechanistic insight into potassium channel modulation. *ACS Chemical Biology*, *12*(5), 1353–1362. https://doi.org/10.1021/acschembio.7b00222.

Woll, K. A., Skinner, K. A., Gianti, E., et al. (2017). Sites contributing to TRPA1 activation by the anesthetic propofol identified by photoaffinity labeling. *Biophysical Journal*, *113*, 2168–2172. https://doi.org/10.1016/J.BPJ.2017.08.040.

Woll, K. A., Weiser, B. P., Liang, Q., et al. (2015). Role for the propofol hydroxyl in anesthetic protein target molecular recognition. *ACS Chemical Neuroscience*, *6*(6), 927–935. https://doi.org/10.1021/acschemneuro.5b00078.

Wu, J., Yan, Z., Li, Z., et al. (2016). Structure of the voltage-gated calcium channel Cav1.1 at 3.6 Å resolution. *Nature*, *537*(7619), 191–196. https://doi.org/10.1038/nature19321.

Yan, Z., Zhou, Q., Wang, L., et al. (2017). Structure of the Nav1.4-β1 complex from electric eel. *Cell*, *170*(3), 470–482.e11. https://doi.org/10.1016/j.cell.2017.06.039.

Yu, F. H., & Catterall, W. A. (2004). The VGL-chanome: A protein superfamily specialized for electrical signaling and ionic homeostasis. *Science STKE: Signal Transduction Knowledge Environment*, *2004*(253), re15. https://doi.org/10.1126/stke.2532004re15.

Zhang, X., Ren, W., DeCaen, P., et al. (2012). Crystal structure of an orthologue of the NaChBac voltage-gated sodium channel. *Nature*, *486*(7401), 130–134. https://doi.org/10.1038/nature11054.

CHAPTER EIGHTEEN

Combining Mutations and Electrophysiology to Map Anesthetic Sites on Ligand-Gated Ion Channels

Stuart A. Forman[1]
Massachusetts General Hospital, Boston, MA, United States
Harvard Medical School, Boston, MA, United States
[1]Corresponding author: e-mail address: saforman@mgh.harvard.edu

Contents

Methods in Enzymology, Volume 602
ISSN 0076-6879
https://doi.org/10.1016/bs.mie.2018.01.014

Abstract

General anesthetics are known to act in part by binding to and altering the function of pentameric ligand-gated ion channels such as nicotinic acetylcholine and γ-aminobutyric acid type A receptors. Combining heterologous expression of the subunits that assemble to form these ion channels, mutagenesis techniques and voltage-clamp electrophysiology have enabled a variety of "structure–function" approaches to questions of where anesthetic binds to these ion channels and how they enhance or inhibit channel function. Here, we review the evolution of concepts and experimental strategies during the last three decades, since molecular biological and electrophysiological tools became widely used. Topics covered include: (1) structural models as interpretive frameworks, (2) various electrophysiological approaches and their limitations, (3) Monod–Wyman–Changeux allosteric models as functional frameworks, (4) structural strategies including chimeras and point mutations, and (5) methods based on cysteine substitution and covalent modification. We discuss in particular depth the experimental design considerations for substituted cysteine modification–protection studies.

ABBREVIATIONS

$GABA_A$ gamma-aminobutyric acid type A
SCAM substituted cysteine accessibility method
SCAMP substituted cysteine modification–protection

1. INTRODUCTION

The parallel development of methods for heterologously expressing and mutating cloned ion channel subunits, and for voltage-clamp electrophysiological assessment of ion channel function has engendered multiple structure–function approaches that contributed to discovery of general anesthetic sites in pentameric ligand-gated ion channels (pLGICs). Here, we review structure–function experimental strategies combining electrophysiology with ion channel subunit chimeras, single-point mutations, and covalent modification of sulfhydryls and their application in identifying general anesthetic sites on nicotinic acetylcholine receptors (nAChRs) and γ-aminobutyric acid type A receptors ($GABA_AR$s). We review detailed methodological considerations for one method that has proven useful in identifying drug–receptor contacts on $GABA_A$ receptors: substituted cysteine modification and protection (SCAMP).

2. CHALLENGES FOR IDENTIFYING GENERAL ANESTHETIC BINDING SITES

Many anesthetic drugs, including alcohols and volatile alkanes and ethers, are low affinity and probably act through multiple molecular targets. More potent anesthetics that display stereoselectivity and greater target specificity include certain barbiturates, etomidate, and alphaxalone. These potent anesthetics present fewer challenges in defining their mechanisms of action and represent bases for developing new drugs with improved clinical utility.

Many anesthetics inhibit the activity of excitatory pLGICs. A well-studied case is inhibition of the muscle-type nAChR, although nAChR activity is also positively modulated by small alcohols (Miller, Firestone, & Forman, 1987). Mechanisms of anesthetic inhibition in nAChRs include channel blockade, desensitization, and negative allosteric modulation (Dilger, Brett, & Lesko, 1992; Dilger, Brett, & Mody, 1993). In contrast, most general anesthetics positively modulate inhibitory pLGICs such as $GABA_A$ receptors and glycine receptors (Harrison, Kugler, Jones, Greenblatt, & Pritchett, 1993). Positive modulation at a subset of $GABA_A$ receptor isotypes is known to mediate major behavioral effects of many general anesthetics, including hypnosis, amnesia, and immobility (Jurd et al., 2003). In addition, high concentrations of some anesthetics directly agonize $GABA_A$ receptors and others also inhibit receptor activity. Thus, there are potentially many types of anesthetic binding sites mediating different anesthetic actions on each pLGIC. Identifying these sites is a first challenge, and a second is demonstrating which anesthetic actions, if any, each binding site mediates. It is also possible that one site can mediate multiple drug effects.

3. pLGIC STRUCTURAL MODELS

Current structural models of heteromeric pLGICs are homology models, typically based on both cryoelectron microscopy and crystallography in channels from bacteria, nematodes, and mammals. The existence of many different subunit isotype genes indicates a very high number of potential structures, whereas a limited number of nAChRs and $GABA_A$ receptors are found in the central nervous system. The subunit arrangement of muscle nAChR is established as $\alpha(\gamma/\varepsilon)\alpha\delta\beta$. The most common synaptic $\alpha\beta\gamma$ $GABA_A$ receptor structure is established as $\beta\alpha\beta\alpha\gamma$ (Fig. 1). All pLGIC subunits include an

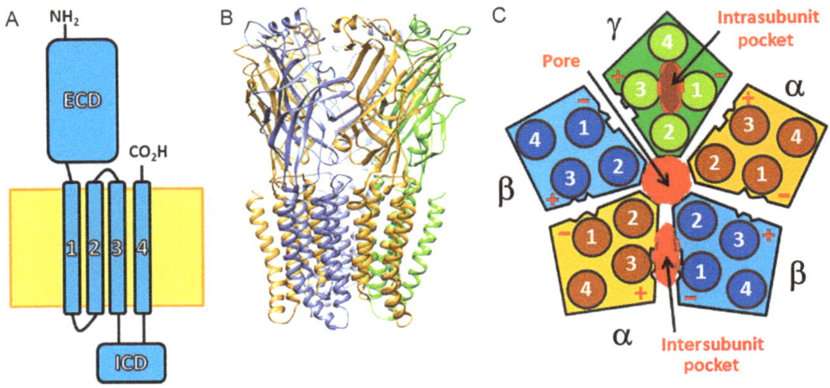

Fig. 1 Pentameric ligand-gated ion channel structure and types of anesthetic binding sites. (A) The canonical subunit structure shared by all pLGICs, including a large N-terminal extracellular domain (ECD), a transmembrane domain consisting of four alpha-helices, and an intracellular domain (ICD) between M3 and M4. The cytoplasmic membrane is depicted as a *yellow zone*. (B) A structural homology model of a synaptic $GABA_A$ receptor consisting of two α subunits (*gold*), two β subunits (*blue*), and one γ subunit (*green*). Peptide backbones are depicted as *ribbons* to highlight secondary structural folding. Subunit ICDs have been removed because their structure is not well defined. (C) A diagram showing a cross section of an αβγ $GABA_A$ receptor through the TMD. Each subunit is *color-coded* as in (B), the arrangement of subunits is detailed, as are the positions of numbered alpha-helices in each subunit. Three types of anesthetic binding pockets in the TMD are highlighted with *red overlays*: the central transmembrane pore, intrasubunit pockets formed by subunit four-helix bundles, and intersubunit pockets. Intersubunit anesthetic binding pockets have also been imaged in the ECD. The faces of each subunit are also labeled "+" or "−" depending on whether they include M3 or M1 helical elements.

N-terminal extracellular domain (ECD) of about 200 amino acids, a transmembrane domain (TMD) consisting of four transmembrane helices (M1–M4), a variable-size intracellular domain (ICD) between M3 and M4, and a short extracellular C-terminus. Sequence alignments in the TMD are well established in M2 helices, where a universally conserved leucine is the 9th residue (L9′) after the positively charged residue defining the onset of the helix. However, sequence alignment of other transmembrane helices is more uncertain due to variation in the length of interhelical linker sequences.

Structure–function studies have established that in muscle nAChRs orthosteric agonists (e.g., ACh and nicotine) bind to α^+/δ^- and $\alpha^+/(\gamma/\varepsilon)^-$ interfaces in the ECD, while in $GABA_ARs$, GABA binds to ECD β^+/α^- interfaces. Thus, nAChR αs can be considered homologs of $GABA_AR$ βs, and in both pLGIC families these archetypal subunits can form functional homopentamers.

Several high-resolution structural studies have focused on anesthetic binding sites in crystallized pLGICs, all of which are homopentamers (Fourati et al., 2017; Laurent et al., 2016; Nury et al., 2011; Pan et al., 2012; Spurny et al., 2013). These confirm that small amphiphilic general anesthetics can fit into multiple types of intrasubunit or intersubunit protein pockets formed in both ECDs and TMDs (Fig. 1C).

4. ELECTROPHYSIOLOGICAL APPROACHES TO ASSESSING pLGIC FUNCTION

4.1 Functional States of pLGICs

Electrophysiological studies have demonstrated that most pLGICs undergo a similar set of functional transitions. Binding of orthosteric agonists to resting state closed receptors stimulates rapid transition to activated or open ion-conductive states that can be observed in voltage-clamp experiments. Prolonged agonist application results in multiple phases of channel desensitization, leading to non-conductive states with high affinity for agonists. After discontinuing agonist application, pLGICs undergo deactivation as open channels return to the resting state. Recovery of desensitized receptors to the resting state may be via "silent" pathways or return to open states, and take up to tens of seconds, depending on the degree of receptor desensitization.

4.2 Cellular Systems for Heterologous Expression and Electrophysiology

The most common cell systems used for heterologous expression and structure–function studies are *Xenopus* oocytes and cultured mammalian cell lines like HEK293. *Xenopus* oocytes are large cells (~1 mm diameter) that can express many receptors on the membrane surface, and under ideal conditions, oocytes can provide stable currents for lengthy experiments lasting several hours. Oocytes are microinjected with messenger RNA or cDNA encoding wild-type or mutant receptor subunits. Oocytes can then be used for whole-oocyte two-electrode voltage-clamp studies or for patch-clamp studies in excised membrane patches (Stuhmer, 1992). Mammalian cells are transfected with cDNA plasmids that direct synthesis of receptor subunits that are assembled and transported to the cytoplasmic membrane and used for whole cell or patch current studies. Posttranslational modification of receptors may vary between expression cell types or depending on cytoplasmic kinase activity. There are examples where either receptor assembly or functional properties differ when expressed in HEK293 cells vs oocytes (Lewis, Harkness, Sivilotti, Colquhoun, & Millar, 1997; Shu et al., 2011).

4.3 Electrophysiological and Solution Exchange Equipment

Commercially available two-microelectrode voltage-clamp amplifiers can typically measure whole-oocyte voltage-clamped currents ranging from under 5 nA to 10 μA. Application of various solutions containing agonists and anesthetics or other chemicals is typically controlled by manual or electronically triggered valves from multiple reservoirs feeding into a manifold with outflow directed at the oocyte in a flow chamber and then to a waste collection system. Special care needs to be used with volatile anesthetics to prevent evaporation and adsorption into flow-system materials. When using very hydrophobic compounds such as potent anesthetics, reservoirs and tubing should be made of impermeable materials like glass, stainless steel, or PTFE. Depending on the rate of flow and the design of the reservoir around the oocyte, solution exchange can be as fast as 10 ms, but is typically around 0.1 s or longer. This is often too slow to observe the fastest rates of desensitization and deactivation observed in nAChRs and synaptic $GABA_A$ receptors. Another important consideration in studying hydrophobic anesthetics is that the large volume of intracellular lipid and protein in oocytes acts as a sink, slowing the rate of equilibration between drug in the oocyte membrane and the surrounding solution. Preequilibrating oocytes with anesthetics before agonist application can reduce errors due to this process. An associated challenge is prolonged washout times to fully remove hydrophobic drugs from oocytes.

Patch-clamp amplifiers can typically measure currents in the 5 pA to 10 μA range. When combined with "artificial synapse" techniques for rapid solution switching, solution exchange times under 1 ms can be achieved at excised membrane patches (Brett, Dilger, Adams, & Lancaster, 1986; Forman, 1999). Solution exchange times in whole HEK cells are typically 2–10 ms using rapid solution switching devices. Equilibration of hydrophobic drugs is fastest in excised patch experiments and can be rate limiting in measuring drug effects in voltage-clamped cells (Akk et al., 2005). These approaches enable studies of state-dependent drug actions with analysis of desensitization and deactivation rates. Single-channel recordings can also be obtained using low-noise patch-clamp techniques.

5. CHIMERIC SUBUNITS

Structure–function studies based on chimeric subunits represent an unbiased approach to identifying peptide sequences within receptor subunits that influence anesthetic sensitivity. Mihic et al. (1997) created a set of chimeric pLGIC subunits containing complementary portions of anesthetic-sensitive α1 glycine receptors and anesthetic-insensitive ρ $GABA_A$ receptors, and used electrophysiology to assess anesthetic modulation in each chimera at its EC20 agonist concentration (the agonist depended on which type of receptor donated the N-terminal domain). The results identified a subunit region including M2 and M3 helices that was both necessary and sufficient to confer positive modulation by volatile anesthetics and alcohols. Subsequent point mutations exchanging the nonhomologous residues in M2 and M3 helices identified two amino acids that influenced anesthetic modulation in both parent receptors: the M2–15′ position, and an M3 residue 21 amino acids away (sometimes called 36′). Mutations at the homologs on $GABA_A$ receptor α and β subunits were subsequently shown to influence sensitivity to both volatile and intravenous anesthetics.

Lambert and colleagues found that $GABA_A$ receptors containing β2 or β3 subunits were sensitive to modulation and direct agonism by the intravenous drug etomidate, while those with β1 subunits were not (Hill-Venning, Belelli, Peters, & Lambert, 1997). Chimeras containing complementary portions of β1 and β2, coexpressed with α and γ subunits and studied electrophysiologically, showed that determinants of etomidate sensitivity were in the TMD (Belelli, Lambert, Peters, Wafford, & Whiting, 1997). Subsequent point mutation experiments showed that swapping β2 and β1 M2–15′ residues determined drug sensitivity. At this position in β2, a N-to-S mutation suppressed etomidate sensitivity, while the complementary β1 S-to-N mutation increased sensitivity to etomidate modulation.

Interpretation of data from chimera experiments is critically dependent on the experimental framework. An important illustration is a study of β3 sequences involved in pentobarbital agonism (Serafini, Bracamontes, & Steinbach, 2000). Chimeric $GABA_A$ β3/ρ subunits were coexpressed with α1 subunits, and pentobarbital vs GABA agonism was assessed electrophysiologically in the resulting receptors. In one chimeric receptor where pentobarbital efficacy was reduced relative to that of GABA, noise and single-channel analyses indicated that the structural modification had increased GABA efficacy, without altering pentobarbital efficacy. Alternative analysis based on deactivation kinetics in the various chimeric receptors suggested that pentobarbital affinity determinants were downstream (toward the C-terminus) from the M2 helix. Another study of pentobarbital actions on $GABA_A$ receptor chimeras containing complementary sequences from γ and δ subunits (Feng & Macdonald, 2010) concluded that sequences upstream from M2 were critical. However, pentobarbital-induced currents were normalized to maximal GABA currents, and this study failed to account for evidence that GABA intrinsic efficacy in αβγ receptors is high, but is very low in αβδ receptors (Feng, Jounaidi, Haburcak, Yang, & Forman, 2014).

6. HYDROPHOBIC POINT MUTATIONS IN nAChRs

6.1 Introduction

Single-point mutations are the simplest hypothesis-based (biased) structure–function approach, because each site chosen for a mutation represents a hypothesis or set of hypotheses. Conceptually, one makes mutations and tests their effects on receptor function and sensitivity to drugs, leading to inferences about drug binding or modulation mechanisms.

6.2 Nicotinic ACh Receptors

Pioneering work by Lester and colleagues focused on the ion-conductive pore of muscle nAChRs, which was thought to be surrounded by M2 helices of the TMD (Leonard, Labarca, Charnet, Davidson, & Lester, 1988). They showed that M2–6′ S-to-A mutations in mouse muscle nAChR alpha subunits reduced sensitivity to QX222, a charged quaternary ammonium analog of an amide local anesthetic thought to block nAChRs via open-channel block. A strength of this study was the use of single-channel kinetics, demonstrating reduced apparent QX222 residence time in its site, as evidence of reduced binding affinity. Voltage-dependent block was also demonstrated, supporting a transmembrane location of the site. Mutations at M2–10′ residues on nAChRs showed that increased hydrophobicity enhanced QX-222 affinity, suggesting that the charged end of the molecule interacted with 6′ residues, while the hydrophobic end interacted with 10′ residues.

Kinetic studies in mouse nAChRs also suggested that general anesthetics might act as open-channel blockers (Dilger, Vidal, Mody, & Liu, 1994). Following the work on QX222, Forman, Miller, and Yellen (1995) tested M2–10′ mutations that enhanced hydrophobicity and found that they dramatically increased mouse nAChR sensitivity to open state-dependent block by both alcohols and volatile anesthetics. Subsequent studies showed that mutations on all five M2–10′ residues affected sensitivity to anesthetic inhibition, suggesting a pore site (Forman, 1997). Scanning mutagenesis studies defined a "hydrophobic patch" in the nAChR channel where anesthetics apparently interacted (Zhou, Zhou, & Forman, 2000).

Unlike long-chain alcohols, ethanol enhances the apparent affinity of ACh for muscle and Torpedo nAChRs. Studies in nAChR mutants that impaired gating demonstrated that ethanol enhances ACh agonist efficacy rather than binding (Forman & Zhou, 1999). However, positive modulation by ethanol is unaffected by M2 mutations that affect alcohol inhibitory potency.

6.3 $GABA_A$ Receptors

In contrast to anesthetic inhibition of excitatory serotonin and nAChRs, many general anesthetics enhance activation of inhibitory glycine and $GABA_A$ receptors (Harrison et al., 1993). Moreover, high concentrations of general anesthetics directly activate (agonize) these receptors in the absence of GABA or other orthosteric agonists, which might indicate the presence of at least two classes of anesthetic sites and presented challenges in defining anesthetic potency for putative sites. A standard approach is to measure anesthetic enhancement of receptor activation by a low agonist concentration that alone generates a small fraction (5%–20%) of maximal receptor-mediated responses.

Based on the findings of chimera studies (Mihic et al., 1997), mutations on various $GABA_A$ receptor subunit M2–15′ and 36′ residues have been tested for effects on anesthetic modulation. These studies showed that βM286W mutations weaken modulation by propofol and related alkylphenols (Krasowski et al., 1998; Krasowski, Nishikawa, Nikolaeva, Lin, & Harrison, 2001) and that the size of the α1 M2–15′ (S270) sidechain determines volatile anesthetic sensitivity (Koltchine et al., 1999). Tryptophan mutant sensitivity has been applied as a scanning method to discover anesthetic contacts (Jenkins, Andreasen, Trudell, & Harrison, 2002; Ueno et al., 2000). Similar approaches have also been used to discern selectivity for multiple $GABA_A$ receptor binding sites by various anesthetics (Maldifassi, Baur, & Sigel, 2016).

6.4 MWC Coagonism as an Interpretive Framework for Anesthetic Actions on $GABA_A$ Receptors

The interpretive framework for mutant function studies is also critically important, because it is difficult, if not impossible, to rule out allosteric effects based on binding studies and functional analysis alone. This is particularly difficult when studying agonist sites (Colquhoun, 1998). Reframing anesthetic actions at $GABA_A$ receptors as allosteric coagonism was based on pharmacological analysis using a formal Monod–Wyman–Changeux (MWC) two-state model (Rüsch, Neumann, Wulf, & Forman, 2012; Rüsch, Zhong, & Forman, 2004; Ziemba & Forman, 2016). These models provide a mechanistic framework for structure–function experiments that accounts for both direct anesthetic agonism and modulation of GABA-evoked responses based on drug interaction with a single class of allosteric agonist sites, providing simplified analysis in many cases. This approach has also proven insightful in explaining results that might otherwise lead to the conclusion that direct activation and modulation are mediated by independent sites. For example, based on azietomidate photolabeling results (Li et al., 2006), we examined the effects of tryptophan mutations at both α1M236 and β2M286 in α1β2γ2L receptors (Stewart, Desai, Cheng, Liu, & Forman, 2008). Oocyte electrophysiology showed that α1M236Wβ2γ2L receptors were spontaneously active and that etomidate potently and efficaciously activated these receptors, whereas etomidate modulation of GABA responses was much weaker than that in wild-type receptors. MWC models predict that spontaneously active mutant channels should display increased sensitivity to agonists (both orthosteric and allosteric), and model-based analysis showed that all of the etomidate-dependent results were consistent with reduced drug efficacy. More importantly, this example illustrates how mutations that alter receptor gating or orthosteric agonist efficacy may produce the appearance of altered anesthetic sensitivity, even when underlying anesthetic site interactions are unaltered.

However, even though MWC mechanisms distinguish binding affinity to resting receptors from agonist efficacy as separate model parameters, we have consistently found that mutations alter apparent efficacy far more than binding affinity, even at residues thought to be important binding contact points. This is likely because agonist efficacy in MWC is defined as the ratio of binding dissociation constants in open vs closed receptors, and molecules sufficiently efficacious to activate receptors bind much more tightly to open state receptors. At the same time, the very high anesthetic concentrations needed to occupy the majority of low-affinity resting state sites usually produce mixed functional effects that include positive modulation and inhibition. Some mutations that produce total insensitivity to anesthetics provide no data for differentiating binding vs efficacy changes (Desai, Rüsch, & Forman, 2009). Thus, even using an analytical approach that differentiates binding and efficacy, identifying binding determinants is a challenge.

6.5 Comparing Mutant Sensitivity and Anesthetic Photolabeling

Given the possibility that mutations allosterically alter apparent anesthetic sensitivity, an important question is how well hydrophobic mutagenesis and functional analysis perform in identifying drug–receptor contacts. We recently compared results of studies based on $GABA_A$ receptor photolabeling with derivatives of etomidate, propofol, mephobarbital, and alphaxalone to tryptophan mutagenesis experiments at two photolabeled amino acids (Nourmahnad et al., 2016). With eight possible interactions (four positive and four negative photolabeling results), Trp mutant studies agreed with photolabeling in six of eight cases, with one false positive and one false negative. This correlation was too weak to support mutant function analysis as a reliable approach to discovering anesthetic contact residues in $GABA_A$ receptors.

7. SUBSTITUTED CYSTEINE ACCESSIBILITY METHOD

7.1 Introduction

Cysteine mutation-based approaches based on covalent sulfhydryl modification are potentially more sensitive to local steric rather than allosteric effects, because implicit in this method is a bimolecular interaction between a chemical probe and the sulfhydryl group. These approaches include the substituted cysteine accessibility method (SCAM), disulfide mapping, and SCAMP. Below, we focus on how these methods have been applied to pLGIC receptor structure and mapping of anesthetic sites.

7.2 Nicotinic ACh Receptors

Akabas and Karlin were first to combine real-time agonist-dependent cysteine modification of pLGICs with electrophysiological analysis in nAChRs (Akabas, Kaufmann, Archdeacon, & Karlin, 1994). Their focus was cysteine accessibility, using a positively charged methanethiosulfonate ethylammonium probe (MTSEA), which they presumed was accessible only to the water-filled cation-conductive pore. Water is usually required for disulfide bond formation, which can also be influenced by the acidity of the targeted –SH group. At the time of these pioneering experiments, it was assumed that water was generally excluded from access to the M1, M3, and M4 helices that surrounded the M2s. Akabas and Karlin assumed that MTSEA modification in the nAChR pore would inhibit cation conduction. Thus, they compared maximal ACh-activated voltage-clamp currents mediated by cysteine-substituted receptors before and after MTSEA exposure. When inhibition was found, they inferred sidechain access from the water-filled ion pore. Other water-soluble reagents have also been used to probe water-accessible portions of pLGIC receptors. Using various modifying reagents at accessible substituted cysteines also enables studies of the effect of size, hydrophobicity, and charge on function.

7.3 $GABA_A$ Receptors

Akabas and colleagues extended the SCAM approach to M2 helices of $GABA_A$ receptors, and using EC_{50} GABA instead of high GABA when testing mutant receptors, found that modification both enhanced receptor activation and inhibited it at various positions (Xu & Akabas, 1996). The finding of enhanced GABA responses after modification meant that inferences regarding accessibility via the central ion pore were uncertain. Accessibility to water remained the sole inferential outcome of these studies. Importantly, however, results in other TM helices showed that water could reach close to the presumed inner ends of TM helices, implying that these helices are flexible and that water intercalates between them, and not just in the ion-conductive pore (Goren, Reeves, & Akabas, 2004; Williams & Akabas, 1999). In short, SCAM revealed water-accessible pockets formed among the various transmembrane helices.

8. DISULFIDE BOND FORMATION USED TO PROBE HELIX–HELIX INTERACTIONS IN $GABA_A$ RECEPTORS

Jansen and Akabas (2006) extended the use of cysteine substitutions in order to gain more information about the orientation of residues on adjacent transmembrane helices in the GABA α1 subunits. In this study, they inserted cysteines at different positions in both α1-M2 and α1-M3 helices, and used oocyte electrophysiology and chemical manipulations to either promote or reverse disulfide bond formation between the two helices. Their results helped constrain structural models and also suggested possible relative movements of M2 and M3 during channel gating.

Bali, Jansen, and Akabas (2009) utilized similar methods but focused on creating disulfide linkages between two transmembrane helical faces across the β^+/α^- intersubunit cleft. Their results further constrained structural models and were consistent with azi-etomidate photolabeling data, suggesting β^+/α^- intersubunit etomidate binding sites based on photolabeling of residues on both β-M3 and α-M1 helices (Li et al., 2006).

9. SUBSTITUTED CYSTEINE MODIFICATION AND PROTECTION

9.1 Introduction

Because covalent modification of a substituted cysteine requires bimolecular interactions between the probe reagent and the engineered sulfhydryl, this reaction can be hindered by noncovalent ligands that bind nearby. This inhibition of cysteine modification or "protection" has been developed as an approach to both complement and supplement other techniques, particularly photolabeling, in identifying anesthetic contacts on $GABA_A$ receptors. The first such study was that of Bali and Akabas (2004), testing whether propofol contacted βN265 (M2–15′) or βM286 (36′) residues that had been identified in chimera and point mutation studies. To do this, they used oocyte electrophysiology to estimate the rates of modification at βN265C or βM286C by the sulfhydryl-reactive probe *p*-chloromercuribenzene sulfonate (*p*CMBS). They then tested whether addition of propofol slowed the rate of modification and found protection at βM286C, but not βN265C.

9.2 SCAMP Experimental Design Considerations

For the SCAMP technique to enable strong inferences of steric interactions, a number of experimental conditions must be met (Forman & Miller, 2016):

9.2.1. The background receptor must be insensitive to probe exposure. This may require removing native cysteines, usually by mutation to alanines or serines.

9.2.2. The cysteine-substituted mutant receptor should retain sensitivity to the anesthetic drug (usually measured as EC5 enhancement by a standard drug concentration), ideally similar to wild-type sensitivity. Loss of drug sensitivity might be caused by the cysteine mutation dramatically reducing drug binding, which could lead to false-negative results. For example, we not only studied βN265C protection by etomidate but also found that the mutation totally eliminated etomidate sensitivity (Stewart, Pierce, Hotta, Stern, & Forman, 2014). Thus, the observed absence of etomidate protection could have been due to the absence of drug binding, rather than binding at a site distant from βN265. It is possible that the lack of propofol protection at βN265C (Bali & Akabas, 2004) was also due to low occupancy of these sites. In cases where drug sensitivity is moderately reduced, protection may still be tested, but using higher drug concentrations.

9.2.3. Modification of the mutant receptor must reliably produce quantifiable functional changes. Multiple probes may be tested to determine which produce the largest functional effects. However, very large probes may increase sensitivity to inhibition by ligands that bind further from the study residue. Electrophysiological signals indicating $GABA_A$ receptor modification can include: (a) spontaneous receptor activity, determined with picrotoxin inhibition; (b) maximal receptor current; or (c) EC5 GABA response, usually normalized to maximal response. In our experience, carefully designed SCAMP studies can lead to valid inferences in cases where receptor modification produces functional changes with magnitudes as small as twofold.

9.2.4. Modification conditions must enable estimation of modification rates. Typically, sequential cycles of receptor modification are separated by repeated electrophysiological assessments of receptor function, with each cycle resulting in approximately 10% of the full modification effect (Fig. 2). With many such cycles, full modification can be achieved and the electrophysiological results fitted to an exponential function to determine a rate parameter. Alternatively, three or four partial modification cycles can be used to establish an initial "apparent" modification rate based on a linear fit to the electrophysiological results. Normalization to each individual cell's maximal current is essential when using "apparent" modification rate analysis.

9.2.5. The distribution of receptor states should be similar in both control modification experiments and in drug protection experiments. This aspect of study design can be challenging when using anesthetics that act as agonists at $GABA_A$ receptors. In our experience, the rate of *p*CMBS modification at cysteines in transmembrane helices of $GABA_A$ receptors is usually faster when *p*CMBS is coapplied with GABA than when applied alone (Nourmahnad et al., 2016; Stewart, Hotta, Desai, & Forman, 2013; Stewart, Hotta, Li, et al., 2013). If adding anesthetics activates receptors, then using a "closed state" modification control is inappropriate, because a fraction of receptors will be activated during protection experiments. To avoid this problem, we use control modification conditions with *p*CMBS + GABA. GABA-bound receptors also bind more avidly to anesthetics (allosteric agonists), resulting in higher anesthetic binding site occupancy. However, GABA may become a partial agonist with introduction of some cysteine mutations, introducing the possibility that the fractions of closed vs open vs desensitized receptors differ in control experiments

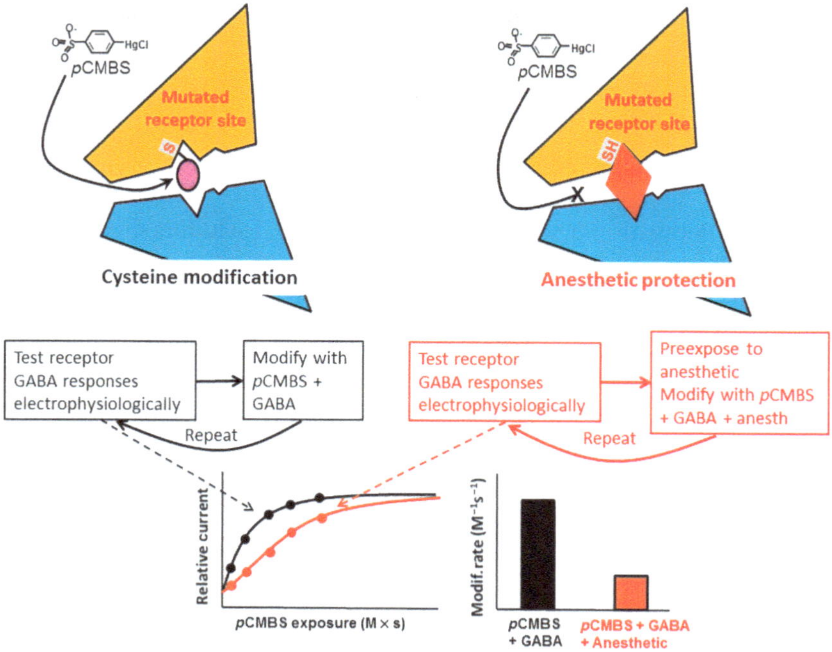

Fig. 2 Substituted cysteine modification protection (SCAMP) experimental approach. The *top and middle left panels* depict a mutant receptor with a cysteine-substitution in an intersubunit anesthetic binding pocket as it is covalently modified by *p*-chloromercuribenzene sulfonate (*p*CMBS). Repeated cycles of partial modification produce functional changes that are measured electrophysiologically. The *top and middle right panels* depict similar experiments on anesthetic-bound receptors, with inhibition of the bimolecular reaction between *p*CMBS and the engineered sulfhydryl group on the receptor. The *bottom left panel* shows how electrophysiological measurements for both control modification (*black dots*) and protection (*red dots*) conditions are plotted against cumulative *p*CMBS exposure and fitted with exponential functions (*black* and *red lines*) to determine modification rates. The *bottom right panel* shows a summary of the analyzed rates indicating that bound anesthetic "protects" this cysteine from modification.

and anesthetic protection experiments. In such cases, addition of another positive modulator that binds to distinct sites may help establish similar distributions of receptor states in control modification and protection experiments (Stewart, Hotta, Desai, et al., 2013). For example, pharmacological and photolabeling evidence indicates that alphaxalone and some barbiturate sites are distinct from those for propofol or etomidate.

9.2.6. Drug site occupancy should correlate with the degree of observed protection. Demonstrating concentration-dependent protection, particularly when anesthetic functional effects parallels protection, supports the presumed local steric hindrance mechanism (Stewart, Hotta, Desai, et al., 2013). By extension, SCAMP will be more sensitive when high anesthetic concentrations are used, leading to higher binding site occupancy. As noted earlier, this is particularly important when studying mutations that are relatively insensitive to modulation by one or more study drugs. Including GABA usually enhances mutant receptor affinity for anesthetics and increases drug site occupancy (see above).

9.2.7. In protection experiments, potent anesthetics should be preapplied before adding modifiers, to assure anesthetic site occupancy *before* modifiers reach their target. This approach is particularly important when studying very hydrophobic anesthetics that slowly equilibrate in oocyte membranes due to intracellular binding to protein and lipid. Higher sulfhydryl probe concentrations will more rapidly reach the target sulfhydryl, so selecting conditions where lower probe concentrations are used (i.e., apply GABA with the modifier) will also reduce this potential artifact.

9.2.8. Even with all these conditions met, it is important to perform additional studies to help rule out allosteric drug inhibition of modification at each substituted cysteine. Allosteric inhibition of modification at a particular cysteine should correlate with receptor functional state, rather than anesthetic site occupancy. Thus, all anesthetics that modulate the mutant receptor should inhibit modification of that cysteine. Our approach has been to compare multiple anesthetics that all produce similar functional effects in $GABA_A$ receptors, but that are also known to bind to distinct sets of sites, based on pharmacological and photolabeling evidence. Sets of such drugs perform as negative controls for each other and lend support to a steric hindrance mechanism in most cases where SCAMP experiments show inhibition.

9.3 SCAMP Limitations

SCAMP does not enable every hypothesized residue to be tested against every drug of interest. Some cysteine mutations prevent functional expression of receptors (e.g., β3Y284C) and some (e.g., α1I239C and α1Q242C) show no functional effects when exposed to sulfhydryl modifiers (Borghese, Hicks, Lapid, Trudell, & Harris, 2014; Stewart, Hotta, Li, et al., 2013). Several cases where cysteine mutants are totally insensitive to drugs of interest have also been reported. In the case of β2N265, we used an indirect SCAMP approach, based on protection of α1M236C by etomidate, to show that β2N265M, which also eliminates etomidate sensitivity, dramatically reduces apparent etomidate affinity for its site in both closed and GABA-activated states. Indirect SCAMP experiments may also be used to distinguish between allosteric and steric effects of other mutations. Despite negative control experiments, we have also encountered cases where anesthetics only partially inhibit sulfhydryl modification. These may represent cases where substituted cysteines are located near the periphery of anesthetic sites, so that steric hindrance is incomplete. This possibility may be tested with different size modifiers that should show size-dependent interactions with bound anesthetic.

9.4 Comparing SCAMP and Anesthetic Photolabeling

We have also formally compared SCAMP results against photolabeling at two positions with four different drugs. The two approaches show 100% agreement for the eight residue-drug pairs we examined, a correlation that is statistically significant and strengthened when other photolabeled residues studied with SCAMP are added. However, in one case, β3H267, SCAMP disagrees with photolabeling, reminding us that both approaches are based on structural variants of either the receptor or the ligand, and therefore potentially inaccurate (Stern & Forman, 2016). The preponderance of evidence supports SCAMP as a technique that is useful both for confirming photolabeling results and for testing potential drug contacts that have not been photolabeled. Indeed, SCAMP experiments have substantially expanded the list of anesthetic contact residues in $GABA_A$ receptors beyond those based on photolabeling alone.

ACKNOWLEDGMENT

This work was supported by the National Institutes of Health (R01-GM089745 and P01-GM058448).

REFERENCES

Akabas, M. H., Kaufmann, C., Archdeacon, P., & Karlin, A. (1994). Identification of acetylcholine receptor channel-lining residues in the entire M2 segment of the alpha subunit. *Neuron*, *13*(4), 919–927.

Akk, G., Shu, H. J., Wang, C., Steinbach, J. H., Zorumski, C. F., Covey, D. F., et al. (2005). Neurosteroid access to the GABAA receptor. *The Journal of Neuroscience*, *25*(50), 11605–11613.

Bali, M., & Akabas, M. H. (2004). Defining the propofol binding site location on the GABAA receptor. *Molecular Pharmacology*, *65*(1), 68–76.

Bali, M., Jansen, M., & Akabas, M. H. (2009). GABA-induced intersubunit conformational movement in the GABAA receptor alpha1M1-beta2M3 transmembrane subunit interface: Experimental basis for homology modeling of an intravenous anesthetic binding site. *The Journal of Neuroscience*, *29*(10), 3083–3092.

Belelli, D., Lambert, J. J., Peters, J. A., Wafford, K., & Whiting, P. J. (1997). The interaction of the general anesthetic etomidate with the gamma-aminobutyric acid type A receptor is influenced by a single amino acid. *Proceedings of the National Academy of Sciences of the United States of America*, *94*(20), 11031–11036.

Borghese, C. M., Hicks, J. A., Lapid, D. J., Trudell, J. R., & Harris, R. A. (2014). GABA(A) receptor transmembrane amino acids are critical for alcohol action: Disulfide cross-linking and alkyl methanethiosulfonate labeling reveal relative location of binding sites. *Journal of Neurochemistry*, *128*(3), 363–375.

Brett, R. S., Dilger, J. P., Adams, P. R., & Lancaster, B. (1986). A method for the rapid exchange of solutions bathing excised membrane patches. *Biophysical Journal*, *50*(5), 987–992.

Colquhoun, D. (1998). Binding, gating, affinity and efficacy: The interpretation of structure-activity relationships for agonists and of the effects of mutating receptors. *British Journal of Pharmacology*, *125*(5), 924–947.

Desai, R., Rüsch, D., & Forman, S. A. (2009). Gamma-amino butyric acid type A receptor mutations at beta2N265 alter etomidate efficacy while preserving basal and agonist-dependent activity. *Anesthesiology*, *111*(4), 774–784.

Dilger, J. P., Brett, R. S., & Lesko, L. A. (1992). Effects of isoflurane on acetylcholine receptor channels. 1. Single-channel currents. *Molecular Pharmacology*, *41*(1), 127–133.

Dilger, J. P., Brett, R. S., & Mody, H. I. (1993). The effects of isoflurane on acetylcholine receptor channels: 2. Currents elicited by rapid perfusion of acetylcholine. *Molecular Pharmacology*, *44*(5), 1056–1063.

Dilger, J. P., Vidal, A. M., Mody, H. I., & Liu, Y. (1994). Evidence for direct actions of general anesthetics on an ion channel protein. A new look at a unified mechanism of action. *Anesthesiology*, *81*(2), 431–442.

Feng, H. J., Jounaidi, Y., Haburcak, M., Yang, X., & Forman, S. A. (2014). Etomidate produces similar allosteric modulation in alpha-1/beta-3/delta and alpha-1/beta-3/gamma2L GABA-A receptors. *British Journal of Pharmacology*, *171*(3), 789–798.

Feng, H. J., & Macdonald, R. L. (2010). Barbiturates require the N terminus and first transmembrane domain of the delta subunit for enhancement of alpha1beta3delta GABAA receptor currents. *The Journal of Biological Chemistry*, *285*(31), 23614–23621.

Forman, S. A. (1997). Homologous mutations on different subunits cause unequal but additive effects on n-alcohol block in the nicotinic receptor pore. *Biophysical Journal*, *72*, 2170–2179.

Forman, S. A. (1999). A hydrophobic photolabel inhibits nicotinic acetylcholine receptors via open-channel block following a slow step. *Biochemistry*, *38*, 14559–14564.

Forman, S. A., & Miller, K. W. (2016). Mapping general anesthetic sites in heteromeric gamma-aminobutyric acid type A receptors reveals a potential for targeting receptor subtypes. *Anesthesia and Analgesia*, *123*(5), 1263–1273.

Forman, S. A., Miller, K. W., & Yellen, G. (1995). A discrete site for general anesthetics on a postsynaptic receptor. *Molecular Pharmaceutics*, *48*(4), 574–581.

Forman, S. A., & Zhou, Q. (1999). Novel modulation of a nicotinic receptor channel mutant reveals that the open state is stabilized by ethanol. *Molecular Pharmaceutics*, *55*(1), 102–108.

Fourati, Z., Ruza, R. R., Laverty, D., Drege, E., Delarue-Cochin, S., Joseph, D., et al. (2017). Barbiturates bind in the GLIC ion channel pore and cause inhibition by stabilizing a closed state. *The Journal of Biological Chemistry*, *292*(5), 1550–1558.

Goren, E. N., Reeves, D. C., & Akabas, M. H. (2004). Loose protein packing around the extracellular half of the GABA(A) receptor beta1 subunit M2 channel-lining segment. *The Journal of Biological Chemistry*, *279*(12), 11198–11205.
Harrison, N. L., Kugler, J. L., Jones, M. V., Greenblatt, E. P., & Pritchett, D. B. (1993). Positive modulation of human gamma-aminobutyric acid type A and glycine receptors by the inhalation anesthetic isoflurane. *Molecular Pharmacology*, *44*(3), 628–632.
Hill-Venning, C., Belelli, D., Peters, J. A., & Lambert, J. J. (1997). Subunit-dependent interaction of the general anaesthetic etomidate with the gamma-aminobutyric acid type A receptor. *British Journal of Pharmacology*, *120*, 749–756.
Jansen, M., & Akabas, M. H. (2006). State-dependent cross-linking of the M2 and M3 segments: Functional basis for the alignment of GABAA and acetylcholine receptor M3 segments. *The Journal of Neuroscience*, *26*(17), 4492–4499.
Jenkins, A., Andreasen, A., Trudell, J. R., & Harrison, N. L. (2002). Tryptophan scanning mutagenesis in TM4 of the GABA(A) receptor alpha1 subunit: Implications for modulation by inhaled anesthetics and ion channel structure. *Neuropharmacology*, *43*(4), 669–678.
Jurd, R., Arras, M., Lambert, S., Drexler, B., Siegwart, R., Crestani, F., et al. (2003). General anesthetic actions in vivo strongly attenuated by a point mutation in the GABA(A) receptor beta3 subunit. *The FASEB Journal*, *17*(2), 250–252.
Koltchine, V. V., Finn, S. E., Jenkins, A., Nikolaeva, N., Lin, A., & Harrison, N. L. (1999). Agonist gating and isoflurane potentiation in the human gamma-aminobutyric acid type A receptor determined by the volume of a second transmembrane domain residue. *Molecular Pharmacology*, *56*, 1087–1093.
Krasowski, M. D., Koltchine, V. V., Rick, C. E., Ye, Q., Finn, S. E., & Harrison, N. L. (1998). Propofol and other intravenous anesthetics have sites of action on the gamma-aminobutyric acid type A receptor distinct from that for isoflurane. *Molecular Pharmacology*, *53*, 530–538.
Krasowski, M. D., Nishikawa, K., Nikolaeva, N., Lin, A., & Harrison, N. L. (2001). Methionine 286 in transmembrane domain 3 of the GABAA receptor beta subunit controls a binding cavity for propofol and other alkylphenol general anesthetics. *Neuropharmacology*, *41*(8), 952–964.
Laurent, B., Murail, S., Shahsavar, A., Sauguet, L., Delarue, M., & Baaden, M. (2016). Sites of anesthetic inhibitory action on a cationic ligand-gated ion channel. *Structure*, *24*(4), 595–605.
Leonard, R. J., Labarca, C. G., Charnet, P., Davidson, N., & Lester, H. A. (1988). Evidence that the M2 membrane-spanning region lines the ion channel pore of the nicotinic receptor. *Science*, *242*(4885), 1578–1581.
Lewis, T. M., Harkness, P. C., Sivilotti, L. G., Colquhoun, D., & Millar, N. S. (1997). The ion channel properties of a rat recombinant neuronal nicotinic receptor are dependent on the host cell type. *The Journal of Physiology*, *505*, 299–306.
Li, G. D., Chiara, D. C., Sawyer, G. W., Husain, S. S., Olsen, R. W., & Cohen, J. B. (2006). Identification of a $GABA_A$ receptor anesthetic binding site at subunit interfaces by photolabeling with an etomidate analog. *The Journal of Neuroscience*, *26*(45), 11599–11605.
Maldifassi, M. C., Baur, R., & Sigel, E. (2016). Functional sites involved in modulation of the GABA receptor channel by the intravenous anesthetics propofol, etomidate and pentobarbital. *Neuropharmacology*, *105*, 207–214.
Mihic, S. J., Ye, Q., Wick, M. J., Koltchine, V. V., Krasowski, M. D., Finn, S. E., et al. (1997). Sites of alcohol and volatile anaesthetic action on GABA(A) and glycine receptors. *Nature*, *389*, 385–389.
Miller, K. W., Firestone, L. L., & Forman, S. A. (1987). General anesthetic and specific effects of ethanol on acetylcholine receptors. *Annals of the New York Academy of Sciences*, *492*, 71–87.

Nourmahnad, A., Stern, A. T., Hotta, M., Stewart, D. S., Ziemba, A. M., Szabo, A., et al. (2016). Tryptophan and cysteine mutations in M1 helices of α1β3γ2L γ-aminobutyric acid type A receptors indicate distinct intersubunit sites for four intravenous anesthetics and one orphan site. *Anesthesiology*, *125*(6), 1144–1158.

Nury, H., Van Renterghem, C., Weng, Y., Tran, A., Baaden, M., Dufresne, V., et al. (2011). X-ray structures of general anaesthetics bound to a pentameric ligand-gated ion channel. *Nature*, *469*(7330), 428–431.

Pan, J., Chen, Q., Willenbring, D., Mowrey, D., Kong, X. P., Cohen, A., et al. (2012). Structure of the pentameric ligand-gated ion channel GLIC bound with anesthetic ketamine. *Structure*, *20*(9), 1463–1469.

Rüsch, D., Neumann, E., Wulf, H., & Forman, S. A. (2012). An allosteric coagonist model for propofol effects on alpha1beta2gamma2L gamma-aminobutyric acid type A receptors. *Anesthesiology*, *116*(1), 47–55.

Rüsch, D., Zhong, H., & Forman, S. A. (2004). Gating allosterism at a single class of etomidate sites on alpha1beta2gamma2L GABA-A receptors accounts for both direct activation and agonist modulation. *The Journal of Biological Chemistry*, *279*(20), 20982–20992.

Serafini, R., Bracamontes, J., & Steinbach, J. H. (2000). Structural domains of the human GABAA receptor beta3 subunit involved in the actions of pentobarbital. *Journal of Physiology*, *524*(Pt. 3), 649–676.

Shu, H. J., Bracamontes, J., Taylor, A., Wu, K., Eaton, M. M., Akk, G., et al. (2011). Characteristics of concatemeric GABA(A) receptors containing alpha4/delta subunits expressed in Xenopus oocytes. *British Journal of Pharmacology*, *165*(7), 2228–2243.

Spurny, R., Billen, B., Howard, R. J., Brams, M., Debaveye, S., Price, K. L., et al. (2013). Multisite binding of a general anesthetic to the prokaryotic pentameric Erwinia chrysanthemi ligand-gated ion channel (ELIC). *The Journal of Biological Chemistry*, *288*(12), 8355–8364.

Stern, A. T., & Forman, S. A. (2016). A cysteine substitution probes beta3H267 interactions with propofol and other potent anesthetics in alpha1beta3gamma2L gamma-aminobutyric acid type A receptors. *Anesthesiology*, *124*(1), 89–100.

Stewart, D. S., Desai, R., Cheng, Q., Liu, A., & Forman, S. A. (2008). Tryptophan mutations at azi-etomidate photo-incorporation sites on α1 or β2 subunits enhance GABAA receptor gating and reduce etomidate modulation. *Molecular Pharmacology*, *74*(6), 1687–1695.

Stewart, D. S., Hotta, M., Desai, R., & Forman, S. A. (2013). State-dependent etomidate occupancy of its allosteric agonist sites measured in a cysteine-substituted GABAA receptor. *Molecular Pharmacology*, *83*(6), 1200–1208.

Stewart, D. S., Hotta, M., Li, G. D., Desai, R., Chiara, D. C., Olsen, R. W., et al. (2013). Cysteine substitutions define etomidate binding and gating linkages in the alpha-M1 domain of gamma-aminobutyric acid type A (GABAA) receptors. *The Journal of Biological Chemistry*, *288*(42), 30373–30386.

Stewart, D. S., Pierce, D. W., Hotta, M., Stern, A. T., & Forman, S. A. (2014). Mutations at beta N265 in gamma-aminobutyric acid type A receptors alter both binding and efficacy of potent anesthetics. *PLoS One*, *9*(10), e111470.

Stuhmer, W. (1992). Electrophysiological recording from Xenopus oocytes. *Methods in Enzymology*, *207*, 319–339.

Ueno, S., Lin, A., Nikolaeva, N., Trudell, J. R., Mihic, S. J., Harris, R. A., et al. (2000). Tryptophan scanning mutagenesis in TM2 of the GABA(A) receptor alpha subunit: Effects on channel gating and regulation by ethanol. *British Journal of Pharmacology*, *131*(2), 296–302.

Williams, D. B., & Akabas, M. H. (1999). Gamma-aminobutyric acid increases the water accessibility of M3 membrane-spanning segment residues in GABA-A receptors. *Biophysical Journal*, 77, 2563–2574.

Xu, M., & Akabas, M. H. (1996). Identification of channel-lining residues in the M2 membrane-spanning segment of the GABA(A) receptor alpha1 subunit. *The Journal of General Physiology*, *107*(2), 195–205.
Zhou, Q. L., Zhou, Q., & Forman, S. A. (2000). The n-alcohol site in the nicotinic receptor pore is a hydrophobic patch. *Biochemistry*, *39*(48), 14920–14926.
Ziemba, A. M., & Forman, S. A. (2016). Correction for inhibition leads to an allosteric co-agonist model for pentobarbital modulation and activation of alpha1beta3gamma2L GABAA receptors. *PLoS One*, *11*(4), e0154031.

CHAPTER NINETEEN

HCN and K_{2P} Channels in Anesthetic Mechanisms Research

Paul M. Riegelhaupt, Gareth R. Tibbs, Peter A. Goldstein[1]
Weill Cornell Medical College, New York, NY, United States
[1]Corresponding author: e-mail address: pag2014@med.cornell.edu

Contents

Abstract

The ability of a diverse group of agents to produce general anesthesia has long been an area of intense speculation and investigation. Over the past century, we have seen a paradigm shift from proposing that the anesthetized state arises from nonspecific interaction of anesthetics with the lipid membrane to the recognition that the function of distinct, and identifiable, membrane-embedded proteins is dramatically altered in the presence of intravenous and inhaled agents. Among proteinaceous targets, metabotropic and ionotropic receptors garnered much of the attention over the last 30 years, and it is only relatively recently that voltage-gated ion channels have clearly and rigorously been shown to be important molecular targets. In this review, we will consider the experimental issues relevant to two important ion channel anesthetic targets, HCN and K_{2P}.

Methods in Enzymology, Volume 602
ISSN 0076-6879
https://doi.org/10.1016/bs.mie.2018.01.015

1. INTRODUCTION

Reviews on the molecular and cellular mechanisms of anesthetic action have often and rightly emphasized the role of $GABA_A$ receptors and inhibitory synaptic transmission (Franks, 2008; Zeller et al., 2008). Despite clear evidence for such modulation, it is now abundantly evident that other ion channels contribute to the anesthetic state (Herold & Hemmings, 2012; Nau, 2008), including hyperpolarization-activated cyclic nucleotide-gated channels (HCN) (Goldstein, 2015) and tandem pore K^+ (K_{2P}) channels (Steinberg, Wafford, Brickley, Franks, & Wisden, 2015). Both classes of channels belong to the potassium channel superfamily, but there are profound differences in their structure and function from the canonical archetype (Gonzalez et al., 2012; He, Chen, Li, & Hu, 2014). Consequently, there are unique considerations in experimental design and data interpretation that one must take into account when studying these channels. We assume that the interested reader is familiar with standard electrophysiologic approaches when recording from neurons and heterologous expression models (HEK 293 cells and oocytes) as well as the basic molecular biologic techniques for heterologous expression, and we will not review the fundamentals of those methods here. Rather, we will provide critical experimental details that one must consider when performing experiments using both native and heterologously expressed channels so as to readily obtain meaningful and interpretable results.

2. HCN CHANNELS

Weakly selective for K^+ over Na^+ and activated by hyperpolarization not depolarization, HCN channels are, nonetheless, members of the K_V superfamily (Biel, Wahl-Schott, Michalakis, & Zong, 2009; Wahl-Schott & Biel, 2009). When activated, the functional channel conducts a mixed cationic current termed (in neurons) I_h; the comparable current in cardiac tissue is referred to as I_f although the molecular basis for the current is identical. Originally described as a current responsible for the Phase 4 spontaneous depolarization that precedes the cardiac action potential, I_h is now known to be widely distributed and to regulate rhythmic oscillations as well as subthreshold excitability in the PNS and CNS through control of the resting membrane potential, input resistance, and synaptic integration (Robinson & Siegelbaum, 2003; Wahl-Schott & Biel, 2009). The channels, most notably HCN1, are sensitive to a wide range of anesthetics at clinically relevant concentrations, and have attracted significant attention over the past decade as important molecular targets with respect to anesthetic action (Goldstein, 2015).

2.1 Physiological Regulation of Channel Gating

An excellent review of the extensive regulatory mechanisms governing HCN channel gating can be found in Lewis, Estep, and Chetkovich (2010). Among the key pathways to consider is the role of cAMP. All HCN channels contain a cyclic nucleotide binding domain (CNBD) located on the cytoplasmic C-terminus distal to the S6 transmembrane domain; cAMP binding to the CNBD relieves tonic inhibition and shifts the $V_{1/2}$ to more depolarized membrane potentials (Wainger, DeGennaro, Santoro, Siegelbaum, & Tibbs, 2001). HCN1 is weakly gated by cAMP, whereas in HCN2 (Wang, Chen, & Siegelbaum, 2001) and HCN4 (Stieber et al., 2003) cAMP gating is pronounced. Despite the presence of the CNBD, HCN3 is insensitive to cAMP (Stieber, Stöckl, Herrmann, Hassfurth, & Hofmann, 2005).

Function of the CNBD in turn is regulated by the tetratricopeptide repeat-containing Rab8b-interacting protein, TRiP8b (Han et al., 2011; Santoro et al., 2011). TRiP8b-dependent regulation of HCN channel gating occurs in a subtype-dependent and mode-specific fashion; thus, cAMP gating of HCN2 and HCN4 is noticeably impaired, while gating by phosphoinositides (PIP) and basal voltage dependence remain unaffected (Zolles et al., 2009). TRiP8 not only regulates cAMP-mediated gating but also contributes to cell-surface expression of HCN (Lewis et al., 2009; Santoro et al., 2009). Multiple TRiP8b splice variants exist, so caution must be used when comparing results obtained from natively and heterologously expressed channels.

HCN channel function is sensitive to the concentrations of Na^+ and K^+ in the extracellular solution (McCormick & Pape, 1990; Moroni et al., 2000). With respect to ion conduction, elevation of external K^+ decreases the channels' modest K^+ selectivity while elevating the inward current amplitude. The elevation of the inward current is not simply due to a depolarizing shift in the reversal potential but involves a K^+ enhancement of the Na^+ flux. In the absence of external K^+, the Na^+ flux is profoundly reduced (Biel et al., 2009). The loss of the inward Na^+ current is somewhat paradoxical given that gating in the absence of external K^+ is actually more favorable (the $V_{1/2}$ is some 10 mV depolarized, open probability is higher, and kinetics are faster) than in the presence of external K^+ (P.A.G. and G.R.T., unpublished observations).

There is also a sensitivity to Cl^- (Mistrík, Pfeifer, & Biel, 2006; Wahl-Schott, Baumann, Zong, & Biel, 2005), but under most recording conditions, external Cl^- is not subject to variation nor consideration.

Another relevant modulator of HCN function is H^+ (Stevens et al., 2001; Zong, Stieber, Ludwig, Hofmann, & Biel, 2001). Acidification of the extracellular environment decreases the slope of the activation curve for I_h and produces a right shift in the $V_{1/2}$ (Stevens et al., 2001). Conversely, acidification of the intracellular environment left shifts the $V_{1/2}$ with variable effects on the slope (Munsch & Pape, 1999; Zong et al., 2001). Internal pH sensitivity is governed by a single, conserved histidine residue (His-321) which is localized to the boundary between the voltage-sensing S4 helix and the S4–S5 linker (Zong et al., 2001). Other important modulators of HCN function include phosphatidic and arachidonic acids (Fogle, Lyashchenko, Turbendian, & Tibbs, 2007) as well as phosphatidylinositol 4,5-bisphosphate (PIP_2) (Pian, Bucchi, Decostanzo, Robinson, & Siegelbaum, 2007; Pian, Bucchi, Robinson, & Siegelbaum, 2006).

2.2 Pharmacological Regulation of Channel Gating

Commercially available pore blockers such as ZD7288 and ivabradine are nonselective with respect to the different isoforms. It is possible to pharmacologically block HCN1 as HCN1-selective presumptive pore blockers have been synthesized (Del Lungo et al., 2012; Melchiorre et al., 2010), and low micromolar concentrations of the alkylphenols propofol (2,6-di-iso-propylphenol) and 2,6-di-*tert*-butylphenol preferentially impair HCN1, but not HCN2–4, channel gating and markedly left shift the $V_{1/2}$ (Tibbs et al., 2013).

Volatile anesthetics (Chen et al., 2005) and propofol (Lyashchenko, Redd, Yang, & Tibbs, 2007) stabilize HCN channels in the closed state; for volatiles, but not propofol, such stabilization is cAMP dependent. Thus, while the effect on the $V_{1/2}$ is similar in both instances (i.e., left-shifted), the mechanistic explanation is quite different. The effect of cAMP on HCN1 and HCN2 saturates at concentrations $\geq 10\ \mu M$ (Wang et al., 2001); consequently, one should consider the need to maintain intracellular cAMP levels when performing experiments where cAMP levels might be dialyzed (as in whole-cell patch clamp) or where adenylyl cyclase activity might be altered.

Nevertheless, the situation is more complicated than this. Wang et al. demonstrated that voltage-dependent changes in HCN2 channel opening could modify the binding of basal levels of cAMP, reflecting allosteric coupling of channel opening and ligand binding (Wang, Chen, Nolan, & Siegelbaum, 2002). cAMP has an exceedingly slow rate of dissociation, which provides I_h with an "activity-dependent memory" that generates a stable channel open state that persists for tens of seconds (Wang et al., 2002). Those results have profound implications with respect to stimulation paradigms when recording I_h-dependent changes in membrane voltage (and hence excitability) from intact cells. Consequently, one must take into account acute (i.e., those occurring within 1–2 s) anesthetic-induced effects vs those that occur on a protracted timescale (tens of seconds) and distinguish them from activity (i.e., hyperpolarization)-induced changes; the use of appropriate control paradigms is of course mandatory.

2.3 Experimental Considerations

With any voltage-sensitive, ion-selective channel, the choices of ionic composition of the recording solutions, holding potential, step durations, and the tail potential are the four most obvious (and generally straightforward) parameters to consider. When recording HCN channels, the decision-making process is atypically problematic.

Superficially, as HCN channels carry a mixed Na^+–K^+ current and display no overt inactivation (but see Shin, Maertens, Proenza, Rothberg, & Yellen, 2004), the simple and obvious choice is to use physiological concentrations of Na^+ and K^+ on both sides of the membrane, a holding potential that is consistent with channels being fully closed but the cells healthy, a tail potential far from the reversal potential (be it at a very negative or positive potential), and activation step durations sufficient to equilibrate gating. This would seem obvious and is, in general terms, what we would advocate, but what does that mean practically and what are our preferred conditions?

When recording from *Xenopus* oocytes, our preferred external solution is 107 m*M* NaCl, 5 m*M* KCl, 1 m*M* $MgCl_2$, 1 m*M* $CaCl_2$ with 10 m*M* HEPES (free acid) adjusted to pH 7.4 with NaOH. We hold cells at −30 mV and excised patches at −40 mV. We attempt to use steps sufficient to fully activate each isoform across the whole range of the activation curve. We routinely measure tails at 0 mV. Why do we advocate for such conditions and what are the problems with these vs others that are often used?

The most obvious effect of elevating external K^+ (while isosmotically lowering Na^+) is to greatly increase the observed macroscopic inward current. As such, raising external K^+ has an attraction when recording poorly expressing mutants of HCN1 or HCN2 channels, or recording HCN3 and HCN4 given that these isoforms tend to express far more poorly in *Xenopus* oocytes than do HCN1 and HCN2. However, isosmotic exchange of Na^+ for K^+ in the extracellular recording solution comes at a considerable price. First, the condition is clearly far from physiological. Second, and far more importantly, as noted earlier, elevated external K^+ markedly slows channel kinetics, shifts the $V_{1/2}$ hyperpolarized, and flattens out the steady-state activation curve. Indeed, while the currents may look "prettier," it is often difficult or impossible to get a fully determined activation curve with the curve showing no clear saturation even at very low voltages—even for the relatively fast isoforms HCN1 and HCN2.

The second parameter to consider is the holding potential. While the true thermodynamic midpoint of gating for all four HCN channels is relatively hyperpolarized (being somewhere between −60 and −80 mV depending on isoform and ionic composition of the recording solutions), HCN channels are not steeply dependent on voltage meaning they can have a not insignificant open probability at potentials as high as −40 mV. Moreover, a well-described mode-shift behavior (Bruening-Wright & Larsson, 2007; Elinder, Männikkö, Pandey, & Larsson, 2006; Männikkö, Pandey, Larsson, & Elinder, 2005) means that the choice of holding potential has the ability to markedly influence gating measurements particularly when inadequate step durations are used. We typically use −30 mV as this is consistent with the health of the cells and ensures that channels are closed under most conditions.

The third experimental issue to deal with when recording HCN channels is to select an activation step duration that yields a relevant measure of gating. While HCN1-mediated currents are relatively fast (Wahl-Schott & Biel, 2009), even at very negative or positive potentials equilibration of gating requires voltage steps of several hundred milliseconds (Santoro et al., 1998). At potentials closer to the $V_{1/2}$ step durations of at least 3 s are required to achieve a thermodynamically sound measure of the $V_{1/2}$. For HCN2 and 3, steps that are at least 10 s long are required (but 30 s is preferable, particularly for HCN3). With HCN4, step durations of 120 s are necessary (Ludwig et al., 1999; Lyashchenko et al., 2007; Seifert et al., 1999). Note that with shorter voltage steps that HCN2- and HCN4-generated currents do not reach stable steady-state values when using protocols with steps of shorter duration (Fig. 1A and B). When one factors in the interpulse interval necessary to allow channels to fully deactivate and recover from mode-shifting, a single, fully equilibrated current–voltage (I–V) relationship for HCN1 may be some 3 min long, while for HCN4 a single I–V may require 30 min. Not only are such long durations limiting in terms of data throughput, the times and voltage excursions are such that few cells tolerate the paradigm well. There are two alternatives. One approach is to use

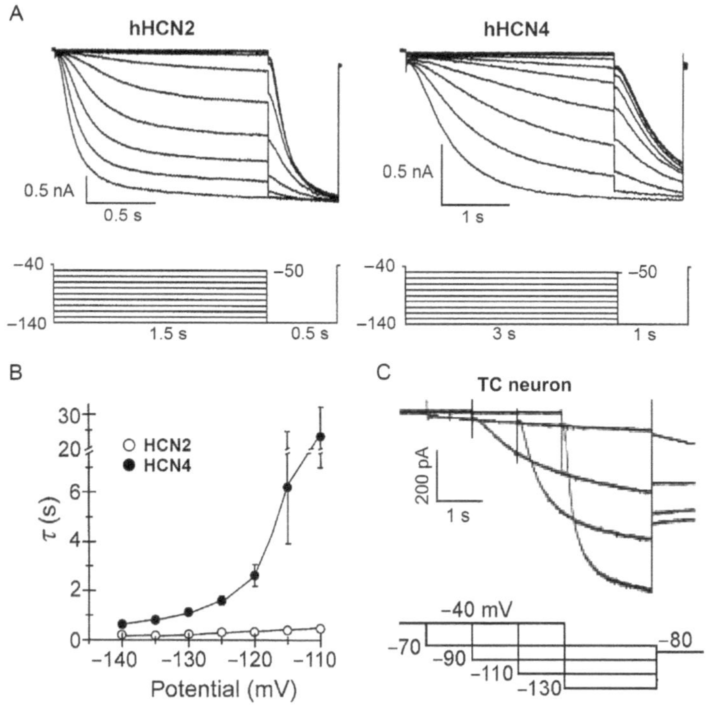

Fig. 1 Voltage-clamp paradigms for recording I_h currents. (A and B) Human (h)HCN2 (*left panel*) and hHCN4 (*right panel*) channels were heterologously expressed in HEK293 cells and whole-cell currents (*top traces*) measured in whole-cell voltage clamp using the voltage-clamp protocols shown (*bottom traces*). (C) Voltage dependence of activation kinetics. Current traces at voltages ranging from −140 to −110 mV from hHCN2- or hHCN4-expressing cells were fitted with a single exponential. The time constants for hHCN2 and hHCN4 activation are plotted against the corresponding potentials. (C) A series of current traces recorded from a mouse thalamocortical (TC) neuron in an acutely prepared brain slice (*top traces*). Neurons were voltage clamped for the times indicated from −40 mV to a range of test potentials using the voltage protocol shown (*bottom*). *Panels (A and B): Modified from Ludwig, A., Zong, X., Stieber, J., Hullin, R., Hofmann, F., & Biel, M. (1999). Two pacemaker channels from human heart with profoundly different activation kinetics. The EMBO Journal, 18(9), 2323–2329, with permission. Panel (C): Modified from Ludwig, A., Budde, T., Stieber, J., Moosmang, S., Wahl, C., Holthoff, K., et al. (2003). Absence epilepsy and sinus dysrhythmia in mice lacking the pacemaker channel HCN2. The EMBO Journal, 22(2), 216–224, with permission.*

variable step durations, shortening the times as the applied voltage moves further from the $V_{1/2}$. The second approach is to use an isochronal paradigm. Both approaches have secondary limitations. First, it is well described that deactivation gating of HCN channels changes as a function of the strength of the preceding activation pulse (this is discussed further below with respect to tail current analysis). It seems likely that so long as gating equilibrates at each potential this "activation strength" effect should not contaminate the measure of the $V_{1/2}$ but the extent to which this may perturb the measure should be borne in mind and checked. Second, while the parameters obtained by fitting the Boltzmann equation to an activation curve ($V_{1/2}$ and slope) are taken as thermodynamic descriptors, when the activation steps are not all equilibrated the parameters are, at best, no more than useful descriptors of channel function. The more abbreviated the activation sweeps, the more kinetically contaminated the $V_{1/2}$ and slope become. It is imperative that when considering how drug- or mutation-perturbation is influencing gating, the effect of step duration is taken into consideration. Has the experimental variable really altered the $V_{1/2}$ or is it actually a kinetic effect that is conflated with thermodynamics due to the choice of a poor (inadequate) step duration? That said, it is important to also ask if these concerns are a reason to always strive for fully equilibrated sweep durations? The answer to that question is "Generally, no." While a desire to cleanly separate thermodynamic and kinetic measures is highly commendable in the face of the technical constraints, the complexities of HCN gating mean that it is not obvious how to combine the $V_{1/2}$ and slope and extract a free energy term.

The fourth challenge in obtaining an activation curve from HCN channels is to select a voltage at which to measure the tail currents. Should this be negative or positive to the reversal potential? How far from the $V_{1/2}$ does it need to be? Outward tail currents measured at extreme negative potentials have the advantage that their amplitude is not blunted by intracellular Mg^{2+} ion block. However, there are technical disadvantages. First, if recording in elevated external K^+ (particularly when recording from the slower isoforms), the failure of activation to fully equilibrate in a reasonable length of time makes determination of a leak-subtracted amplitude problematic. Second, imposing additional time at hyperpolarized potentials can be problematic with respect to cell survival. The obvious alternative is to measure tails at depolarized potentials. Under the quasi-physiological conditions we advocated above, 0 mV is a good choice. Deactivation is relatively fast but is slow enough to be readily separated from both uncompensated capacitance and the fast block by internal Mg^{2+}. There is, however, an additional caveat to measuring tails at depolarized potentials that is not present when tails are measured at hyperpolarized potentials—conversion of tail currents from exponential to sigmoidal decay as a function of strength of activation. A simple explanation of this behavior is that basic HCN gating is well described by a cyclic allosteric model with 5 closed and 5 open states each representing activation of 0–4 voltage sensors. At modest potentials channels open when incompletely activated, whereas following activation at a very negative potential channels transit back through multiple open states before closing. Irrespective of the biophysical origin, this behavior makes the choice of cursor positions for determination of amplitude a less than objective issue. Place the cursors too close to the beginning of the tail and risk picking up capacity and block transients. Place the cursors further to the right and underestimate the current following sweeps at intermediate potentials. While one could measure tails at intermediate potentials, as this is where channels gate their slowest, having sweeps long enough to allow for gating to equilibrate, and hence allowing for leak subtraction, greatly exacerbates the data collection time problem. We opt to measure tails at 0 mV and are careful to be consistent with the position of the cursors.

2.3.1 Heterologous Expression of HCN Channels in Xenopus Oocytes

HCN2 expression in *Xenopus* oocytes is straightforward. The clone is easy to handle in bacteria, and DNA and RNA synthesis and recovery poses no problems (we aim to recover RNA at 1–5 μg/μL) and HCN2 expresses at very high levels within 24–48 h postinjection (30–100 nL/cell). Accordingly, even when using the quasi-physiological external concentrations of Na^+ and K^+ recommended earlier, macroscopic currents are readily recordable in TEVC. In IOPC, large macroscopic inward currents can also be consistently observed without the need to increase external K^+. Not all these conditions are met with HCN1, 3, or 4.

Full-length HCN1 is deleterious to bacteria. In our hands, use of standard cells, LB plates, and LB broth results in poor growth and very low yields of DNA. We have found that STBL2 cells grown at 30°C on CIRCLEGROW plates supplemented with glucose and in CIRCLEGROW media are far preferable. As a routine, we use this combination for all four HCN isoforms. The exception to this is when we use QuikChange mutagenesis. However, with respect to HCN1, we typically only perform that manipulation on a minimal HCN1 channel wherein both the first half of the N-terminus and all of the C-terminus have been deleted (HCN1–ΔNvΔC).

Like HCN2, HCN1 expresses to very high levels. However, in contrast to HCN2, HCN1 is poorly amenable to excised inside-out patching due to channel clustering. If it is necessary to record full-length HCN1 in IOPC, patience is required as most patch will have little or no currents—but some will. If acceptable within the design of the experiment, working with deletion constructs can and will overcome this problem. For example, HCN1–Δ NvΔC does not cluster and is readily recorded in IOPC.

HCN3 and HCN4 express considerably more poorly. Indeed, the combination of low expression and slow gating means obtaining acceptable levels of current and current free of contamination from the endogenous hyperpolarization-activated current is consistently problematic. Successful data collection will often involve sampling cells from many donor frogs with a low expectation of successful recording from many of them.

In inside-out recording it is commonplace to include Mg^{2+} for the first 3–8 min as this promotes depletion of 4,5-PIP_2 from the membrane and allows channel gating to stabilize at its run-down state. Following Mg^{2+} perfusion, we recommend exchanging the internal solution to one that is divalent free so as to eliminate open-channel block of the tail currents, thereby maximizing tail amplitude. For oocytes, we recommend using a Mg^{2+}-containing solution ((in m*M*) 112 KCl, 1 $MgCl_2$, 1 EGTA, and 10 HEPES) and a Mg^{2+}-free solution (112 KCl, 1 EDTA, 1 EGTA, and 10 HEPES, with EDTA, EGTA, and HEPES all added as their free acid forms with the pH subsequently adjusted to 7.4 with KOH). It should be noted that the run-down of gating that occurs on patch excision is accompanied by a hyperpolarizing shift in gating. The need to go to yet more hyperpolarized potentials, combined with the slow gating, places a demand on patches that few tolerate. When recording I_h from *Xenopus* oocytes in which HCN is heterologously expressed, close inspection of the current traces is required to avoid analyzing currents contaminated by an endogenous hyperpolarization-activated Cl^- current, which is also present in these cells (Chao & Katayama, 1991; Thiemann, Gründer, Pusch, & Jentsch, 1992).

2.3.2 Study of HCN Channels in Mammalian Cells

In mammalian cells, both those used for heterologous expression and native cells, similar considerations apply though the actual ionic compositions will vary. With respect to voltage paradigms, HEK293 (and other cells used for heterologous expression) as well as healthy neurons will readily tolerate short

(1–3 s) excursions to very hyperpolarized potentials (i.e., <−110 mV). However, extended multisecond voltage steps to such negative potentials are difficult to maintain, and variable-duration voltage step protocols can be considered (Fig. 1C). But as discussed earlier, the possibility that the steps may still be insufficient to allow gating to approach or attain equilibrium must be considered.

HCN channel isoforms are widely and variably expressed throughout the central nervous system (Notomi & Shigemoto, 2004). When measuring neuronal I_h in the voltage-clamp recording configuration, one must keep in mind that the channels will form functional homo- and heterotetramers, with only the HCN2/HCN3 combination disfavored (Much et al., 2003). Homomeric channels display marked differences in kinetics and nuanced differences in terms of voltage dependence. They also exhibit significant differences with respect to the recording milieu including biologically controlled factors (such as cAMP) and technically influenced factors (such as ion identities, concentrations, and pH). Heteromerization can, and does, add to this diversity (Altomare et al., 2003; Much et al., 2003; Ulens & Tytgat, 2001; Wang et al., 2001). The total current measured, then, will reflect the sum of the conductances arising from each channel configuration present, and one cannot necessarily ascribe the measured current to any one particular channel configuration.

Trafficking of assembled channels to the cell membrane is regulated by both glycosylation (Much et al., 2003) and the auxiliary protein, TRiP8b (Lewis et al., 2009; Santoro et al., 2009; Santoro, Wainger, & Siegelbaum, 2004; Zolles et al., 2009), of which numerous splice variants have been identified (Lewis et al., 2009; Santoro et al., 2009). Trafficking effects also gives rise to nonlinear distribution of functional channels such that HCN1 density is significantly higher in distal dendrites (Lörincz, Notomi, Tamás, Shigemoto, & Nusser, 2002), while HCN2 is preferentially located in dendritic spines of thalamic reticular neurons (Abbas, Ying, & Goldstein, 2006; Ying et al., 2007). Such sequestration in electronically distant sites can result in insignificant current measurements at the soma despite the presence of appreciable functional protein; one approach to resolve this apparent dichotomy is the use of an "I_h isolation" extracellular solution in which other major voltage- and ligand-gated channel conductances are blocked (Abbas et al., 2006; Rateau & Ropert, 2006).

Further complicating the story are regional (Notomi & Shigemoto, 2004) and age-dependent (Bender et al., 2001; Kanyshkova et al., 2009) differences in isoform expression. Thus, what may be true in early postnatal

(P2–P7) life with respect to relative channel expression in the hippocampus vs that in the thalamus (or elsewhere) may not be true during adolescence or adulthood, and appropriate thought given when comparing results obtained from neurons acquired from different brain regions and different aged animals.

One can assess the *relative* contribution of a specific isoform to the observed phenomenon using a suitable gene deletion model, as knockout models exist for HCN1 (Nolan et al., 2003), HCN2 (Chung et al., 2009; Stieber, Wieland, Stockl, Ludwig, & Hofmann, 2006), and HCN3 (Fenske et al., 2011); global HCN4 deletion is embryonic lethal, but it is possible to generate conditional HCN4 deletions using an appropriate Cre/loxP system (Herrmann, Stieber, Stockl, Hofmann, & Ludwig, 2007).

3. K_{2P} CHANNELS

The tandem pore domain (K_{2P}) potassium channel family includes 14 individual channel members, divided into six subgroups based on sequence identity and functional properties. K_{2P} activity is typically referred to as background or "leak" potassium current and is responsible for establishing an electronegative resting membrane potential in biological membranes. K_{2P} channels exhibit only weak voltage or time dependence (with some differences in these properties based on the subtype being studied), but generally approximate the Goodman, Hodgkin, and Katz constant field theory for an open rectifying potassium selective pore (Enyedi & Czirjak, 2010; Honoré, 2007). A unique shared structural architecture also distinguishes K_{2P} channels from other potassium channels classes. While nearly all other potassium channels are formed by a tetrameric assembly of pore-forming subunits, K_{2P}'s are dimers. The canonical tetrameric arrangement of the ion-conducting path is retained because each subunit contributes two potassium channel signature sequence motifs. A number of the functional properties unique to K_{2P} channels are derived in part from this distinctive structural arrangement and the interested reader is directed to excellent reviews of K_{2P} structure/function (Feliciangeli, Chatelain, Bichet, & Lesage, 2015; Renigunta, Schlichthörl, & Daut, 2015).

3.1 Physiological Modulation of K_{2P} Channel Activity

K_{2P} channel members are modulated by numerous physiochemical stimuli including temperature, mechanical stretch, and alteration in pH (Lopes, 2001; Maingret et al., 2000; Sandoz, Douguet, Chatelain, Lazdunski, &

Lesage, 2009). Many K_{2P}'s are also regulated by cellular factors, including second messenger signaling systems and membrane lipid composition (Bockenhauer, Zilberberg, & Goldstein, 2001; Chemin et al., 2005). The intracellular C-terminal domain (CTD) region of the K_{2P} protein has been demonstrated to play an important role in sensing or transducing these modulatory effects, as truncation of this domain attenuates or eliminates temperature sensitivity, mechanosensitivity, lipids modulation, and other gating modulatory cues (Honoré, Maingret, Lazdunski, & Patel, 2002; Kim, Bang, Gnatenco, & Kim, 2001; Patel et al., 1999). While it remains possible that the CTD region is the sensor for all of these diverse signals, there is an abundance of evidence to suggest that gating of K_{2P} channels occurs far from the CTD at the potassium selectivity filter, via a process termed pore gating or C-type gating (Bagriantsev, Peyronnet, Clark, Honoré, & Minor, 2011; Cohen, Ben-Abu, Hen, & Zilberberg, 2008; Piechotta et al., 2011). A likely explanation for the role of the CTD in so many distinct modulatory gating cues is that the CTD acts as an allosteric modulator of the pore gate. The complex biophysical interplay between modulatory inputs to K_{2P} channels and their effects of channel gating continues to be an area of intense focus in this field.

3.2 Pharmacological Modulation of K_{2P} Channel Activity

Standard potassium channel pore-blocking agents, including barium ions, sulfonylureas, 4-aminopyridine, or the quaternary ammonium (QA) cation tetraethylammonium, are unable to block K_{2P}'s when applied to the external face of the channel (Honoré, 2007). This observation has been ascribed to the presence of a unique cap domain situated above the K_{2P} pore, thought to prevent entry of blockers into the channel. Bypassing the cap domain by applying QA derivatives to the intracellular face of the channel does effectively block K_{2P} channels, albeit only with larger derivatives including tetrapentylammonium or tetrahexylammonium (Piechotta et al., 2011; Rapedius et al., 2012). The resistance of K_{2P} channels to the extracellular application of standard potassium channel blockers is a useful feature when studying K_{2P}'s in native tissue, as K_{2P}-mediated potassium current can be readily distinguished from other endogenous potassium channels (Alloui et al., 2006; Franks & Lieb, 1988).

A number of membrane-permeable K_{2P}-specific blocking agents have been identified, although none exhibit affinities below the micromolar range and few completely inhibit channel activity even when applied at saturating concentrations. Readers interested in K_{2P} pharmacologic modulators (known activators and inhibitors) should reference an excellent review by Feliciangeli et al. (2015) and take note of a number of more recently identified K_{2P} modulatory agents (Lolicato et al., 2017; Luo et al., 2017; Su, Brown, Wang, & Mackinnon, 2016).

The mechanism by which volatile anesthetics (VAs) modulate K_{2P} channels is relatively poorly understood. VAs only modulate a subset of K_{2P} channels (TREK1, TREK2, TASK1, TASK3, TRESK, THIK) and are inactive at others (TRAAK, TALK, TWIK) (Patel et al., 1999; Steinberg et al., 2015). Application of a VA typically causes K_{2P} channel activation, although THIK channels are inhibited by the VA halothane. There is also significant variability between channel subtypes in responsiveness to a given anesthetic agent such that not all anesthetics activate every anesthetic-responsive K_{2P}. VA modulation of the TREK1 and TASK1 K_{2P} channels is eliminated by deletion of the CTD region, a finding that suggests VAs may coopt the normal K_{2P} gating machinery. In addition to the CTD region, the third and fourth transmembrane domains of TASK1 and TASK3 channels have been shown to play a role in both VA modulation and stereoselectivity (Andres-Enguix et al., 2007; Conway & Cotten, 2012), raising the possibility that VAs may bind at these transmembrane regions.

Given the broad set of physiochemical and cellular inputs for many of the anesthetic-responsive K_{2P} channels, there is opportunity to understand the mechanism of action of VAs on K_{2P} channels by examining the interplay between anesthetics and other modulatory K_{2P} signals. However, failure to control for variation in the biophysical or physiochemical environments that are inherent to alternative electrophysiological approaches can lead to variability between experimental systems and subsequent errors of interpretation. In the following sections, we will describe the most common approaches used to study K_{2P} channel physiology and address the impact of experimental approach on K_{2P} channel behavior.

3.3 Experimental Considerations

Simple voltage ramp protocols from −140 to +60 mV are sufficient to grossly demonstrate the outward-rectifying behavior of the anesthetic-sensitive members of the K_{2P} channel family. As a general rule, the reversal potential of a cell heterologously expressing a K_{2P} channel and bathed in a physiological extracellular solution will approximate the Nernst equilibrium potential for potassium (−90 mV). Of note, ramp protocols may be inadequate in evaluating the reversal potential of K_{2P} channels, as the lack of significant inwardly directed K_{2P} current leads to only negligible K_{2P} activity at voltages near the reversal potential. Simple voltage step or tail current protocols are more useful when a detailed examination of channel reversal potential is required. Alternatively, the use of buffers containing isosmotic substitutions of potassium for sodium while assaying for changes in reversal potential and inwardly directed current has been an effective method to more accurately assess ionic selectivity and permeability (Thomas, Plant, Wilkens, McCrossan, & Goldstein, 2008). The process of C-type gating thought to regulate many K_{2P} channels is characterized by an increase in sodium permeability and detailed examination of shifts in reversal potential can be mechanistically important findings. For example, modulation of TREK1 currents by extracellular acidification is characterized by a shift in the reversal potential toward more depolarized values (Bagriantsev et al., 2011) and a number of K_{2P} channel mutations exhibit alterations in ionic selectivity, characterized by shifted reversal potentials (Decher et al., 2017; Ma, Zhang, & Chen, 2011; Ma, Zhang, Zhou, & Chen, 2012).

When studying wild-type K_{2P} channels, a significant deviation of the reversal potential in the depolarizing direction (reversal at values more positive than −70 mV) is indicative of cells with significant non-K_{2P}-mediated leak, likely due to poor viability. Recordings from these cells should be interpreted with extreme caution. Differentiating K_{2P}-mediated leak current from background leak can be difficult, given the relatively few pharmacologic agents that block K_{2P} channels with high affinity.

By convention, we, and many groups, use the recorded current at 0 mV as a standard metric for consideration of channel activity. It should be noted that holding oocytes at 0 mV while assaying K_{2P} gating or anesthetic responsiveness is inadvisable. Prolonged depolarizing potentials are known to modulate *Xenopus* oocyte second messenger systems and have been demonstrated to diminish TREK1 channel activity via protein kinase-mediated phosphorylation (Cohen & Zilberberg, 2006). As such, oocytes should be held at hyperpolarizing potentials and ramp or step protocols should be relatively short to avoid this confounding effect.

While TEVC recording from oocytes is a rapid and convenient way to analyze K_{2P} function, IOPC or whole-cell patch clamp should be considered. Use of such approaches that allow for control of the chemical composition on both faces of the membrane may be particularly important for studies of K_{2P} gating, as K_{2P}'s have been shown to exhibit "flux gating" properties, in which movements of permeant ions through the pore region play a key role in modulating channel activity (Schewe et al., 2016).

3.3.1 Heterologous Expression of K_{2P} Channels in Xenopus Oocytes

K_{2P}'s can be readily expressed in isolated *Xenopus* oocytes. For all of the anesthetic-responsive K_{2P} channels, injection of 25–50 nL of cRNA at 50 pg/nL concentration produces outwardly rectifying K_{2P} currents 24–48 h postinjection. We typically aim to record from oocytes that produce 1 μA currents at a holding potential of 0 mV. By achieving this level of channel activity, we are able to accurately record 10-fold decreases in channel activity above background noise and 10-fold increases in channel activity without saturating the recording amplifier. While characterizing K_{2P} channel mutations, it is often useful to first inject at a standard concentration of 50 ng/nL to assess for changes in channel density. Once this result has been established, it is important to adjust the injected cRNA concentration such that channel gating and modulation can be assessed at a current density that matches wild-type channels. We find that gain-of-function mutations often require 10- to 50-fold dilution of injected cRNA to avoid inadvertent interpretation of a saturated signal as loss of channel gating properties.

3.3.2 Study of K_{2P} Channels in Mammalian Cells

For the study of K_{2P}s in mammalian cells, we use a specialized PcDNA3.1 vector containing an internal ribosome entry site (IRES) sequence between the K_{2P} to be studied and a GFP open reading frame (pcDNA3.1(+)IRES GFP, addgene plasmid #51406). Mammalian cells transiently transfected with the pcDNA3.1(+)IRES GFP vector express both the K_{2P} of interest and cytoplasmic GFP such that transiently transfected cells are easily identified with a fluorescence microscope. HEK293 or CHO cells are the most commonly used heterologous expression systems for the study of K_{2P}s in the mammalian background. HEK293 cells are very easy to maintain in culture and are relatively easier than CHO cells to transfect and patch. CHO cells are particularly useful for the study of K_{2P} mechanosensitivity, as they do not express significant mechanically activated current, whereas HEK293 cells express endogenous Piezo1 (Dubin et al., 2017). In either expression system, transient transfection with ~100 ng of K_{2P} wild-type containing expression vector will produce nanoamp currents appropriate for whole-cell patch clamp experiments at 24–48 h posttransfection. If using the pcDNA3.1 (+)IRES GFP vector, the degree of GFP fluorescence roughly correlates with K_{2P} expression level, allowing the experimenter to avoid patching cells that either under or overexpress the K_{2P} of interest.

A number of groups have studied K_{2P} activity in isolated dorsal root ganglion cells and in other native tissues (Alloui et al., 2006; Duprat, Lauritzen, Patel, & Honoré, 2007). These studies can be complicated to interpret due to cell-type variation in K_{2P} modulatory factors (PIP2, arachidonic acid, PKA, and PKC activity) known to alter channel activity and coexpression of multiple K_{2P} subtypes in a single cell. Further complicating the interpretation of in vivo recordings is the demonstrated ability of K_{2P} subtypes to heterodimerize and form channels that exhibit properties distinct from either parent subtype (Czirják & Enyedi, 2002; Lengyel, Czirják, & Enyedi, 2016; Levitz, Royal, Comoglio, Wdziekonski, & Schaub, 2016). Genetic knockout or siRNA techniques have been proven to be effective in isolating the role of a given K_{2P} subtype, with reported global genetic knockouts of many isoforms including TASK1 and TASK3 and TREK1, TREK2, and TRAAK channels (as well as double and triple combination knockouts).

3.4 Modality-Specific Considerations

3.4.1 Temperature Gating

The temperature-responsive K_{2P} channels TREK1, TREK2, and TRAAK are modulated by temperatures in a range from 15°C to 40°C, with Q10 estimated to be six- to eightfold in the maximally responsive range (Kang, Choe, & Kim, 2005; Maingret et al., 2000). TASK and TRESK K_{2P}'s are not temperature sensitive. While assaying K_{2P} temperature sensitivity in *Xenopus* oocytes, it should be noted that cooling to 15°C is well tolerated but incubations for prolonged periods at temperatures above 30° C can cause a rapid decrease in oocyte viability. K_{2P} modulation by temperature can only be assayed in whole-cell configurations, as temperature sensitivity has been suggested to be dependent on an unknown cellular factor and is not clearly an intrinsic property of the K_{2P} channel itself. Patch excision fully eliminates temperature-dependent gating, whereas other gating modalities including activation by membrane stretch or application of bioactive lipids such at arachidonic acid remain intact in excised patches (Maingret et al., 2000).

Operationally, it is important to note that in-line temperature controllers are calibrated based on a set flow rate and thus are prone to error with changes in flow rate. It had been our observation that the kinetics and extent of K_{2P} modulation by a range of pharmacologically active agents are also altered by changes in perfusion flow rate. Reproducibility is thus significantly improved by carefully calibrating the perfusion flow rate on the recording rig.

3.4.2 Mechanosensitivity

Unlike temperature gating, mechanosensitivity of the TREK1, TREK2, and TRAAK channels is thought to be an intrinsic channel property and is maintained even in recombinantly expressed and purified K_{2P} channels reconstituted in a lipid bilayer or proteoliposome (Brohawn, Su, & MacKinnon, 2014). The study of K_{2P} mechanosensitivity is possible in *Xenopus* oocyte inside-out patches (Patel et al., 1998), although the presence of an endogenous mechanosensitive channel that becomes activated with membrane stretch is a confounding factor. While the endogenous mechanosensitive channel is blocked by application of lanthanides (gadolinium completely blocks at 10 μM), these agents also inhibit K2P activity (Maingret, Fosset, Lesage, Lazdunski, & Honore, 1999). For this reason, nearly all studies of K_{2P} mechanosensitivity have been performed in membrane patches from mammalian heterologous systems. Perturbation by stretch has been assayed in both the cell-attached and excited patch configurations. However, while cell-attached recording is done at essentially zero membrane tension, the act of excising a patch induces a 0.5–4 mN/m tension (Opsahl & Webb, 1994), one possible explanation for the higher intrinsic K_{2P} activity observed in excised patch over whole-cell recordings. High-speed pressure clamp devices are useful to impart additional membrane tension in either configuration and are used to study K_{2P} gating by pressure pulses. Care should be taken to optimize recording pipette shape, aperture, and resistance for a given experimental design, as changes in any of these parameters may alter the degree with which pressure pulses imparted by the pressure clamp are transmitted to the membrane patch. This poorly controlled variable may explain some differences in the K_{2P} literature with regard to the absolute pressure responsiveness of K_{2P} channels.

4. GENERAL METHODOLOGICAL DETAILS AND CONSIDERATIONS

4.1 Expressing HCN and K_{2P} Channels in Oocytes

The relative ease of introducing genetic modifications into K_{2P} and HCN genes of study prior to in vitro RNA transcription makes this system ideally suited for mutagenesis studies. TEVC recordings in *Xenopus* oocytes are robust and easily achieved, allowing for rapid evaluation of the effects of mutations on basic channel properties and pharmacology.

To study HCN and K_{2P} channels in oocytes, we use a modified version of the pGEMHE expression vector (Liman, Tytgat, & Hess, 1992) suitable for in vitro transcription of 5′-capped messenger RNA. This vector features a T7 or SP6 RNA polymerase translation initiation site, 5′- and 3′-untranslated regions (UTRs) from the *Xenopus laevis* β-globin gene flanking the K_{2P} gene to be transcribed (to increase cRNA stability), and a 3′ poly-alanine sequence. The small size of the pGEMHE vector (3 kDa) makes it very amenable to molecular biological techniques including simple mutagenesis or insertions. The linearization sequence can be readily changed if the restriction site selection is contraindicated by the inserted gene—such is the case for HCN1 vs HCN2. Alternative expression systems include the pCDNA3.1 or pXOOM vectors, both of which also contain a CMV promoter and are thus uniquely suited for both in vitro cRNA production for injection into oocytes or transient transfection of mammalian cells (Jespersen, Grunnet, Angelo, Klaerke, & Olesen, 2002).

4.2 Biochemistry

A confounding factor in interpretation of any whole-cell response data is that changes in channel activity are only descriptive. Mutations that alter protein expression or plasma membrane targeting can appear indistinguishable from mutations that alter fundamental electrophysiological properties including single-channel conductance or open probability.

Western blot analysis of protein extracts can assess for changes in membrane protein expression. Importantly, commercially available monoclonal antibodies against all of the anesthetic-sensitive K_{2P} variants and all HCN isoforms are readily available. Protocols for surface biotinylation and subsequent streptavidin isolation of oocyte membrane proteins are well described and are more appropriate than whole-cell protein preparations when attempting to assess for changes in ion channel surface expression levels (Riegelhaupt, Frame, & Akabas, 2010).

4.3 Working With Lipophilic and Volatile Agents

Experiments that utilize volatile anesthetic agents require care to ensure minimal loss of anesthetic to the perfusion rig or the environment. Glass syringes and polytetrafluoroethylene (PFTE; Teflon®) tubing and valves are ideal for preventing loss of anesthetic agent to the perfusion system.

Use of a sealed or pressurized perfusion pump is favored over gravity flow perfusion, to prevent loss of anesthetic to the environment. The concentration of soluble volatile anesthetic in a saturated solution has been empirically determined for all of the modern anesthetics and allows for dilution of saturated solutions to produce reproducible working concentrations (Franks & Lieb, 1991). When not working with volatile agents, a gravity flow perfusion system is an acceptable alternative to a pump or pressure-driven system.

REFERENCES

Abbas, S. Y., Ying, S. W., & Goldstein, P. A. (2006). Compartmental distribution of hyperpolarization-activated cyclic-nucleotide-gated channel 2 and hyperpolarization-activated cyclic-nucleotide-gated channel 4 in thalamic reticular and thalamocortical relay neurons. *Neuroscience*, *141*(4), 1811–1825.

Alloui, A., Zimmermann, K., Mamet, J., Duprat, F., Noël, J., Chemin, J., et al. (2006). TREK-1, a K+ channel involved in polymodal pain perception. *The EMBO Journal*, *25*(11), 2368–2376.

Altomare, C., Terragni, B., Brioschi, C., Milanesi, R., Pagliuca, C., Viscomi, C., et al. (2003). Heteromeric HCN1-HCN4 channels: A comparison with native pacemaker channels from the rabbit sinoatrial node. *The Journal of Physiology*, *549*(Pt. 2), 347–359.

Andres-Enguix, I., Caley, A., Yustos, R., Schumacher, M. A., Spanu, P. D., Dickinson, R., et al. (2007). Determinants of the anesthetic sensitivity of two-pore domain acid-sensitive potassium channels: Molecular cloning of an anesthetic-activated potassium channel from Lymnaea stagnalis. *Journal of Biological Chemistry*, *282*(29), 20977–20990.

Bagriantsev, S. N., Peyronnet, R., Clark, K. A., Honoré, E., & Minor, D. L. (2011). Multiple modalities converge on a common gate to control K_{2P} channel function. *The EMBO Journal*, *30*(17), 3594–3606.

Bender, R. A., Brewster, A., Santoro, B., Ludwig, A., Hofmann, F., Biel, M., et al. (2001). Differential and age-dependent expression of hyperpolarization-activated, cyclic nucleotide-gated cation channel isoforms 1-4 suggests evolving roles in the developing rat hippocampus. *Neuroscience*, *106*(4), 689–698.

Biel, M., Wahl-Schott, C., Michalakis, S., & Zong, X. (2009). Hyperpolarization-activated cation channels: From genes to function. *Physiological Reviews*, *89*(3), 847–885.

Bockenhauer, D., Zilberberg, N., & Goldstein, S. A. N. (2001). KCNK2: Reversible conversion of a hippocampal potassium leak into a voltage-dependent channel. *Nature Neuroscience*, *4*(5), 486–491.

Brohawn, S. G., Su, Z., & MacKinnon, R. (2014). Mechanosensitivity is mediated directly by the lipid membrane in TRAAK and TREK1 K+ channels. *Proceedings of the National Academy of Sciences of the United States of America*, *111*(9), 3614–3619.

Bruening-Wright, A., & Larsson, H. P. (2007). Slow conformational changes of the voltage sensor during the mode shift in hyperpolarization-activated cyclic-nucleotide-gated channels. *The Journal of Neuroscience*, *27*(2), 270–278.

Chao, A. C., & Katayama, Y. (1991). Regulation of endogenous chloride conductance in Xenopus oocytes. *Biochemical and Biophysical Research Communications*, *180*(3), 1377–1382.

Chemin, J., Patel, A. J., Duprat, F., Lauritzen, I., Lazdunski, M., & Honoré, E. (2005). A phospholipid sensor controls mechanogating of the K+ channel TREK-1. *The EMBO Journal*, *24*(1), 44–53.

Chen, X., Sirois, J. E., Lei, Q., Talley, E. M., Lynch, C., 3rd, & Bayliss, D. A. (2005). HCN subunit-specific and cAMP-modulated effects of anesthetics on neuronal pacemaker currents. *The Journal of Neuroscience*, *25*(24), 5803–5814.
Chung, W. K., Shin, M., Jaramillo, T. C., Leibel, R. L., LeDuc, C. A., Fischer, S. G., et al. (2009). Absence epilepsy in apathetic, a spontaneous mutant mouse lacking the h channel subunit, HCN2. *Neurobiology of Disease*, *33*(3), 499–508.
Cohen, A., Ben-Abu, Y., Hen, S., & Zilberberg, N. (2008). A novel mechanism for human K2P2.1 channel gating. *Journal of Biological Chemistry*, *283*(28), 19448–19455.
Cohen, A., & Zilberberg, N. (2006). Fluctuations in Xenopus oocytes protein phosphorylation levels during two-electrode voltage clamp measurements. *Journal of Neuroscience Methods*, *153*, 62–70.
Conway, K. E., & Cotten, J. F. (2012). Covalent modification of a volatile anesthetic regulatory site activates TASK-3 (KCNK9) tandem-pore potassium channels. *Molecular Pharmacology*, *81*(3), 393–400.
Czirják, G., & Enyedi, P. (2002). Formation of functional heterodimers between the TASK-1 and TASK-3 two-pore domain potassium channel subunits. *Journal of Biological Chemistry*, *277*(7), 5426–5432.
Decher, N., Ortiz-bonnin, B., Friedrich, C., Schewe, M., Kiper, A. K., Rinné, S., et al. (2017). Sodium permeable and "hypersensitive" TREK-1 channels cause ventricular tachycardia. *EMBO Molecular Medicine*, *9*(4), 403–414.
Del Lungo, M., Melchiorre, M., Guandalini, L., Sartiani, L., Mugelli, A., Koncz, I., et al. (2012). Novel blockers of hyperpolarization-activated current with isoform selectivity in recombinant cells and native tissue. *British Journal of Pharmacology*, *166*(2), 602–616.
Dubin, A. E., Murthy, S., Lewis, A. H., Brosse, L., Cahalan, S. M., Grandl, J., et al. (2017). Endogenous Piezo1 can confound mechanically activated channel identification and characterization. *Neuron*, *94*(2), 266–270.e3.
Duprat, F., Lauritzen, I., Patel, A., & Honoré, E. (2007). The TASK background K2P channels: Chemo- and nutrient sensors. *Trends in Neurosciences*, *30*(11), 573–580.
Elinder, F., Männikkö, R., Pandey, S., & Larsson, H. P. (2006). Mode shifts in the voltage gating of the mouse and human HCN2 and HCN4 channels. *The Journal of Physiology*, *575*(2), 417–431.
Enyedi, P., & Czirjak, G. (2010). Molecular background of leak K+ currents: Two-pore domain potassium channels. *Physiological Reviews*, *90*, 559–605.
Feliciangeli, S., Chatelain, F. C., Bichet, D., & Lesage, F. (2015). The family of K2P channels: Salient structural and functional properties. *The Journal of Physiology*, *593*(12), 2587–2603.
Fenske, S., Mader, R., Scharr, A., Paparizos, C., Cao-Ehlker, X., Michalakis, S., et al. (2011). HCN3 contributes to the ventricular action potential waveform in the murine heart. *Circulation Research*, *109*(9), 1015–1023.
Fogle, K. J., Lyashchenko, A. K., Turbendian, H. K., & Tibbs, G. R. (2007). HCN pacemaker channel activation is controlled by acidic lipids downstream of diacylglycerol kinase and phospholipase A2. *The Journal of Neuroscience*, *27*(11), 2802–2814.
Franks, N. P. (2008). General anaesthesia: From molecular targets to neuronal pathways of sleep and arousal. *Nature Reviews. Neuroscience*, *9*(5), 370–386. https://doi.org/10.1038/nrn2372.
Franks, N. P., & Lieb, W. R. (1988). Volatile general anesthetics activate a novel neuronal K + current. *Nature*, *333*, 662–664.
Franks, N. P., & Lieb, W. R. (1991). Stereospecific effects of inhalational general anesthetic optical isomers on nerve ion channels. *Science*, *254*(5030), 427–430.
Goldstein, P. A. (2015). HCN1 channels as targets for volatile anesthetics: Coming to the fore. *Anesthesia and Analgesia*, *121*(3), 594–596.

Gonzalez, C., Baez-Nieto, D., Valencia, I., Oyarzun, I., Rojas, P., Naranjo, D., et al. (2012). K+ channels: Function-structural overview. *Comprehensive Physiology, 2*(3), 2087–2149.

Han, Y., Noam, Y., Lewis, A. S., Gallagher, J. J., Wadman, W. J., Baram, T. Z., et al. (2011). Trafficking and gating of hyperpolarization-activated cyclic nucleotide-gated channels are regulated by interaction with tetratricopeptide repeat-containing Rab8b-interacting protein (TRIP8b) and cyclic AMP at distinct sites. *Journal of Biological Chemistry, 286*(23), 20823–20834.

He, C., Chen, F., Li, B., & Hu, Z. (2014). Neurophysiology of HCN channels: From cellular functions to multiple regulations. *Progress in Neurobiology, 112*, 1–23.

Herold, K. F., & Hemmings, H. C. (2012). Sodium channels as targets for volatile anesthetics. *Frontiers in Pharmacology, 3*, 1–7.

Herrmann, S., Stieber, J., Stockl, G., Hofmann, F., & Ludwig, A. (2007). HCN4 provides a "depolarization reserve" and is not required for heart rate acceleration in mice. *The EMBO Journal, 26*(21), 4423–4432.

Honoré, E. (2007). The neuronal background K2P channels: Focus on TREK1. *Nature Reviews. Neuroscience, 8*(4), 251–261.

Honoré, E., Maingret, F., Lazdunski, M., & Patel, A. J. (2002). An intracellular proton sensor commands lipid- and mechano-gating of the K(+) channel TREK-1. *The EMBO Journal, 21*(12), 2968–2976.

Jespersen, T., Grunnet, M., Angelo, K., Klaerke, D., & Olesen, S. (2002). Dual-function vector for protein expression in both mammalian cells and Xenopus laevis oocytes. *BioTechniques, 32*(3), 536–538.

Kang, D., Choe, C., & Kim, D. (2005). Thermosensitivity of the two-pore domain K+ channels TREK-2 and TRAAK. *The Journal of Physiology, 564*(1), 103–116.

Kanyshkova, T., Pawlowski, M., Meuth, P., Dube, C., Bender, R. A., Brewster, A. L., et al. (2009). Postnatal expression pattern of HCN channel isoforms in thalamic neurons: Relationship to maturation of thalamocortical oscillations. *The Journal of Neuroscience, 29*(27), 8847–8857.

Kim, Y., Bang, H., Gnatenco, C., & Kim, D. (2001). Synergistic interaction and the role of C-terminus in the activation of TRAAK K+ channels by pressure, free fatty acids and alkali. *Pflügers Archiv: European Journal of Physiology, 442*(1), 64–72.

Lengyel, M., Czirják, G., & Enyedi, P. (2016). Formation of functional heterodimers by TREK-1 and TREK-2 two-pore domain potassium channel subunits. *Journal of Biological Chemistry, 291*(26), 13649–13661.

Levitz, J., Royal, P., Comoglio, Y., Wdziekonski, B., & Schaub, S. (2016). Heterodimerization within the TREK channel subfamily produces a diverse family of highly regulated potassium channels. *Proceedings of the National Academy of Sciences of the United States of America, 113*(15), 1–6.

Lewis, A. S., Estep, C., & Chetkovich, D. M. (2010). The fast and slow ups and downs of HCN channel regulation. *Channels (Austin, Tex.), 4*(3), 215–231.

Lewis, A. S., Schwartz, E., Chan, C. S., Noam, Y., Shin, M., Wadman, W. J., et al. (2009). Alternatively spliced isoforms of TRIP8b differentially control h channel trafficking and function. *The Journal of Neuroscience, 29*(19), 6250–6265.

Liman, E. R., Tytgat, J., & Hess, P. (1992). Subunit stoichiometry of a mammalian K+ channel determined by construction of multimeric cDNAs. *Neuron, 9*(5), 861–871.

Lolicato, M., Arrigoni, C., Mori, T., Sekioka, Y., Bryant, C., Clark, K. A., et al. (2017). K2P2.1 (TREK-1)-activator complexes reveal a cryptic selectivity filter binding site. *Nature, 547*(7663), 364–368.

Lopes, C. M. B. (2001). Block of Kcnk3 by protons. Evidence that 2-p-domain potassium channel subunits function as homodimers. *Journal of Biological Chemistry, 276*(27), 24449–24452.

Lörincz, A., Notomi, T., Tamás, G., Shigemoto, R., & Nusser, Z. (2002). Polarized and compartment-dependent distribution of HCN1 in pyramidal cell dendrites. *Nature Neuroscience*, *5*(11), 1185–1193.
Ludwig, A., Zong, X., Stieber, J., Hullin, R., Hofmann, F., & Biel, M. (1999). Two pacemaker channels from human heart with profoundly different activation kinetics. *The EMBO Journal*, *18*(9), 2323–2329.
Luo, Q., Chen, L., Cheng, X., Ma, Y., Li, X., Zhang, B., et al. (2017). An allosteric ligand-binding site in the extracellular cap of K2P channels. *Nature Communications*, *8*(378), 1–9.
Lyashchenko, A. K., Redd, K. J., Yang, J., & Tibbs, G. R. (2007). Propofol inhibits HCN1 pacemaker channels by selective association with the closed states of the membrane embedded channel core. *The Journal of Physiology*, *583*(Pt. 1), 37–56.
Ma, L., Zhang, X., & Chen, H. (2011). TWIK-1 two-pore domain potassium channels change ion selectivity and conduct inward leak sodium currents in hypokalemia. *Science Signaling*, *4*(176), ra37.
Ma, L., Zhang, X., Zhou, M., & Chen, H. (2012). Acid-sensitive TWIK and TASK two-pore domain potassium channels change ion selectivity and become permeable to sodium in extracellular acidification. *Journal of Biological Chemistry*, *287*(44), 37145–37153.
Maingret, F., Fosset, M., Lesage, F., Lazdunski, M., & Honoré, E. (1999). TRAAK is a mammalian neuronal mechano-gated K+ channel. *Journal of Biological Chemistry*, *274*(3), 1381–1387.
Maingret, F., Lauritzen, I., Patel, A. J., Heurteaux, C., Reyes, R., Lesage, F., et al. (2000). TREK-1 is a heat-activated background K(+) channel. *The EMBO Journal*, *19*(11), 2483–2491.
Männikkö, R., Pandey, S., Larsson, H. P., & Elinder, F. (2005). Hysteresis in the voltage dependence of HCN channels. *The Journal of General Physiology*, *125*(3), 305–326.
McCormick, D. A., & Pape, H. C. (1990). Properties of a hyperpolarization-activated cation current and its role in rhythmic oscillation in thalamic relay neurones. *The Journal of Physiology*, *431*(1), 291–318.
Melchiorre, M., Del Lungo, M., Guandalini, L., Martini, E., Dei, S., Manetti, D., et al. (2010). Design, synthesis, and preliminary biological evaluation of new isoform-selective f-current blockers. *Journal of Medicinal Chemistry*, *53*(18), 6773–6777.
Mistrík, P., Pfeifer, A., & Biel, M. (2006). The enhancement of HCN channel instantaneous current facilitated by slow deactivation is regulated by intracellular chloride concentration. *Pflügers Archiv: European Journal of Physiology*, *452*(6), 718–727.
Moroni, A., Barbuti, A., Altomare, C., Viscomi, C., Morgan, J., Baruscotti, M., et al. (2000). Kinetic and ionic properties of the human HCN2 pacemaker channel. *Pflügers Archiv: European Journal of Physiology*, *439*(5), 618–626.
Much, B., Wahl-Schott, C., Zong, X., Schneider, A., Baumann, L., Moosmang, S., et al. (2003). Role of subunit heteromerization and N-linked glycosylation in the formation of functional hyperpolarization-activated cyclic nucleotide-gated channels. *The Journal of Biological Chemistry*, *278*(44), 43781–43786.
Munsch, T., & Pape, H. C. (1999). Modulation of the hyperpolarization-activated cation current of rat thalamic relay neurones by intracellular pH. *The Journal of Physiology*, *519*(Pt. 2), 493–504.
Nau, C. (2008). Voltage-gated ion channels. *Handbook of Experimental Pharmacology*, *182*, 85–92.
Nolan, M. F., Malleret, G., Lee, K. H., Gibbs, E., Dudman, J. T., Santoro, B., et al. (2003). The hyperpolarization-activated HCN1 channel is important for motor learning and neuronal integration by cerebellar Purkinje cells. *Cell*, *115*(5), 551–564.
Notomi, T., & Shigemoto, R. (2004). Immunohistochemical localization of Ih channel subunits, HCN1-4, in the rat brain. *The Journal of Comparative Neurology*, *471*(3), 241–276.

Opsahl, L. R., & Webb, W. W. (1994). Lipid-glass adhesion in giga-sealed patch-clamped membranes. *Biophysical Journal*, *66*(1), 75–79.

Patel, A. J., Honoré, E., Lesage, F., Fink, M., Romey, G., & Lazdunski, M. (1999). Inhalational anesthetics activate two-pore-domain background K+ channels. *Nature Neuroscience*, *2*(5), 422–426.

Patel, A. J., Honoré, E., Maingret, F., Lesage, F., Fink, M., Duprat, F., et al. (1998). A mammalian two pore domain mechano-gated S-like K+ channel. *The EMBO Journal*, *17*(15), 4283–4290.

Pian, P., Bucchi, A., Decostanzo, A., Robinson, R. B., & Siegelbaum, S. A. (2007). Modulation of cyclic nucleotide-regulated HCN channels by PIP2 and receptors coupled to phospholipase C. *Pflügers Archiv: European Journal of Physiology*, *455*(1), 125–145.

Pian, P., Bucchi, A., Robinson, R. B., & Siegelbaum, S. A. (2006). Regulation of gating and rundown of HCN hyperpolarization-activated channels by exogenous and endogenous PIP2. *The Journal of General Physiology*, *128*(5), 593–604.

Piechotta, P. L., Rapedius, M., Stansfeld, P. J., Bollepalli, M. K., Erhlich, G., Andres-Enguix, I., et al. (2011). The pore structure and gating mechanism of K2P channels. *The EMBO Journal*, *30*(17), 3607–3619.

Rapedius, M., Schmidt, M. R., Sharma, C., Stansfeld, P. J., Sansom, M. S. P., Baukrowitz, T., et al. (2012). State-independent intracellular access of quaternary ammonium blockers to the pore of TREK-1. *Channels (Austin, Tex.)*, *6*(6), 473–478.

Rateau, Y., & Ropert, N. (2006). Expression of a functional hyperpolarization-activated current (Ih) in the mouse nucleus reticularis thalami. *Journal of Neurophysiology*, *95*(5), 3073–3085.

Renigunta, V., Schlichthörl, G., & Daut, J. (2015). Much more than a leak: Structure and function of K_2p-channels. *Pflügers Archiv: European Journal of Physiology*, *467*, 867–894.

Riegelhaupt, P. M., Frame, I. J., & Akabas, M. H. (2010). Transmembrane segment 11 appears to line the purine permeation pathway of the Plasmodium falciparum equilibrative nucleoside transporter 1 (PfENT1). *The Journal of Biological Chemistry*, *285*(22), 17001–17010.

Robinson, R. B., & Siegelbaum, S. A. (2003). Hyperpolarization-activated cation currents: From molecules to physiological function. *Annual Review of Physiology*, *65*, 453–480.

Sandoz, G., Douguet, D., Chatelain, F., Lazdunski, M., & Lesage, F. (2009). Extracellular acidification exerts opposite actions on TREK1 and TREK2 potassium channels via a single conserved histidine residue. *Proceedings of the National Academy of Sciences of the United States of America*, *106*(34), 14628–14633.

Santoro, B., Hu, L., Liu, H., Saponaro, A., Pian, P., Piskorowski, R. A., et al. (2011). TRIP8b regulates HCN1 channel trafficking and gating through two distinct C-terminal interaction sites. *Journal of Neuroscience*, *31*(11), 4074–4086.

Santoro, B., Liu, D. T., Yao, H., Bartsch, D., Kandel, E. R., Siegelbaum, S. A., et al. (1998). Identification of a gene encoding a hyperpolarization-activated pacemaker channel of brain. *Cell*, *93*(5), 717–729.

Santoro, B., Piskorowski, R. A., Pian, P., Hu, L., Liu, H., & Siegelbaum, S. A. (2009). TRIP8b splice variants form a family of auxiliary subunits that regulate gating and trafficking of HCN channels in the brain. *Neuron*, *62*(6), 802–813.

Santoro, B., Wainger, B. J., & Siegelbaum, S. A. (2004). Regulation of HCN channel surface expression by a novel C-terminal protein-protein interaction. *The Journal of Neuroscience*, *24*(47), 10750–10762.

Schewe, M., Nematian-ardestani, E., Sun, H., Tucker, S. J., Rapedius, M., & Bucci, G. (2016). A non-canonical voltage-sensing mechanism controls gating in K2P K(+) channels. *Cell*, *164*, 937–949.

Seifert, R., Scholten, A., Gauss, R., Mincheva, A., Lichter, P., et al. (1999). Molecular characterization of a slowly gating human hyperpolarization-activated channel

predominantly expressed in thalamus, heart, and testis. *Proceedings of the National Academy of Sciences of the United States of America, 96*(16), 9391–9396.
Shin, K. S., Maertens, C., Proenza, C., Rothberg, B. S., & Yellen, G. (2004). Inactivation in HCN channels results from reclosure of the activation gate: Desensitization to voltage. *Neuron, 41*(5), 737–744.
Steinberg, E. A., Wafford, K. A., Brickley, S. G., Franks, N. P., & Wisden, W. (2015). The role of K2P channels in anaesthesia and sleep. *Pflügers Archiv: European Journal of Physiology, 467*(5), 907–916.
Stevens, D. R., Seifert, R., Bufe, B., Muller, F., Kremmer, E., Gauss, R., et al. (2001). Hyperpolarization-activated channels HCN1 and HCN4 mediate responses to sour stimuli. *Nature, 413*(6856), 631–635.
Stieber, J., Stöckl, G., Herrmann, S., Hassfurth, B., & Hofmann, F. (2005). Functional expression of the human HCN3 channel. *Journal of Biological Chemistry, 280*(41), 34635–34643.
Stieber, J., Thomer, A., Much, B., Schneider, A., Biel, M., & Hofmann, F. (2003). Molecular basis for the different activation kinetics of the pacemaker channels HCN2 and HCN4. *Journal of Biological Chemistry, 278*(36), 33672–33680.
Stieber, J., Wieland, K., Stockl, G., Ludwig, A., & Hofmann, F. (2006). Bradycardic and proarrhythmic properties of sinus node inhibitors. *Molecular Pharmacology, 69*(4), 1328–1337.
Su, Z., Brown, E. C., Wang, W., & Mackinnon, R. (2016). Novel cell-free high-throughput screening method for pharmacological tools targeting K+ channels. *Proceedings of the National Academy of Sciences of the United States of America, 113*(20), 5748–5753.
Thiemann, A., Gründer, S., Pusch, M., & Jentsch, T. J. (1992). A chloride channel widely expressed in epithelial and non-epithelial cells. *Nature, 359*, 150–152.
Thomas, D., Plant, L. D., Wilkens, C. M., McCrossan, Z. A., & Goldstein, S. A. (2008). Alternative translation initiation in rat brain yields K2P2.1 potassium channels permeable to sodium. *Neuron, 58*(6), 859–870.
Tibbs, G. R., Rowley, T. J., Sanford, R. L., Herold, K. F., Proekt, A., Hemmings, H. C., et al. (2013). HCN1 channels as targets for anesthetic and nonanesthetic propofol analogs in the amelioration of mechanical and thermal hyperalgesia in a mouse model of neuropathic pain. *The Journal of Pharmacology and Experimental Therapeutics, 345*(3), 363–373.
Ulens, C., & Tytgat, J. (2001). Functional heteromerization of HCN1 and HCN2 pacemaker channels. *The Journal of Biological Chemistry, 276*(9), 6069–6072.
Wahl-Schott, C., Baumann, L., Zong, X., & Biel, M. (2005). An arginine residue in the pore region is a key determinant of chloride dependence in cardiac pacemaker channels. *Journal of Biological Chemistry, 280*(14), 13694–13700.
Wahl-Schott, C., & Biel, M. (2009). HCN channels: Structure, cellular regulation and physiological function. *Cellular and Molecular Life Sciences, 66*(3), 470–494.
Wainger, B. J., DeGennaro, M., Santoro, B., Siegelbaum, S. A., & Tibbs, G. R. (2001). Molecular mechanism of cAMP modulation of HCN pacemaker channels. *Nature, 411*(6839), 805–810.
Wang, J., Chen, S., Nolan, M. F., & Siegelbaum, S. A. (2002). Activity-dependent regulation of HCN pacemaker channels by cyclic AMP: Signaling through dynamic allosteric coupling. *Neuron, 36*(3), 451–461.
Wang, J., Chen, S., & Siegelbaum, S. A. (2001). Regulation of hyperpolarization-activated HCN channel gating and cAMP modulation due to interactions of COOH terminus and core transmembrane regions. *The Journal of General Physiology, 118*(3), 237–250.
Ying, S. W., Jia, F., Abbas, S. Y., Hofmann, F., Ludwig, A., & Goldstein, P. A. (2007). Dendritic HCN2 channels constrain glutamate-driven excitability in reticular thalamic neurons. *The Journal of Neuroscience, 27*(32), 8719–8732.

Zeller, A., Jurd, R., Lambert, S., Arras, M., Drexler, B., Grashoff, C., et al. (2008). Inhibitory ligand-gated ion channels as substrates for general anesthetic actions. *Handbook of Experimental Pharmacology*, *182*, 31–51.

Zolles, G., Wenzel, D., Bildl, W., Schulte, U., Hofmann, A., Muller, C. S., et al. (2009). Association with the auxiliary subunit PEX5R/Trip8b controls responsiveness of HCN channels to cAMP and adrenergic stimulation. *Neuron*, *62*(6), 814–825.

Zong, X., Stieber, J., Ludwig, A., Hofmann, F., & Biel, M. (2001). A single histidine residue determines the pH sensitivity of the pacemaker channel HCN2. *The Journal of Biological Chemistry*, *276*(9), 6313–6319.

FSC
www.fsc.org
MIX
Papier | Fördert gute Waldnutzung
FSC® C083411

Kolibri 360 GmbH
Ferdinand-Jühlke-Straße 7
99095 Erfurt, Deutschland
produktsicherheit@kolibri360.de